Rebreathers in Diving Science

This book covers investigations on the diving apparatus operational features, including research investigations, basics of measuring methods, their technical realization, elaboration, and discussion of the results. It contains analyses of research reports prepared in leading research diving centers to formulate opinions when comparing the methods used and equipment presented, including the accuracy of experiments, complexity analysis, laboratory expertise, metrology features of the used instruments, and correctness of the calibration procedures.

Features:

- Presents a novel comprehensive approach to the design of semi–closed circuit diving apparatuses.
- Provides a methodically documented approach to the modelling and validation processes.
- Replaces statistical, empirical, or semi-empirical models with deterministic models, for which all parameters have physical interpretation.
- Includes flexible procedures at one of the highest technology readiness levels.
- Discusses the reasons for using artificial breathing media in special UBAs.

This book is aimed at researchers, professionals, and graduate students in life support system design, diving submarine safety, and ventilation.

Diving Sciences

Series Editor: Ryszard Kłos

The series is aimed at encouraging scientists from around the world to publish research related to diving and hyperbaric exposures pertaining to the technology readiness levels (TRL), with a particular focus on system prototype demonstration in an operational environment, system complete and qualified, and actual system proven in operational environment. The intended audience includes researchers in diving science; military and civilian divers; diving contractors; navies, defense, and civilian diving schools; and related academic audiences.

Oxygen Diving
Ryszard Kłos

Rebreathers in Diving Science
Semi-Closed and Closed Circuit Diving Apparatuses
Ryszard Kłos

For more information about this series, please visit: www.routledge.com/Diving-Sciences/book-series/CRCDS.

Rebreathers in Diving Science

Semi-Closed and Closed Circuit Diving Apparatuses

Ryszard Kłos

CRC Press
Taylor & Francis Group
Boca Raton London New York

CRC Press is an imprint of the
Taylor & Francis Group, an **informa** business

Designed cover image: Ryszard Kłos

First edition published 2025
by CRC Press
2385 NW Executive Center Drive, Suite 320, Boca Raton FL 33431

and by CRC Press
4 Park Square, Milton Park, Abingdon, Oxon, OX14 4RN

CRC Press is an imprint of Taylor & Francis Group, LLC

Library of Congress Cataloging-in-Publication Data
Names: Kłos, Ryszard, author.
Title: Rebreathers in diving science : semi-closed- and closed-circuit diving apparatuses / Ryszard Kłos.
Description: First edition. | Boca Raton, FL : CRC Press, 2025. | Series: Diving sciences | Includes bibliographical references and index.
Identifiers: LCCN 2024055879 (print) | LCCN 2024055880 (ebook) | ISBN 9781032596082 (hardback) | ISBN 9781032649191 (paperback) | ISBN 9781032649207 (ebook)
Subjects: LCSH: Scuba apparatus.
Classification: LCC GV838.672 .K56 2025 (print) | LCC GV838.672 (ebook) | DDC 627/.720284—dc23/eng/20250222
LC record available at https://lccn.loc.gov/2024055879
LC ebook record available at https://lccn.loc.gov/2024055880

ISBN: 978-1-032-59608-2 (hbk)
ISBN: 978-1-032-64919-1 (pbk)
ISBN: 978-1-032-64920-7 (ebk)

DOI: 10.1201/9781032649207

Typeset in Times
by Apex CoVantage, LLC

My family
wife, Yvonne
daughter, Alice
and son, Tom

Contents

Preface

The book is one of the publications in the series *Diving Science* aimed to publish research on diving and hyperbaric exposures pertaining to the technology readiness levels (TRL). It is especially focused on a system prototype demonstration in an operational environment, a system complete and qualified, and a system proven in an operational environment. The intended audience includes researchers in diving science; military and civilian divers; diving contractors; navies, defense, and civilian diving schools; and related academic audience.

This book will be of use to diving scientists, as they must assess the risk, and it will also be useful for special operations forces tactics specialists, because:

- Navies expect progress in technologies of diving dedicated to use in military operations
- Defense diving schools expect materials for training in modernized technologies of diving
- Military divers expect aided materials in the process of self-education

From the date of its establishment (1976), evaluation and study of diving equipment and decompression systems have been the primary task carried out by the Department of Underwater Work Technology[1] of the Naval Academy. The projects referred to required proper research stations. We mostly built them ourselves or acquired them special through orders. At the very beginning, the special purpose team of technicians was established for this task. They were able to produce technology demonstrators[2] as well as prototypes of diving and scientific instruments. Experimental Deepwater Diving Complex *DGKN* – 120 was supposed to be a universal test bench and was designed to provide comprehensive research capabilities for the diving apparatus. At that time, the initial concepts of research stations were designed by the first head, late Capt. Medard Przylipiak, MSc Eng. The ideas of conducting a comprehensive study of diving apparatus were common with his successors, the late Cmdr. Marian Pleszewski, MSc Eng, and Capt. Stanislaw Skrzyński, PhD Eng, which did not prevent them from making other scientific achievements.[3] One of the persons who played the leading role as captain was Prof. Tadeusz Doboszyński, PhD, DSc Med, from the Military Medical Academy.

It was Capt. Stanislaw Skrzyński, PhD Eng, who became an inspiring leader in the development and research on new types of diving equipment. The results of this work have been largely wasted due to the failure to promote research focused on developing the original Polish diving apparatus and decompression systems.

Bearing in mind the old concepts and the undeniable achievements, efforts were made to raise funds to continue the work done by their predecessors. Initially, the purchases of foreign diving equipment facilitated this task, at the same time creating a need to start the implementation work.[4] The effective stage of research was reached with modelling breathing in *CCR*[5] and *SCR*[6] systems, which made it possible to connect the diving apparatus model with the model of decompression diving. Another completed project made it possible to develop a statistical approach to the safety of diving. The research into saturation diving and decompression systems allowed us to extend our laboratory facilities and expand the filed specific knowledge. Using predecessors' ideas and not-so-small funds, we managed to build a material base and theoretical knowledge to carry out research on diving apparatuses and decompression systems. The research facilities are also used for training and as an emergency center. It is hoped that these efforts will not be wasted and Poland will come back to and continue to be present among the countries developing diving technologies.

This report presents the results of research carried out to develop new semi–closed circuit rebreathers, *SCRs*.

Just as diving requires teamwork, this report could not have been made without efficient assistance and help provided by my friends at the Polish Naval Academy, the Department of Diving Gear and Underwater Work Technology, the Military Medical Academy, and the navy units. My special thanks also to my friend Capt. (Navy) Stanisław Skrzyński, PhD Eng.

Putting new underwater breathing apparatuses, *UBAs*, into use is inseparably dependent on manned experiments; therefore, I would like to thank my colleagues, Polish navy divers, for their trust and their wish to undertake the risk. As this type of research is hazardous and diving accidents can occur, I wish to direct my sincere thanks to my friends diving medical officers for their medical care, which has also allowed me to continue to be an efficient experimental diver.

I wish to extend my gratitude to all people of goodwill, whose names are not specified here, for their assistance.

NOTES

1 Former: Department of Diving Gear and Underwater Work Technology.
2 By *technology demonstrator* meant is the practical demonstration of technical solutions aimed at the development of a particular system, demonstrating the possibility of achieving the required parameters for this system.
3 For example, the implementation of saturation dives.
4 But giving only evidence of the possibility of using domestic supplies.
5 Closed circuit rebreather.
6 Semi–closed circuit rebreather.

Acknowledgments

Apart from financing, the presented research required access to unique combat equipment. I express my gratitude to the Navy Command of the Republic of Poland and the 3rd Ship Flotilla for the trust that I received.

The research included experiments on humans, which required consent of the Scientific Research Ethics Committee. In addition to appreciation to the predecessors, gratitude must be expressed for the contribution given by the research staff of the department and the cooperating medical doctors from the Military Medical Institute. Of them, Cmdr. Maciej Konarski, PhD Med, has significantly contributed to assembling the experimental data described in this monograph.

Thanks to the confidence demonstrated by the divers and professionals from military units, police, antiterrorist groups, and others who have supported this work, and owing to the support rendered by the Ministry of Defense, the Ministry of Science and Higher Education, and the National Research and Development, who provided a considerable amount of funds invested in research carried out in the department, the desired and expected results have been attained.

In particular, I would like to express my gratitude to Kazimierz Szczepański, PhD, who reviewed the entire manuscript.

Finally, in appreciation of her assistance and perseverance, I wish to express my love and sincere thanks to my dear wife, Yvonne, who, through the years, has supported and aided me in my efforts.

About the Author

Ryszard Kłos graduated from the Technical University of Wrocław with a MSc degree in physicochemistry. He started his professional career in the Institute of Low Temperature and Structural Research, Polish Academy of Sciences, Wrocław.

In 1985, he attended military training required of university graduates and graduated from Chemical Warfare College. The following year, he joined the Polish Navy as platoon leader in the Military Diving School. In 1988, he was transferred to the Naval Academy, to the Department of Diving Gear and Underwater Work Technology, where he earned his PhD in the field of design and operation of machines. He completed a postdoctoral fellowship at the DR-DC Toronto (former: Defence and Civil Institute of Environmental Medicine). He defended his habilitation thesis (higher-PhD thesis) in the field of machine operation at Gdańsk University of Technology.

The author is a retired Polish Navy captain. He is currently employed as an associate professor at the Polish Naval Academy, where he is responsible for various diving and physicochemical research projects and programs. He has written nine books and many scientific articles. From 2012 to 2015, he served as the deputy of Rector for Scientific Research at the Polish Naval Academy. From 2004 to 2007, he was a president of the Polish Hyperbaric Medical and Technical Society.

List of Important Definitions

Item	Description
batynautics	the totality of knowledge about a man's underwater activities
bradycardia	the state when heart rate decreases upon exposure of the body to hyperbaric conditions compared to the heart in normobaric conditions
buffer	a solution whose pH value after the addition of small amounts of strong acids or alkali, as well as after dilution with water, is relatively constant
catalysts	substances that increase the rate of reaction and remain the same, unchanged, after the reaction
catalytic reaction	*catalysis*; a chemical reaction in which a change in reaction rate occurs under the influence of a catalyst
chemical equation	*reaction equation*; a quick record of the chemical reaction using, located on the left side, the formulas of the starting materials (substrates), and on the right side formulas of the substances obtained in the reaction (reaction products)
chemical reaction yield	efficiency of a chemical reaction; the ratio between the amount of reacted material and the amount of matter that could theoretically react by chemical reaction
cyclo-ergometer	a device with a structure similar to a bicycle, controlled by inhibiting the pedal effort at a controlled rate of cycling
decompression	with respect to the dive, a gradual lowering of pressure applied to the diver at the exit of the hyperbaric environment carried out so that there are no signs of disease
diffusion	the spontaneous mixing of the components of the system as a result of the chaotic motion of particles
dissimulation	understood as a presence manifested in an attempt to meet the norms or hide the actual situation and feelings, impulses, behaviors, etc.
endothermic reaction	chemical reaction that occurs with heat being absorbed
enzyme	mostly protein; macromolecular chemical compound regulating life processes
equilibrium constant[1]	number given by the ratio of the product of the molar concentrations of the products to the product of the molar concentration of substrates; molar concentrations should be expressed in molar fractions and raised to the power such as results from the reaction's stoichiometric ratio
exothermic reaction	a chemical reaction that proceeds with the liberation of heat
hypocapnia	also hypocarbia; a state of reduced partial pressure of CO_2 in the blood below normal values
hypoxia	the state of oxygen deficiency in the tissue resulting from reduced oxygen diffusion in the lungs (hypoxemia) or disorder of the blood oxygen transport to the tissues (ischemia)
hypercapnia	a state of elevated partial pressure of the CO_2 in the blood above $pCO_2 > 45\,mmHg$, referred to here as the symptoms of CO_2 poisoning

(*Continued*)

(*Continued*)

Item	Description
hypoxemia	a reduction of the diffusion of the oxygen in the lungs
hyperbaric oxygen therapy	a therapeutic procedure often used for general and local tissue oxygenation against anaerobes, after gas poisoning, to secure the graft, frostbite, burns, radioactive damage, osteomyelitis, slow-healing wounds, sudden deafness, etc.
hyperoxia	a state of breathing respiratory medium wherein oxygen partial pressure exceeds 21 kPa; used herein for cases of breathing with medium with partial pressure of oxygen, which creates a toxic threat
homeostasis	the ability of a living organism, by proper coordination and regulation of life processes, to maintain a relatively constant state of equilibrium, for example, blood composition, temperature, etc.
ischemia	a disorder of oxygen transport by blood to the tissues
metabolite	an organic or inorganic metabolic product
mathematical model	a mathematical description of the behavior of the modelled system, usually expressed in the form of algebraic mathematical equations
model	a system whose function is to imitate the distinguished features of another system, known as the original
mol	a basic unit in the SI system that specifies the amount of substance in the system[2]
paresthesia	a deficit in sensorial perception involving the wrong location of the stimulus and warped fillings described as tingling, numbness, etc.
premix	a mixture of two or more components standardized prior to use[3]
pre-oxygenation	flushing the body with oxygen[4]
radical	a group of atoms, generally incapable of independent existence, having unpaired electrons[5]
reaction kinetics	reaction duration/mechanism as a function of time
reaction mechanism	a mechanism created during chemical reaction via intermediate stages, which further results in final products, as summarized in the reaction equation, illustrating transition from the starting particles into particles which are reaction products
respiratory dead space volume	lung space volume in which the ventilation does not occur or mass transfer occurs only to a limited extent
saturated solution	a solution which, under specified conditions, is/can be in equilibrium with excess of the solute
semi-empirical methods	methods describing the behavior of the object according to criterion equation[6] typically containing one or several constants determined empirically for a model; do not have constant physical interpretation and are often dependent on the units of measurement[7]
somatic	concerns the body; carnal, physical
solution concentration	the amount of solute in a given amount of solvent; concentration of the solution most often expressed using a fraction or percentage concentration[8]
system	distinguished from the reality collection of elements with links between them

(*Continued*)

(*Continued*)

Item	Description
target strength	most commonly expressed in decibels $[dB@1m]$ as the ratio of the intensity of the wave $[W \cdot m^{-2}]$ reflected from the target toward the receiver at a distance from its center and an intensity of the acoustic wave $[W \cdot m^{-2}]$ incident on the object from the receiver

NOTES

1 If the reaction reaches equilibrium, it can be characterized by the equilibrium constant, which depends only on the type of reaction and temperature.
2 One mole of the substance occurs when the number of particles is equal to the number of atoms contained in the carbon isotope ^{12}C with mass 0.012 kg.
3 This term is used here to emphasize that nitrox used in a diving apparatus must be premixed, seasoned, tested, certified, and operationally tested before use.
4 In normal tissues, dissolved nitrogen is in equilibrium with atmospheric air, but during the initial phase of diving, nitrogen is purged from the tissues, causing delay in the dissolution of nitrogen in the tissues during the phase of nitrox diving, thus allowing a slight shortening of the decompression process without increasing the risk of DCS.
5 Free valences.
6 Received, for example, via dimensional analysis.
7 Algebraic model thus obtained is not adequate when you change the units of measurement, and due to the approximate nature of the relationship (consisting of approximately a function of the unknown laws of physics), semi-empirical and empirical models are valid only to the extent of their determination and allow in this field interpolation, and every extrapolation is always risky.
8 There are several kinds of concentration expressed using a fraction or percentage concentration: mole (normal), mass, volume, etc.

List of More Important Acronyms and Abbreviations

Item	Description
CCR SCUBA	closed circuit rebreather self-contained breathing apparatus
CNSyn	oxygen toxicity impact on the central nervous system, called the Paul Bert effect; acronym *CNS*, for *central nervous syndrome*, also used; for the purpose of this book, to distinguish it from the acronym used to define the central nervous system (*CNS*), acronym *CNSyn* will be used
CO_2	carbon dioxide
DCS	decompression sickness
fsw	feet seawater
HBO	hyperbaric medicine (*HBOT*: hyperbaric oxygen therapy); the use of pressurized oxygen for medical purposes; in addition to treatment of decompression sickness pressure, also a good method for the treatment of gangrene, diabetic foot, exhaust gases, and carbon monoxide poisoning; helps in the healing of wounds after frostbite and transplant of skin; also reported for nerve tissue regeneration, especially auditory nerve
IR	infrared
MoD	Ministry of Defense
Nx	*Nitrox*: nitrogen–oxygen breathing mixtures
NOAA	National Oceanographic and Atmosphere Administration
OTT	oxygen tolerance test
pH	the exponent of multiplicity of hydrogen ions,[1] which is regarded as a physical quantity and a quantitative measure of the activity of hydrogen ions in solution
ppm	parts per million
premix	a fixed-content gas mixture stored in supply bottles integrated with a self-contained diving apparatus or delivered via a supply hose
SEV	surface equivalent value
SCR SCUBA	semi-closed self-contained breathing apparatus
SCUBA	self-contained breathing apparatus
STANAG	NATO standardization agreement
STP	standard temperature and pressure: $T = 273\,K$ and $p = 101.325\,kPa$
TT	treatment table
Tx	*Trimix*: helium–nitrogen–oxygen breathing gas mixture
UBA	underwater breathing apparatus

NOTE

1 Negative logarithm of the hydrogen ion activity; pH of water and neutral solution: pH = 7.0, acidic solutions: pH <7; alkaline: pH >7.

List of Main Symbols

Item	Description
$a,b,c,A,B,C\ldots$	constant values
α	limit probability, or the level of significance
$\forall$	universal quantification (for all . . .)
$\exists$	existential quantification (there exists . . .)
∞	infinity
$\triangleq$	corresponds to or proportional to
$\gg$	considerably greater ($\ll$ means considerably lesser)
$const$	constant
δ	relative value
Δ	difference
$\frac{\partial}{\partial x}$	partial derivative with respect to x
exp	exponent
ε	breathing module
ε_k	apparatus design module
ε_p	pressure module
$\int f(x)\cdot dx$	indefinite integral whose derivative is the given function $f(x)$ with respect to x
$f(x)$	function f of the variable x
H	depth of water
$idem$	the same
L	level of confidence
ln	natural logarithm
$\dot{m}$	mass flow rate
$(N\ n)$	Newton's symbol; the number of n combinations from a given set of N elements (N choose n)
n_X	number of moles of substance X
p	pressure
P	probability
p_X	partial pressure of substance X
π	half of the circumference of the elementary radius
R	universal gas constant
t	temperature in the centigrade scale
τ	time
T	absolute thermodynamic temperature
$\sum$	sum
ω	angular frequency
W	work

(*Continued*)

(Continued)

Item	Description
V	volume
$\dot{V}$	volumetric flow rate
$\dot{v}$	volumetric rate of oxygen consumption
$x_n\ (X)$	mole fraction of substance X
$x_v\ (X)$	volume fraction of substance X

1 Introduction

The history of the special underwater breathing apparatus *UBA* – often known as semi–closed circuit rebreather *SCR*, or closed circuit rebreather *CCR*, goes back to Fleuss's design from 1879. The *oxy-CCR* designed by Fleuss was the first *CCR*. It was immediately put into practical use as the first combat, engineering, and recreational underwater breathing apparatus *UBA*. Further, *CCRs* were supplied by gas mixtures too.

1.1 HISTORY

The practical use of *SCRs* started in 1915, when Drägerwerk AG Lübeck introduced the *DM* 40 *UBA* (Figure 1.1).

The operational depth limit of the *DM* 40 *UBA* was $40\,mH_2O$. It was supplied from three gas cylinders connected to the classic diving helmet. One of the gas cylinders was filled with air, and the second one with pure oxygen. These two gases were mixed in the third gas cylinder that was situated between the other two – the breathing gas mixture was prepared during diving by dynamic gas mixing and created a nitrogen–oxygen breathing gas mixture *Nx*.[1] *SCRs* were used during World War II by the Netherlands, Norway, Belgium, Denmark, and the United Kingdom.

In 1943, the British developed a self-contained underwater breathing apparatus *SCUBA*[2] similar to the *DM* 40 (Figure 1.2). Its bottles were filled with the nitrogen–oxygen breathing gas mixture *Nx*, prepared before diving[3] (Donald, 1992).

In the 1960s, a lot of efforts were made to develop *SCRs*, especially by Drägerwerk AG Lübeck, which built two types of *SCRs*: *SCR FGG III UBA*[4] and *SCR FGT I / D SCUBA*[5] (Haux, 1968; Figure 1.3). Some of the *SCRs* developed in the 1960s have never been used in practice, for example, *SCR SMS I UBA* (Figure 1.4).

Other countries worked on other types of *UBA*, for example, *SCR Fenzy* – 68 *SCUBA* developed in France (Figure 1.5; Przylipiak, 1971).

Following the rapid development of open circuit *UBA* in the 1980s, it seemed that designers had forgotten *SCRs*. Only some of them continued to work on improving these types of *SCRs*, for example, the French have improved the types *SCR DC* – 55 *SCUBA* to *SCR CRABE SCUBA*, and Sweden, alternatively closed and semi–closed circuit ACSC type *SCR / CCR AGA SCUBA* to *SCR IS – mix SCUBA*, respectively (Figure 1.6 and Figure 1.7).

In the 1970s in Poland, work on new *SCRs* was begun. Two types of *SCRs* were designed: *SCR APW* – 3 *Czajka SCUBA*, *APW* – 6 *Kszyk* and *APW* – 6*M SCUBA* (Figure 1.8). Although they have never been put into use, they were the only *SCR* experimental preproduction batch.

In the 1990s, the *SCRs* philosophy was revived, and *UBAs* began to be developed again. Now it seems that it is their second youth. These types of *UBAs* are used in

DOI: 10.1201/9781032649207-1

FIGURE 1.1 *DM* 40 *UBA* manufactured by Drägerwerk AG Lübeck.

Source: Kłos (2008a).

FIGURE 1.2 Davis–Hill $Nx - SCUBA$ designed in 1943.

Source: Donald (1992).

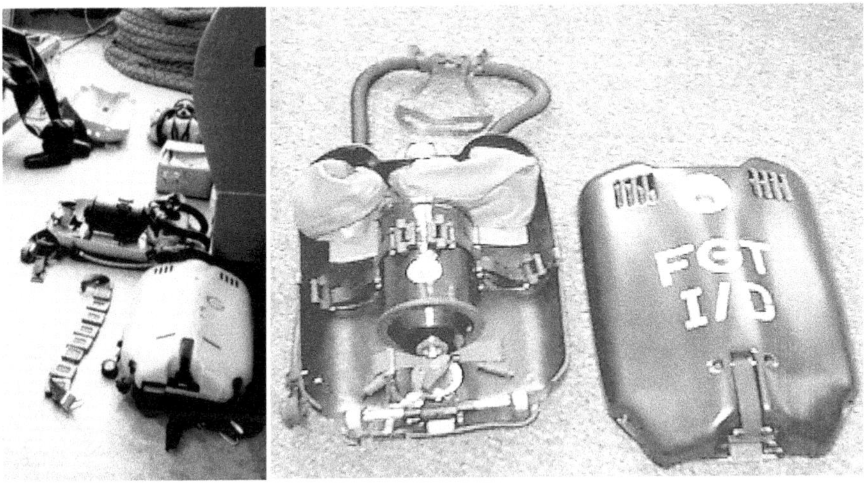

FIGURE 1.3 Examples of Drägerwerk AG Lübeck *SCRs*: *left*, *FGG III UBA* being prepared for diving; *right*, *FGT I / D SCUBA*.

Source: Kłos (2000).

FIGURE 1.4 Diving apparatus type *SCR SMS I UBA*.

Source: Technical information from Drägerwerk AG Lübeck.

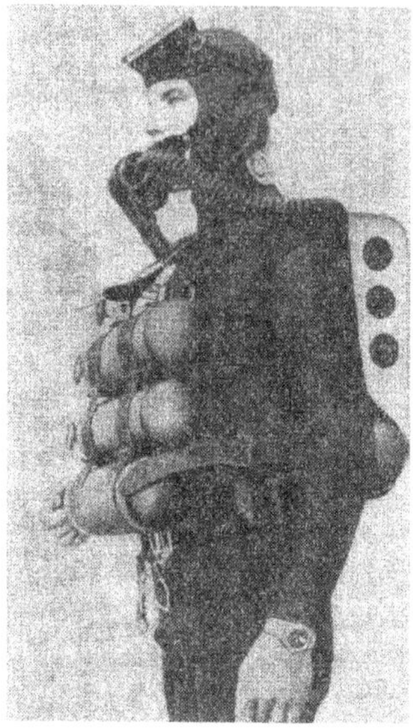

FIGURE 1.5 Diving apparatus type *SCR Fenzy – 68 SCUBA*.

Source: Przylipiak (1971).

FIGURE 1.6 Diving apparatuses: *top*, type *SCR DC – 55 SCUBA*; *bottom*, type *SCR CRABE SCUBA*.

Source: *Top*, Kłos (2008b); *bottom*, author's own photos.

FIGURE 1.7 Diving apparatuses: *left*, type *SCR / CCR ACSC AGA SCUBA*; *right*, type *SCR IS – MIX SCUBA*.

Source: Servicing instruction Interspiro AB.

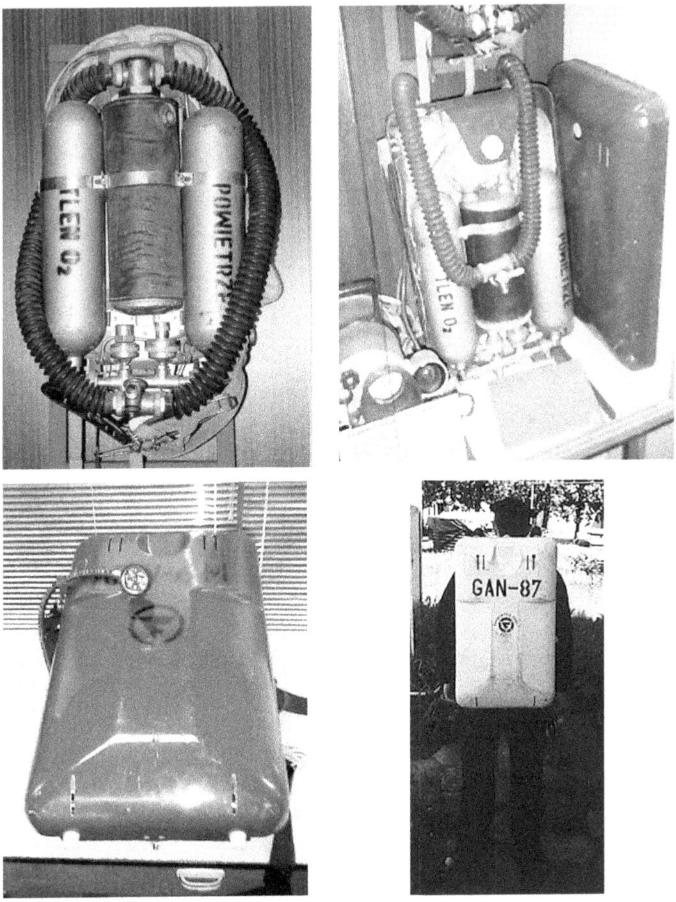

FIGURE 1.8 Diving apparatuses: *top left*, type *SCR APW – 3 Czajka SCUBA*; *top right*, type *SCR APW – 6 Kszyk SCUBA*; *bottom left*, type *SCR APW – 6M SCUBA*'; *bottom right*, type *SCR GAN UBA*.

Source: Kłos (2008b).

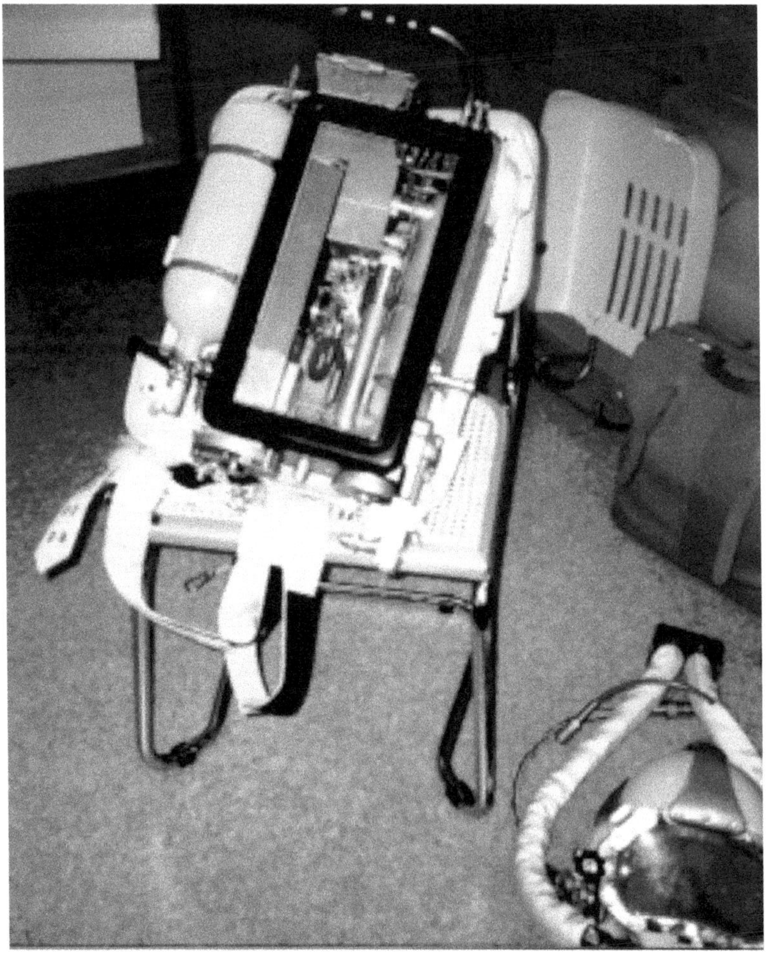

FIGURE 1.9 Diving apparatus type $CCR OBS-90-400 UBA$.

Source: Kłos (2000).

deep-saturation diving systems, for example, $OBS-UBA-90-400-$ (Figure 1.9; Ottestad, 1992).

Information concerning the rapid increase in application of *SCRs* for recreational diving can be found on the internet.

1.2 POLISH SCR PERSPECTIVE

In the early 1970s, special *SCRs*-focused research was commenced at the Polish Naval Academy, the Diving Gear and Underwater Work Technology Department, and the Military Medical Academy's Marine and Tropical Medicine Department. The goal of the research was to develop and then introduce special *SCRs* to the Polish Navy.[6]

It was assumed that the *FASER* company, located at Tarnowskie Góry in Poland, would be the manufacturer of these special *SCRs* – (Figure 1.8). The investigations were carried out until the 1990s. As a result of the investigations carried out within the framework of the research, diving systems enabling $1h$ exposures at the operating depth of $120\,mH_2O$ on oxygen, nitrogen, and helium gas mixture Tx,[7] with use in diving apparatus type $SCR - GAN - 87\,UBA$, and $30mH_2O$ on oxygen and nitrogen gas mixture Nx, with use in diving apparatuses $SCR\,APW - 3\,Czajka\,SCUBA$, $SCR\,APW - 6\,Kszyk\,SCUBA$, and $SCR\,APW - 6M\,SCUBA$, have been developed and put into use.

The same deep diving system dedicated for $SCR - GAN - 87\,UBA$ has been used many times in the course of oil extraction and technical dives performed at oil fields exploited by the ORLEN SA, former PP Petrobaltic, Gdańsk, with the use of $SCR\,FGG\,III\,UBA$. The diving system was used to install underwater exploitation heads, the central oil field collector, the central manifold, and for the connection of the system by means of underwater flow lines (Figure 1.10).

The work on the new *SCRs* continued in the Polish Naval Academy up to the late 1990s. At that time, the Polish Naval Academy was accredited by the Polish Centre for Testing and Certification and registered in the United Europe testing laboratory, equipped with laboratory stands especially designed for the purpose of experimental investigations of *UBA* prototypes for various tasks.

1.3 FACILITIES

The Laboratory of Hyperbaric Technologies is a specialized research center of the Polish Naval Academy made available to the scientific community. The Hyperbaric Technologies laboratory was established as part of Poland's efforts to specialize in the field of diving technology in the times of COMECON[8] and has been used to implement new technologies of diving, diving equipment, and equipment for underwater works. The most important achievements in this area include the development and implementation of deep diving technology in the ORLEN SA, a Polish-made diving complex in the National Rescue Hyperbaric Centre in Gdynia, Redłowo,[9] among others.

In the initial period, the deep diving technology was intended to be used only for delivery and acceptance trials of diving complexes manufactured in Poland for drilling rigs operated by the former Soviet Union. It was at that time that the technology of saturation diving, unique in the world, using *Tx* as a breathing mix was developed. Later, the technology of operational deepwater *Tx* dives outside the saturation zone, used to provide support services for Polish offshore platforms, was developed. It was used to assemble the field for underwater oil extraction – it was the only operation in the world at a depth of approximately $80\,mH_2O$ performed without the use of the saturation technology.

In the times of the Polish People's Republic, Poland's own strategic defense technologies were developed as part of the defense policy pursued by the state. For this reason, advanced diving technologies[10] had to be, for the most part, supported with domestically produced equipment. The Polish Naval Academy was the place where

FIGURE 1.10 The Polish Navy supports drilling activity at the Baltic Sea: *top*, the central oil field manifold before its underwater installation; *bottom*, the Polish Navy diving vessel with equipment designed for deep dives.

Source: Kłos (2000).

tests were performed, and it developed guidelines used to design changes in the diving equipment. The *UBA* laboratory, established to carry out unmanned and manned experiments, is a specialized research center of the Polish Naval Academy. It was also used as a design laboratory for new diving equipment.

Many solutions have been implemented, including several prototypes of the diving apparatus. After the production of domestic diving apparatuses was suspended, the Hyperbaric Technologies laboratory began to carry out tasks to implement and integrate foreign equipment with the domestic system. Currently, the Polish Naval Academy is recognized by the Polish Register of Shipping, and it is also recognized by the Polish Accreditation Centre and notified in the European Union as a certification body in the field of testing for compliance with Directive 89/686/EEC on personal protective equipment[11] and is designated for the certification of diving equipment under defense and security.

The Polish Naval Academy also has its own capacity to produce prototypes and models and is capable of experimental small-lot production. The most important achievements in this area include life-saving systems for hyperbaric[12] and normobaric[13] facilities, experimental diving equipment as well as research and measuring equipment, containerized hyperbaric systems for rescue of submarine crews,[14] and so on (Figure 1.11).

The experiments concerning mathematical ventilation models were performed during manned and unmanned experimental diving aimed at the verification of the

FIGURE 1.11 Containerized mobile hyperbaric system under the code name SERCÓWKA.

Source: Author's own photo.

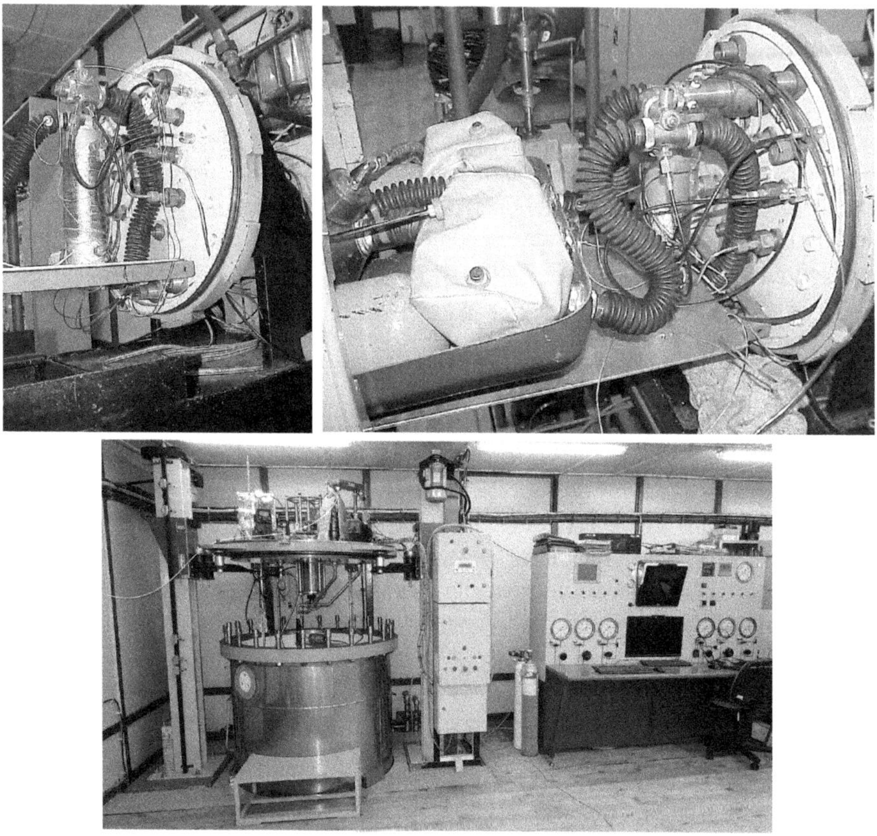

FIGURE 1.12 Stand of combined simulators – respiratory, metabolic, and hyperbaric – at the Polish Naval Academy: *upper left*, the old one metabolic reactor; *upper right*, the diving apparatus in the hyperbaric chamber connected to a metabolic simulator and the respiratory simulator visible behind the hyperbaric chamber; *down*, the newest complex of simulators, which includes respiratory, metabolic, and hyperbaric simulators.

Source: Author's own photos.

decompression procedure proposed. The experiments were carried out for diving apparatuses. The tested diving apparatuses should feature the same parameters as the parameters assumed to derive the mathematical model. It is the main condition necessary to perform experiments with a metabolic simulator (Figure 1.12; Kłos, 2021).

The adequacy of the assumed mathematical model at limits of oxygen partial pressure is tested during pressure tests with the use of diving apparatuses. The experiments were carried out in parallel with verification experiments of the decompression procedure based on the experiments carried out in the hyperbaric swimming simulator (Kłos, 2019, 2021; Figure 1.13).

There were more similar trials carried out, but the system was modified. That is why the results obtained are not sufficient to draw conclusions concerning the

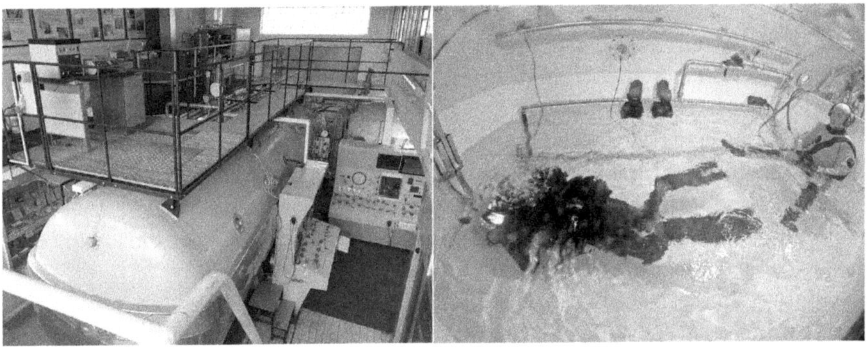

FIGURE 1.13 The land-based hyperbaric diving complex *DGKN* – 120 at the Polish Naval Academy: *left*, main view; *right*, the hyperbaric swimming simulator inside the hyperbaric complex.

Source: Author's own photos.

adequacy of the mathematical ventilation model. The probability of the proposed mathematical model's acceptance was not further maximized because of the high experimental costs.

The adequacy of the assumed mathematical model at limits of oxygen partial pressures is tested during pressure tests with the use of diving apparatuses. The experiments were carried out in parallel with verification experiments of the decompression procedure based on the experiments carried out in the hyperbaric swimming simulator (Kłos, 2019, 2021; Figure 1.13).

There were more similar trials carried out, but the system was modified. That is why the results obtained are not sufficient to draw conclusions concerning the adequacy of the mathematical ventilation model. The probability of the proposed mathematical model's acceptance was not further maximized because of the high experimental costs.

It seems that the best method for experimentation is to apply a breathing machine simulating a breathing action with simultaneous oxygen sampling from the breathing space and carbon dioxide emission for unmanned tests (Figure 1.12). As in this case the carbon dioxide emission is maintained at the proper level, there is no necessity to carry out manned experiments, and the experimental results are repeatable with known accuracy of reproduction. It can be used to carry out less-expensive experiments and get the results without the necessity to carry out manned experiments.

The laboratory of hyperbaric chambers with the experimental land-based diving complex type *DGKN* – 120 is the base for experimental research (Figure 1.13). It also includes facilities for manned and unmanned tests of the diving equipment. The land-based complex was used in diving experiments with human subjects lasting up to four weeks. During the experiments, two subjects were placed inside an ecologically confined atmosphere complex. In order to carry out long-duration experiments with the subjects, the diving complex is suitably equipped and includes gas analyzers supplied by Polish and worldwide manufacturers.

The diving complex atmosphere can be modified[15] and regenerated according to the specific requirements. The diving complex is equipped with a swimming pool having water temperature regulation (Figure 1.13, *right*). It can be used to simulate natural diving conditions.

The diving complex is owned by the Polish Naval Academy, and its personnel are highly qualified and hold certificates issued by the Main Marine Board, authorizing them to perform the technical inspections, high-pressure operations, and diving supervision of activities appointed. The Polish Naval Academy personnel are certified to carry out medical experiments with human subjects.

For over 25 years, the physiochemical laboratory has provided support for experiments with the participation of people in a hyperbaric ecologically closed atmosphere and for research on diving equipment. The laboratory is equipped with specialist equipment, inter alia, for precise measurements of the composition of breathing mixtures (Figure 1.14). The Polish Naval Academy employs specialists trained in the field of metrology. The operating personnel have the appropriate technical qualifications.

Since the 1960s, the breathing mix production laboratory has made breathing mixes used in diving works carried out by the Polish Navy and ORLEN SA. The breathing mix production laboratory is equipped with appropriate stations (Figure 1.15). The operating personnel have the appropriate qualifications, both technical[16] and specialist.[17]

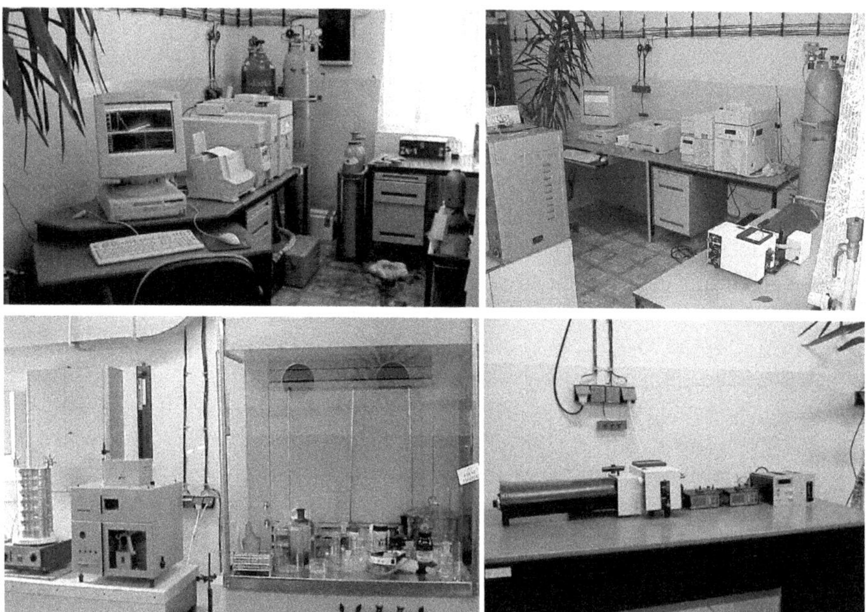

FIGURE 1.14 Examples of laboratory equipment at the Polish Naval Academy: *top left and right*, gas chromatograph stand with mass spectrum detector; *bottom left*, laboratory stands; *bottom right*, IR spectrometer and electrochemical and paramagnetic gas analyzers.

Source: Author's own photos.

FIGURE 1.15 The breathing mix production laboratory at the Polish Naval Academy: *top left*, stand designed for obtaining small amounts of breathing mixtures; *top right*, control panel of the stand designed for obtaining large amounts of breathing mixtures; *bottom left*, a high-efficiency breathing gas pumping stand; *bottom right*, pressure tanks for breathing mixtures and clean gases.

Source: Author's own photos.

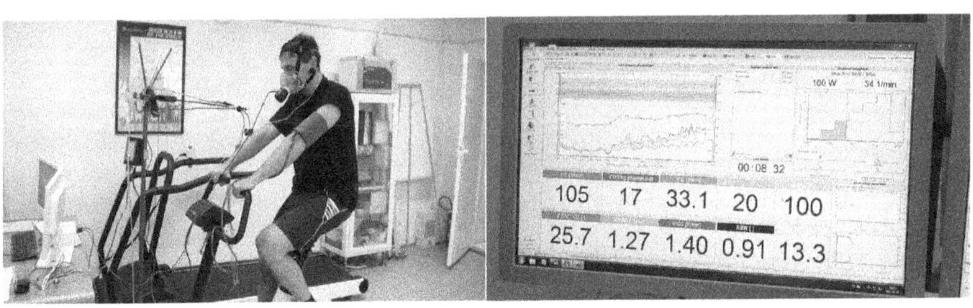

FIGURE 1.16 The stress laboratory at the Polish Naval Academy: *left*, a cyclo-ergometer stand for stress tests, behind which is a treadmill stand for stress tests; *right*, a parameter monitor screen during test (ECG parameters are displayed on a separate screen).

Source: Author's own photos.

The stress stand is equipped with measuring equipment designed to assess training based on the direct measurement of oxygen consumption in relation to spirometry parameters[18] and physiological parameters[19] or, in a simplified manner, based on the PWC_{170} test[20] stations (Figure 1.16).

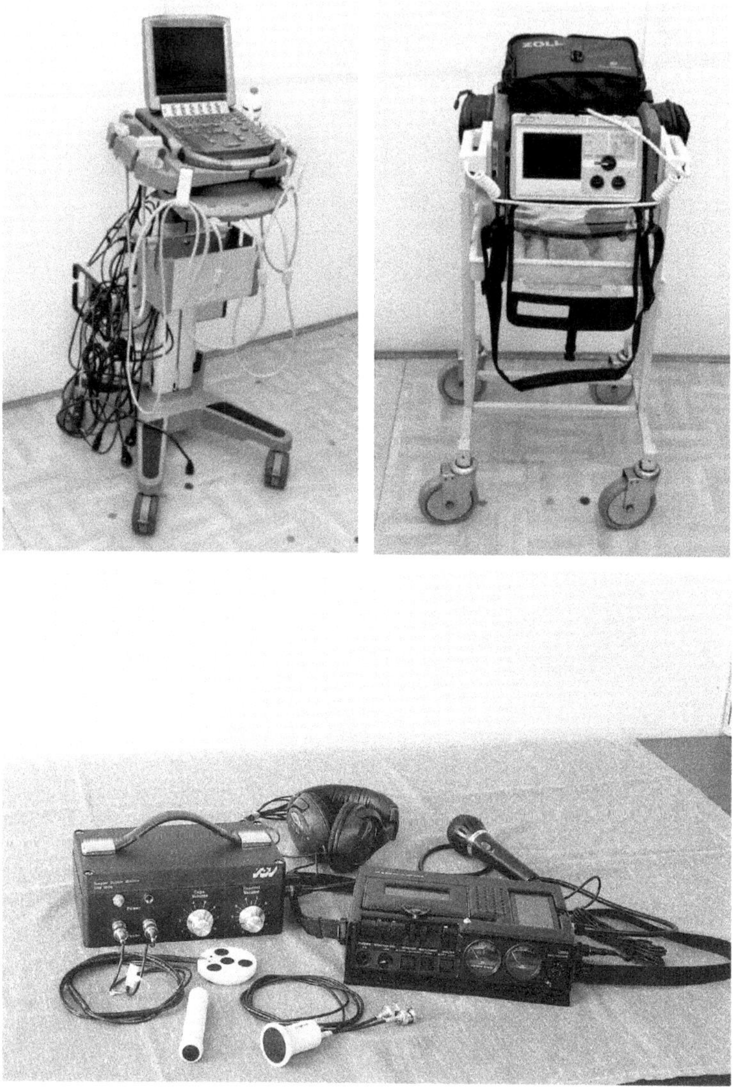

FIGURE 1.17 The stress laboratory at the Polish Naval Academy: *top left*, ultrasound scanner; *top right*, defibrillator; *down*, free gas phase in veins detector: Doppler Bubble Monitor.

Source: Author's own photos.

The laboratory is also equipped with devices for emergency conditions[21] and free gas phase detection[22] (Figure 1.17).

The Naval Academy has implemented a quality system for research and development projects carried out at the Maritime Technology Centre, which is a specialized, licensed unit of the Naval Academy. The center was established to carry out projects related to the country's defense needs, and a management control system has been implemented at the academy. The following are chosen completed diving projects:

1. Title: Validation of decompression tables/schedules
   No: NATO Standardization Agency
   Realization time: 2003–2005
   Sponsor: Statutory annual budget
   Place: The Naval Academy/Defense and Civil Institute of Environmental Medicine, Toronto
   Role: Leader, primary contractor
   Budget: PLN 0.1M (actual rate ca. €22k)
2. Title: Containerized rescue and treatment hyperbaric system
   No: DPZ/4/TM/W/3.5.3.1/2004/WR
   Realization time: 2004–2008
   Sponsor: Armament Policy Department, Ministry of Defense
   Place: The Naval Academy
   Role: Leader, primary contractor
   Budget: PLN 1.8M (actual rate, ca. €0.4M)
3. Title: Saturation diving methodology
   No: R00-O0014/3
   Realization time: 2007–2009
   Sponsor: National Research and Development Centre, Ministry of Scientific Research and Higher Education
   Place: The Naval Academy
   Role: Leader, primary contractor
   Publications:
   Budget: PLN 2.98M (actual rate, ca. €0.66M)
4. Title: Planning of decompression schedules for combat diving mission
   No: O R00 001 08
   Realization time: first, 2009–2011
   Sponsor: National Research and Development Centre, Ministry of Scientific Research and Higher Education
   Place: The Naval Academy
   Role: Leader, primary contractor
   Budget: PLN 2.5M (actual rate, ca. €0.55M)
5. Title: Detection and countermeasure against terrorist divers
   No: O R 0000 98 1
   Realization time: 2010–2012
   Sponsor: National Research and Development Centre, Ministry of Scientific Research and Higher Education
   Place: The Naval Academy
   Role: Primary contractor
   Budget: PLN 0.9M (actual rate, ca. €0.2M)
6. Title: Planning of decompression schedules for MCM divers
   No: O ROB 0047 03 001
   Realization time: first, 2012–2015
   Sponsor: National Research and Development Centre, Ministry of Scientific Research and Higher Education
   Place: The Naval Academy

Role: Leader, primary contractor
Budget: PLN 2.8M (actual rate, ca. €0.6M)

7. Title: Planning of decompression schedules for MCM divers II
No: DOB-BI08/09/01/2016
Realization time: first, 2016–2021
Sponsor: National Research and Development Centre, Ministry of Scientific Research and Higher Education
Place: The Naval Academy
Role: Leader, primary contractor
Budget: PLN 6.0M (actual rate, ca. €1.3M)

1.4 PRESENTED MATERIAL

The aim of the book concerns the main tasks specified in Table 1.1.

The investigations on ventilation in hyperbaric facilities, and particularly on diving apparatus operational features, are the main aim of exchanging scientific reports between the Naval Academy and the following:

- DRDC Toronto,[23] Canada
- Norwegian Underwater Technical Centre,[24] Ytre Låksevag, Norway
- US Navy Experimental Diving Unit, Panama City, USA

TABLE 1.1
The Objectives

Theme	Semi–Closed and Closed Circuit Diving Apparatuses
Cognitive tasks	• Literature review of the results of the research on selected operational features of special *SCRs* obtained worldwide. • Consultations with other countries on the achievements in the field of research of selected operational features of special *SCRs*. • Information exchange and its analysis with leading research centers dealing with the operational features of the special *SCRs*.
Operational tasks	• Elaboration of the Polish research methods used to investigate operational features of special *SCRs*. • Research on operational features of special *SCRs* focused on their usefulness in carrying out combat diving tasks by Polish Navy divers.
Educational tasks	• Elaboration of the early history of Polish achievements in the field of special *SCR* design and research.
Final standard	• Elaboration of the initial design for research on the operational features of special *SCRs*.
Methodical directions	• Research on selected operational features of *SCRs* should be based on domestic capabilities. Particularly, the existing experimental base of the diving gear and underwater work technology of the Naval Academy should be used and developed. The capabilities of the other Polish Navy laboratories and establishments should be employed as well. • The operational features of *SCRs* should be presented as a comprehensive collection so that possible uses of special *SCRs* in the Polish Navy can be determined.

This book consists of eight chapters. The beginning of each chapter offers a review of the available literature. This chapter analyzes research reports prepared in leading research diving centers. It has been assumed that such research reports are more complete when compared to books or articles and easier for additional interpretation. Moreover, these reports facilitate making opinions on the methods and equipment used, research reliability, completion of the research objectives, accuracy of experiments, analysis of complexity approach to the problem, laboratory expertise, metrology features of the used instruments, correctness of the calibration procedures, etc.

These features make research reports more convenient to use than generally accessible articles or set books. In most cases, the latter are controversial in the interpretation of experimental results. Often, this is related to the limitations imposed by the publisher, for example, related to the report volume. Most research reports are confidential – some of them are for internal use only. Access to such a report may not be useful, as its content is not user-oriented. Some limitations are also placed in order to ensure security or protection of inventions. Literature reviews emphasize the Polish achievements in the field of new underwater breathing apparatus *UBA* design and new diving technologies.

The successive chapters, concerning research investigations, deal with the theoretical fundamentals of measuring methods, their technical realization, elaboration, and a discussion of the results. Initially, the idea to provide the whole results of the research was considered. However, it was concluded that not all the research results obtained during a long period[25] might not be significant. The research difficulties are of a twofold nature.

Firstly, research is very expensive. The costs and problems characteristic of aeronautics research are similar; therefore, diving and aeronautics research centers try to collaborate. Sometimes, diving studies are financially supported within the framework of aviation projects. Owing to diverless technology, financial support provided for diving research decreased to an indispensable minimum.

As diving research is complex, interdisciplinary science, that is, the knowledge and technologies used, therefore requires large teams of specialists. Therefore, successful research would have been impossible without the cooperation of a large group of workers from the Polish Naval Academy, the Military Medical Institute,[26] and officers and civilians from Polish Navy units.

The second set of difficulties involves the necessity for experimental divers to participate in the experiments. However, according to the international conventions signed by Poland, approval for human subjects research is difficult to obtain. Approval for such research should be given by the responsible authorities, for example, in Poland by the Committee of Ethics in Scientific Research.

Chapter 1 presents the short world and Polish history of the special underwater breathing apparatus *UBA*, often known as semi–closed circuit rebreather *SCR* or closed circuit rebreather *CCR*. It also presents the Polish *UBA* laboratory established to carry out unmanned and manned experiments. The laboratory was also used to design new diving equipment. The chapter also presents the aim of the book and the main tasks.

Chapter 2 presents the general scheme of investigations related to the use of *UBA*s. The procedure described here has been verified for many years. The chapter includes generalizations and summaries of the investigations performed. The assumed final standard will be used to perform the tasks specified in this book. The procedures used to describe the basic operational features of a *UBA* can be employed to determine the *UBA*'s practical suitability and do the planning required to perform special and experimental dives.

Chapter 3 presents the characteristics, specification, and construction of *UBA*s. Most of the details were obtained during consultations with specialists and experts from foreign countries. As construction details of some *UBA*s are confidential and cannot be published, a detailed analysis of *UBA*s is not possible. The section on the development of *UBA*s allows collecting, systemizing, and analyzing the tendencies in the construction development of *UBA*s. Although this section is not perfect, it is still the most detailed description of the issues concerning this subject in the literature. Its practical usefulness has been confirmed in the course of systemizing and data analysis. As it follows from the analysis of tendencies in the development of *UBA*s based on the preceding section, interest in *SCR*s in Europe dates back to many years before the large *SCR* market emerged. Simultaneous application of three *UBA* division criteria was an innovative solution, because most specifications of *UBA*s presented in the literature are only criteria-dependent constructions.

Chapter 4 discusses the reasons for using artificial breathing media in special *UBA*s. It briefly describes the history of Polish work on special *SCR*s and the world's research on artificial breathing media. The book has also been revised to include recent Polish achievements. The role of some institutions and teams employed in the research has been accentuated here. It is worthy of note that it also contains a detailed section on the breathing media used in bathynautics.

The chapter also discusses the issues related to the preparation of gas mixtures used in diving. The problems concerning human reactions to gases in the hyperbaric environment are completely different as compared to those in hypobaric or normobaric environments. Human reaction to oxygen and gases known as inert in normobaric atmosphere and contaminants depends upon the partial pressure of these gases. The statement that oxygen, the basic component of all gaseous mixtures, is toxic in the hyperbaric environment is a surprise for common people. Some problems can occur during dives due to oxygen toxicity. All gases known as inert in the normal environment can be toxic in the hyperbaric environment. Also, the content of contaminants known as safe and permissible in the normobaric environment exceeds the threshold of toxicity limits in the hyperbaric environment, and so they are not allowed in the hyperbaric environment. As it follows from the previous, the breathing medium for deep diving should be of special purity and especially prepared. Its composition should be carefully checked. Special measuring instruments and procedures ensuring high measurement accuracy should be employed. Due to fire hazard, special precautions to ensure oxygen purity for all technical equipment should be taken whenever gas mixtures are used. In most situations, gas mixtures used for deep diving are hypoxic in normal conditions and can be lethal for a human being if breathed on the surface.

Moreover, Chapter 4 discusses the selection of methods used to measure gas mixture composition. It also deals with the issues related to gas composition and pollution contents. The measuring methods were experimentally verified during experimental saturation dives and on research on the diving systems' prototypes. The chapter describes the methods used to prepare gas mixtures. Additionally, it includes some practical recommendations based on the experiments conducted. It refers to the verified special methods developed for preparing gas mixtures, using prepared gas mixtures, using distribution installations and recommended consumables. Finally, it deals with fire hazard issues. The special *UBA*s are characterized with closed or semi-closed breathing loop. As it follows from the preceding, the breathing gas must be supplemented with oxygen and regenerated because of the occurrence of contaminants.

Chapter 5 introduces the mathematical models for ventilation of breathing space in different types of *SCR*s. The mathematical models presented fulfil the boundary conditions and take into account the influence of partial errors of the measured component values on the final value. The problem of breathing space ventilation in breathing apparatuses is inherent in decompression safety and generally in diving work. For this reason, it is the basic utilitarian feature of *SCR*. The other parameters having effect on diving safety are stabilization rate and flushing effectiveness of the ecologically confined or semiconfined loop in breathing apparatuses.

Chapter 6 deals with carbon dioxide removal. It is a very important problem due to the protection time of *SCR*s and *CCR*s. The breathing loop volume is small; therefore, variation of gas mixture composition occurs rapidly. In one minute, the composition of the breathing gas may suddenly change, leading even to a diver's death.

Chapter 7 describes the measuring methods used in the most important operational parameters of special *UBA*s.

1.5 ACKNOWLEDGMENT

This scientific monograph is the result of work and research programs in which the author was a participant:

- Evaluation of diving apparatuses – research sponsored by the Polish Navy (1989–1990)
- Legal, technical, and medical problems of saturation dives – research sponsored by the Polish government's central R&D project purpose: 35 carried out in 1989–1991
- Diving systems for Polish shelf underwater work – research sponsored by the State Committee for Research project no. 9937794 C/1993, carried out in 1993–1994
- Principles of interrupted and continuous diving complex ventilation during air expositions – research sponsored by the State Committee for Research project no. 7 T07C 034 19, carried out in 2001–2002
- Detecting and counteracting terrorist underwater threats performed by divers – research sponsored by the State Committee for Research project no. O R 0000 98 1, carried out in 2010–2012

As well as projects conducted by the author:

- Diving apparatus for special forces – research sponsored by the Polish Navy, carried out in 1989–1992 (Kłos, 2000)
- Evaluation of the semi–closed circuit diving apparatus for deep diving – research sponsored by the Polish Navy, carried out in 1990–1992 (Kłos, 2000)
- Operational examinations of diving equipment – research sponsored by the Polish Navy, carried out in 1993–1996
- Oxygen diving technology – research sponsored by the State Committee for Research project no. 148–101/C-T00/96, carried out in 1996–1998 (Kłos, 2000)
- Principles of ventilation in diving complexes during air expositions by means of oxygen inhalators for decompression and treatment – project sponsored by the Armament Policy Department, Ministry of Defense, carried out in 1998 (Kłos, 2021)
- New generation of analytical sensors for submarine atmosphere monitoring – purposeful project sponsored by the Polish Scientific Research Committee contract no. 148 181/C-T00/98, carried out in 1998–2001 (Kłos, 2008c)
- The mathematical models of diving apparatus atmosphere ventilation with partial regeneration of the breathing medium – research sponsored by the State Committee for Research project no. 0T00A 072 18, carried out in 2000–2002 (Kłos, 2007, 2021)
- Submarine life support systems – project sponsored by the Armament Policy Department, Ministry of Defense, carried out in 2003–2004
- Submarine rescue and therapeutic containerized hyperbaric set – project no. DPZ/4/TM/W/3.5.3.1/2004/WR, sponsored by the Armament Policy Department, Ministry of Defense, carried out in 2004–2008 (Kłos, 2008c)
- Chemisorption of carbon dioxide in military applications – project no. 148–414/C-T00/2004, sponsored by the State Committee for Research, carried out in 2004–2008 (Kłos and Kłos, 2004; Kłos, 2008c, 2009)
- Protection of miners through the use of escape chambers in the event a harmful breathing atmosphere appears – project sponsored by KGHM Polska Miedź SA in 2007–2008 (Kłos, 2021)
- The method of saturation dives – project no. R00-O0014/3, sponsored by the State Committee for Research, carried out in 2007–2009 (Kłos, 2014)
- A new generation of breathing simulators – grant no. O N504 497734, sponsored by the State Committee for Research, carried out in 2007–2010
- Decompression design of in-combat missions – project no. O R00 001 08, sponsored by the State Committee for Research, carried out in 2009–2011 (Kłos, 2011, 2012, 2023)
- Decompression design for MCM dives – project agreement no. DOBR/0047/R/ID1/2012/03, sponsored by the State Committee for Research, carried out in 2012–2015 (Kłos, 2016)
- Decompression design for MCM/EOD dives II – project agreement no. DOB-BIO8/09/01/2016, sponsored by the State Committee for Research

1.6 THANKS

Apart from financing, the presented research required access to unique combat equipment. I express my gratitude to the Navy Command of the Republic of Poland and the 3rd Ship Flotilla for the trust that I received.

The research included experiments on humans, which required consent of the Scientific Research Ethics Committee. In addition to appreciation to the predecessors, gratitude must be expressed for the contribution given by the research staff of the department and the cooperating medical doctors from the Military Medical Institute. Of them, Cmdr. Maciej Konarski, PhD Med, has significantly contributed to assembling the experimental data described in this monograph.

Thanks to the confidence demonstrated by the divers and professionals from military units, police, anti-terrorist groups, and others who have supported this work, and owing to the support rendered by the Ministry of Defense, the Ministry of Science and Higher Education, and the National Research and Development, who provided a considerable amount of funds invested in research carried out in the department, the desired and expected results have been attained.

In particular, I would like to express my gratitude to prof. Anna Mjchrzycka, PhD, and Kazimierz Szczepański, PhD, who reviewed the entire manuscript.

NOTES

1. *Nitrox*.
2. The Davis–Hill unit.
3. Premix.
4. For underwater work to maximum operational depth of $200\,mH_2O$.
5. For mine countermeasure operations – *MCM SCR* to maximum operational depth of $50\,mH_2O$.
6. Mine countermeasure (*MCM*) units.
7. Trimix.
8. The Council for Mutual Economic Assistance – Russian: Совéт Экономи́ческой Взаимопо́мощи, СЭВ.
9. Now the property of Gdańsk Medical University.
10. Like the already-mentioned *Tx* deep-diving technology.
11. Diving apparatuses, diving suits, buoyancy compensators, etc.
12. For example, decompression chambers, oxybarotherapy chambers, diving bells, etc.
13. For example, for submarines.
14. Containerized mobile hyperbaric system under the code name SERCÓWKA.
15. The composition of the atmosphere, humidity, and temperature.
16. Authorizations of the technical inspection staff for servicing pressure equipment; authorizations of the technical inspection staff for the operating pressure and energy equipment.
17. Authorizations of diving works managers issued by the Maritime Office.
18. For example, lung vital capacity, first second forced expiratory volume, forced lung vital capacity, peak expiratory flow, forced expiratory flow calculated between 25 and 75% of lung vital capacity, etc.
19. For example, heart rate, blood pressure, blood oxygen saturation, other parameters of heart function based on 12-channel ECG, etc.
20. Load measurement at which the heart rate of the respondents during the PWC_{170} test reaches the value of $HR = 170\ min^{-1}$.

21 Such as defibrillator, ECG, blood oxygen saturation device, and medical equipment for first aid.
22 Ultrasonic portable detectors and a laboratory-clinical ultrasound stand.
23 Formerly Defence and Civil Institute of Environmental Medicine.
24 Formerly Norwegian Underwater Institute.
25 Even that not presented here.
26 Formerly the Military Medical Academy.

REFERENCES

Donald K.W. 1992. *Oxygen and the diver*. Harley Swan: The SPA Ltd. ISBN: 1-85421-176-5.

Haux G. 1968. *Diving technics*. Lübeck: Drägerwerk AG.

Kłos I., Kłos R. 2004. *Polish soda lime in military applications*. Oświęcim: Chemical Company DWORY S.A. ISBN: 83-920272-0-5.

Kłos R. 2000. *Aparaty Nurkowe z regeneracją czynnika oddechowego*. Poznań: COOPgraf. ISBN: 83-909187-2-2.

Kłos R. 2007. *Mathematical modelling of the normobaric and hyperbaric facilities ventilation*. Gdynia: Wydawnictwo Polskiego Towarzystwa Medycyny i Techniki Hiperbarycznej. ISBN: 978-83-924989-0-2.

Kłos R. 2008a. Classification of the underwater diving equipment. *Polish Maritime Research*. 55, 80–85. DOI: 10.2478/v10012-007-0056-3.

Kłos R. 2008b. Principles of work of different types of underwater breathing apparatus. *Polish Maritime Research*. 58, 72–84. DOI: 10.2478/v10012-007-0100-3.

Kłos R. 2008c. *Systemy podtrzymania życia na okręcie podwodnym*. Gdynia: Polskie Towarzystwo Medycyny i Techniki Hiperbarycznej. ISBN: 978-83-924989-4-0.

Kłos R. 2009. *Wapno sodowane w zastosowaniach wojskowych*. Gdynia: Polskie Towarzystwo Medycyny i Techniki Hiperbarycznej. ISBN: 978-83-924989-5-7.

Kłos R. 2011. *Możliwości doboru dekompresji dla aparatu nurkowego typu CRABE*. Gdynia: Polskie Towarzystwo Medycyny i Techniki Hiperbarycznej. ISBN: 978-83-924989-4-0.

Kłos R. 2012. *Możliwości doboru ekspozycji tlenowo-nitroksowych dla aparatu nurkowego typu AMPHORA—założenia do nurkowań standardowych i eksperymentalnych*. Gdynia: Polskie Towarzystwo Medycyny i Techniki Hiperbarycznej. ISBN: 978-83-924989-8-8.

Kłos R. 2014. *Helioksowe nurkowania saturowane—podstawy teoretyczne do prowadzenia nurkowań i szkolenia*. Gdynia: Polskie Towarzystwo Medycyny i Techniki Hipernbarycznej. ISBN: 978-83-938322-1-7.

Kłos R. 2016. *System trymiksowej dekompresji dla aparatu nurkowego typu CRABE*. Gdynia: Polskie Towarzystwo Medycyny i Techniki Hiperbarycznej. ISBN: 978-83-938322-5-5.

Kłos R. 2019. Modelling of the nnormobaric and hyperbaric facilities ventilation. *International Journal of Mechanical Engineering and Applications*. 7, 26–33. DOI: 10.11648/j.ijmea.20190701.14.

Kłos R. 2021. *Ventilation of normobaric and hyperbaric objects*. Boca Raton, FL: CRC Press (Taylor & Francis Group, LLC). ISBN: 978-0-367-67523-3 (hbk); ISBN: 978-0-367-67524-0 (pbk); ISBN: 978-1-003-13164-9 (ebk).

Kłos R. 2023. *Oxygen diving*. Boca Raton, FL: CRC Press (Taylor & Francis Group, LLC). ISBN: 978-1-032-31389-4 (hbk); ISBN: 978-1-032-31390-0 (pbk); ISBN: 978-1-003-30950-5 (ebk). DOI: 10.1201/9781003309505.

Ottestad N.T. 1992. Pneumatically assisted semi-closed breathing apparatus. In *Lung physiology and divers' breathing apparatus*, ed. Flook V., Brubakk A.O., 419–494. Ballater: BPCC-AUP Aberdeen Limited. Proceedings from the International Workshop, 1–4 November 1991, Ballater, Scotland. ISBN-10: 8259571250; ISBN-13: 978-8259571250.

Przylipiak M. 1971. Sprzęt nurkowy i technika nurkowania. In *Prace podwodne*, ed. Praca zbiorowa (group work), 71–296. Gdańsk: Wydawnictwo Morskie.

2 Development

For many years, the Naval Academy carried out investigations of underwater breathing apparatuses *UBAs* and their possible uses in the Polish Navy. The investigations were financed by the Central Development Projects, the Scientific Research Committee, the Polish Navy, and industrial companies. The Ministry of Internal Affairs made use of the investigation results in post-accident procedures. Some professional, sport, and recreational diving organizations were involved as well. The investigations presented were used to devise an activity sequence plan for testing *UBAs*, with particular respect to regenerating *UBAs* and their application in diving technology. Under the proposed research on implementation of new *UBAs*, a sequence of activities to be undertaken[1] is devised. The aims of the implementation research on the new types of *UBAs* are as follows:

- To develop the initial preliminary design
- To evaluate the preliminary initial design through unmanned laboratory tests, manned normobaric laboratory tests, manned hyperbaric laboratory tests, and tests conducted at worksite
- To maintain engineering supervision of the *UBAs*

The aforementioned activities must be part of the implementation research on a diving system composed of the following:

- A decompression system
- Systems of decompression protection
- A diver's transport system
- Breathing mixture supply systems

The implementation research on new *UBAs* should progress simultaneously with the investigations on the decompression system. The other specified elements of the diving system can be tested separately. However, in some cases, a diving system will not include some elements at all. Hence, the technical and constructional activities in the implementation research on *UBAs* should be in conformity with safety of ascent, staying at depth, and first of all, further safe decompression of divers.

2.1 SYSTEM AND PROCESS

A specific[2] *system* consists of a reasonably minimal[3] set of elements together with the synergic[4] links between them, guaranteeing the implementation of the defined processes. *A process* is a series of actions occurring at a given time aimed at achieving a specific goal. The conditions influencing the process constitute its internal and external *context*.[5]

DOI: 10.1201/9781032649207-2

A *system* is a series of elements organized according to strictly defined rules. A *relationship* is any kind of dependency between two or more elements of a system. The system and relationships make up the structure of the system. A *process* is a set of activities distinguished in the system and arranged in causal order which occur in a specific time, constituting a study of development of something (Figure 2.1).

The basic[6] *system* is a reasonably minimal set of elements together with the synergic links between them, ensuring implementation of the defined *process*.

The concepts of *system* and *process* are fundamental in system engineering, which seeks synergic optimal solutions for technical and economic *systems*,[7] capable of undisturbed maintenance of established *processes* throughout the system's life cycle.

Systems engineering currently seems to be the best approach to solving scientific problems, both technology- and organization-related.[8] This is an analytical method and is therefore based on breaking down a problematic situation into its components. The starting point should be the basic *processes* that the created *system* must support. In the case of diving technology, these will be scenarios of diving operations. The risk analyses of such operations will form the basis for the development of a database of emergency scenarios. An analysis of the main and emergency *processes* of the adopted technology of diving operations can be used to develop an optimal engineering and organizational *system* in which the required *processes* can be safely carried out.

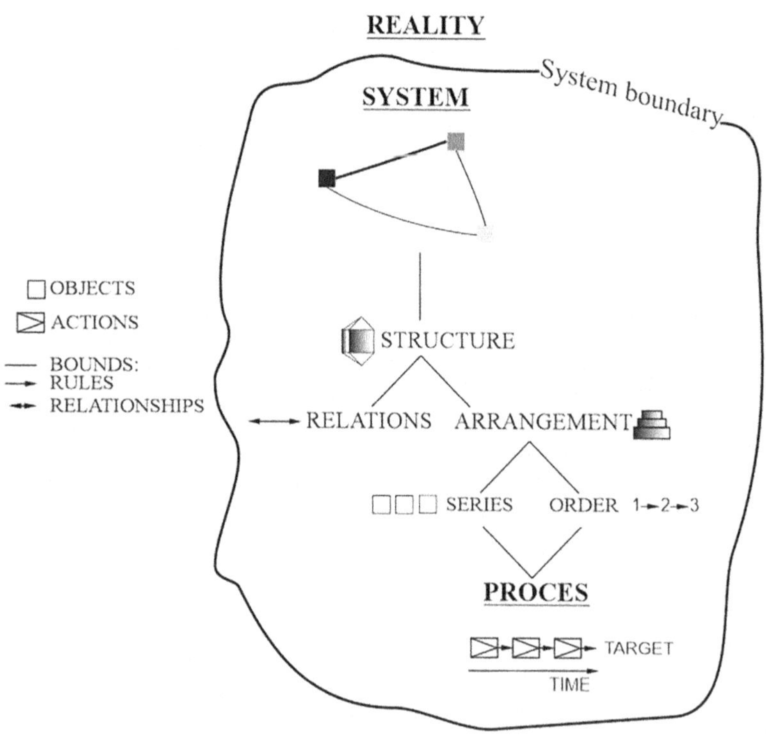

FIGURE 2.1 The *process–system* relationship.

2.1.1 Initial Design

Preliminary initial design of the *UBA* should be preceded with an answer concerning the scope of the diving *system*, the tasks and role of the *UBA*, and the external conditions that make up the context. There are several typical diving systems and technologies employed in underwater works. If the experience shows that effective diving activity is possible when a specified unique diving system is used, then the diving system could be designed[9] only for the project that is to be carried out (e.g., Bøe J. and Hartung, 1983). A solution applicable to small projects may not always be cost-effective. However, it is an unquestionable fact that there are very few standard diving systems in the world. Hence, the preceding premise indirectly proves that we have to adopt an individual approach if we need to have an effective diving system. An individual approach to each of the diving problems is needed especially in the case of extreme diving,[10] which requires specific knowledge and technical skills. The traditional elements of diving tasks are based on the depth of diving. The elements of the diving tasks proposed by the author are given in what follows. In most cases, the diving depths are conventional and depend upon the local obligatory standards.

The diving tasks used in this book are related the depth of diving. Diving operations, divided according to depth, can be as follows:

- Shallow-water diving: $12\,mH_2O$ to $6mH_2O-7mH_2O$ for oxygen
- Medium-depth diving: $12\,mH_2O-30\,mH_2O$
- Great-depth diving: $30\,mH_2O-50\,mH_2O$
- Typical large-depth diving: $50\,mH_2O-200\,mH_2O$
- Typical very large-depth diving: $200\,mH_2O-300\,mH_2O$
- Very large-depth diving: up to $300\,mH_2O-470\,mH_2O$
- Experimental diving at very large depth: $470-600\,mH_2O$
- Experimental breaking diving record at very large depth to $600+mH_2O$[11]

Typical air diving ranges are as follows: shallow-, medium-, and great-depth diving. The normal limit for shallow diving is based on the standard oxygen diving rules. The maximum limit is based on the use of air without exceeding the no-decompression limit. There is a great variety of diving systems.

For diving at great and large depths ($200\,mH_2O$ and $300\,mH_2O$), standard diving systems authorized by classification societies are most often used for the depth. Only a few diving companies have decided to carry out deep-sea diving operations without scientific personnel. In the most situations, deep-sea diving operations are performed for scientific research. The limit of experimental deep-sea operational dives has been determined in real experiments carried out within the framework of the *Janus IV Project* (Gussmann, 1984). The depth record was established within the framework of the *Hydra Project* in simulated saturation diving performed in the wet pot of the diving system[12] by Comex (Hydra 10 – a 701 msw onshore record dive using, 1993).

According to the diving tasks, the diving operations are subdivided into thus:

- Recreational dives
- Sport/technical dives

- Commercial dives
- Military dives
- Rescue dives
- Other dives, for example, experimental dives

Recreational dives are mainly air dives[13] at shallow depths. Shallow air diving is a hobby of groups of divers interested in amusement, tourism, recreation, etc. Decompression procedures are standard in recreational dives.[14] There are a lot of decompression tables[15] for recreational dives (Lippmann, 1990; Lippmann and Mitchell, 2009). In most cases, the differences between them are small. As a rule, no decompression is recommended by most diving federations, and every decompression dive schedule is treated as not standard[16] or emergency.[17]

Sport/technical dives are usually record-breaking ones. They are carried out without support from a diving installation. The world's deepest sport dive with an open circuit scuba is $332.35 mH_2O$ (1,090 ft). It was attempted by Ahmed Gabr in Dahab in the Red Sea on September 18/19, 2014, after nearly a decade of preparation. In sport dives, divers usually use mixed gas. Most often, *SCUBAs*[18] are adapted for the particular attempt. Like in experimental diving, special decompression procedures are individually elaborated almost for each of the diving operations.

Commercial dives involve most of the standard diving technologies. As regards diving technologies, diving procedures should cover a wide scope of the diving depth and emergency situations. Standard decompression procedures are applicable to commercial diving. However, each diving company has their own set of procedures.

Military dives are divided here into three groups: combat, *MCM*,[19] and rescue dives. Usually, combat dives are carried out with the use of $oxy-CCR$. Other *UBAs* can also be employed. In combat dives, the *UBA* is used for water infiltration and/or execution of combat tasks without enemy detection. For *MCM dives*, most often, special *UBAs*, which differ from open circuit apparatuses, are used. The diving apparatus tactical features are improved, owing to the minimized magnetic field, the *UBA* -generated noise in the increased gas dissipation rate. *MCM* dives are employed to perform clearance diving operations. Rescue divers carry out a broad scope of diving tasks, sometimes at a great depth, including both deep bounce and saturation diving. Military rescue dives can differ from normal rescue missions. The basic differences may include objects being rescued, but not exclusively. The following tasks can be performed in the course of rescue dives: recovering or destroying documents or equipment from sunken naval vessels to deny access to them by the enemy. In spite of the prevailing view, destroying own submarine in distress can be recognized as a rescue operation when a lifting operation is impossible to carry out, and the enemy should be denied access to the submarine's equipment. Special individual decompression procedures are adhered to in the course of combat dives.

UBAs offer three main functions in a diving system. They are as follows:

- Basic function
- Bailout
- Emergency

The basic function of the *UBA* is to provide breathing gas to the diver during routine work of the diving system. In order to perform that function, the *UBA* can operate as self-contained or surface-supplied. Before a dive begins, *UBAs* can be prepared for the particular kind of work. Usually, when the independent supply is selected, it cannot be changed during the dive.[20] When the breathing system is surface-supplied, only some elements of the *UBA* can be used, for example, carbon dioxide absorber, hoses, mouthpiece assembly, etc. After the external supply has been cut off[21] and self-contained breathing is activated, the *UBA* can work independently.[22]

The bailout function of the *UBA* is used when the basic *UBA* system fails, creating a hazardous situation.[23] In order to enable its immediate use, the *UBA* should be accessible to the diver.[24] A *Bail – out UBA* is usually a short functioning *SCUBA* that enables the diver to escape from the worksite to a place where the diver will be relatively safe.[25] It can be used in case of emergency.[26] Hose-supplied *UBAs* are used in most cases of underwater works. These *UBAs* also play the role of an independent *bail – out SCUBA* supplied with emergency breathing gas stored in an integral cylinder set. In many underwater systems, emergency gas can be supplied through the basic or emergency hose.[27] At present, small cylinders[28] with a mouthpiece assembly attached or a simplified *SCR* connected are used. The cylinders are filled up with compressed gas, most often air or nitrox.

Emergency *UBAs*/systems are different from the *bail – out UBAs*. The emergency function of the *UBA* is based on the long-lasting gas supply to the diver when the diving system fails.[29] Emergency *UBAs* double an emergency function of the basic breathing.[30] Owing to this, the diver can remain under extreme conditions, but as a rule, he/she cannot work.

2.1.2 Conclusion

The presented observations are a result of analyses of diving accidents[31] and of the experience earned in the course of developing new diving systems and modernizing old ones in response to orders placed by different organizations concerned. It is also a result of work done in different commissions involved in evaluating diving systems, as well as research on the problem referred to. The functions of *UBAs* proposed earlier can be used to design the whole system without disregarding its essential elements. The analysis of diving accidents provides a lot of evidence in support of the thesis that sometimes knowledge of people who are responsible for project/preparation of underwater work is not sufficient enough. Sometimes, divers' death is the highest price for the errors that have been made in the course of designing an underwater system.

Basic working parameters have been precisely specified for a new *UBA* in order to make its short initial design. As it follows from the earlier text, it is necessary to take into consideration very important ergonomic parameters and human factors as well. Some parameters concerning human factors are available in the literature, but most of them are usually obtained through experimental studies. Mathematical simulation of *UBA* functioning is often performed at this research stage. The simulation is based on mathematical models of decompression and breathing loop ventilation.[32] In order to determine the probability of *UBA* failure and reliability parameters, mathematical modelling is also performed (e.g., Stone, 1995).

As a rule, the procedures that have been mentioned earlier are not available in the literature because they are confidential. However, there are some companies which publish general information on their research and its final results.[33] For these reasons, at this stage of research, it is necessary to take full advantage of one's own research that has been already carried out. Some models and procedures used to design the *SCR* as well as constant dosage of breathing gas will be discussed further.

2.2 UNMANNED TESTS

This stage of experiments is relatively well described in world literature. There are two research centers in particular which have published research reports, for example:

- US Navy Experimental Diving Unit – NEDU Reports
- Norwegian Underwater Technical Centre, former Norwegian Underwater Institute – NUTEC/NUI Reports

The statutory activities of the aforementioned research centers are aimed at investigating different types of diving equipment. The reports prepared by these organizations are available in catalogs or on the internet. There are also standards concerning the basic parameters of the diving equipment (e.g., European Committee for Standardization, 2014; Frånberg, 2015). Among many organizations, classification societies[34] issue diving equipment regulations. Large companies involved in underwater activity also issue rules and regulations concerning the investigations of *UBAs*/diving systems, for example, the *Polish Navy*, the *US Navy* (NEDU, 1994; Kłos, 2000, 2021), etc. These publications are helpful in preparing a brief initial design and verifying it in unmanned experiments (Figure 1.12). Commercial reports and advertisement materials concerning diving and measuring equipment used in laboratory stands and diving research are useful as well. Professional meetings, conferences, symposia, and personal contacts are of utmost importance. Information from the internet can be useful as well.

This stage of research includes experimental verification of assumptions and models accepted for creating a brief initial design. Satisfactory research results enable commencing manned laboratory tests.

Various laboratory sets are used for unmanned laboratory verification of the brief initial design. As is evident from the author's own research, the following parameters should be tested at this stage of investigations:

- *Respiratory parameters*: positive-pressure breathing, negative-pressure breathing, respiratory work, changes in respiratory resistance due to hydrostatic influences, protection time of the *UBA*
- *Stability parameters*: weight of the *UBA*, weight distribution and displacement of elements, *UBA* buoyancy, bending moments while center of buoyancy does not cover the center of *UBA* gravity, head resistance

- *Technological parameters*: dimensions and weight of the *UBA* elements, level of generated noise and magnetic field, amount of gas vented to water, corrosion resistance and shock resistance, respiratory protection duration related to margin of breathing gas consumption and absorption of carbon dioxide
- *Ergonomic parameters*: valves accessibility and handling comfort, elimination of mistakes in valve handling, facility of manometers observation, facility of communication, mask visibility, fit and form, selection of materials
- *Fatigue and reliability parameters*: resistance of the UBA, harness configuration to damage while jumping into/immersing in water, resistance to corrosion and fire with respect to high-percentage oxygen mixture, valve resistance to handling and resistance of breathing regulator to freezing, strength of gas cylinders
- *Work parameters:* most important constructional elements of a *UBA*, that is, investigation of the regeneration system and output of breathing regulator or entrance nozzle, measuring elements, investigation of the breathing gas contamination, investigation of the dead space ventilation, and investigation of the inspired gas composition at fixed emission parameters of carbon dioxide

2.2.1 Conclusion

Unmanned laboratory verification of the brief initial design is one of the most important stages of research on *UBAs* and diving technology. Each of the problems should be examined separately. For this reason, at this stage of research, the only one specific procedure algorithm cannot be presented. This stage of research will be discussed further.

2.3 NORMOBARIC MANNED LABORATORY TESTS

Normobaric manned laboratory verification of the brief initial design is performed without immersing or immersing in shallow depths (Figure 2.2).

Research investigations carried out in immersion are preceded by surface investigations. Most often, a bicycle ergometer is used to perform Figure 1.16 and Figure 2.2, *left*. Research in immersion is performed in a shallow swimming pool (Figure 2.3).

When research is performed in a swimming pool, the diver is swimming at a given distance at a fixed speed[35] (Figure 2.4, *right*).

To simulate depth inside a dry chamber, the diver is exposed only to gas pressure; in the case of wet pot, the diver is exposed to water and gas pressure of inside the chamber (Figure 1.13). The swimming simulator is the most convenient device for simulating diving at depths. The diver can exercise by being pressed against the wall while connected to a dynamometer measuring the force exerted[36] (Figure 2.5).

FIGURE 2.2 Divers during investigations of *UBAs* on surface in a wet pot and a swimming pool.[37]

Source: Kłos (2000).

FIGURE 2.3 A swimming simulator with adjustable water stream, Karlskrona Diving and Maritime Medicine Center.

Source: Author's own photos, courtesy of the Swedish Armed Forces.

The swimming simulator for shallow-water depth can be employed to establish the workload for the swimming simulator. It keeps the subject stationary and allows him/her to carefully monitor the *UBA* and the medical parameters. A simulator of swimming is the most convenient device used to fix the diver's workload. In a full-diameter wet pot, to perform the research, an underwater bicycle ergometer is used (Figure 2.6 and Figure 2.7).

When there is a lack of such a simulator, other measures are taken to carry out, that is, balancing with a fixed load or diving against an expander. Other kinds of tasks can be performed by the diver, for example, a specific number of revolutions.

FIGURE 2.4 A swimming simulator: *left*, hyperbaric push-on ergometer into wet pot; *right*, 60 meter circumference pool.

Source: Lundgren and Pendergast (2010).

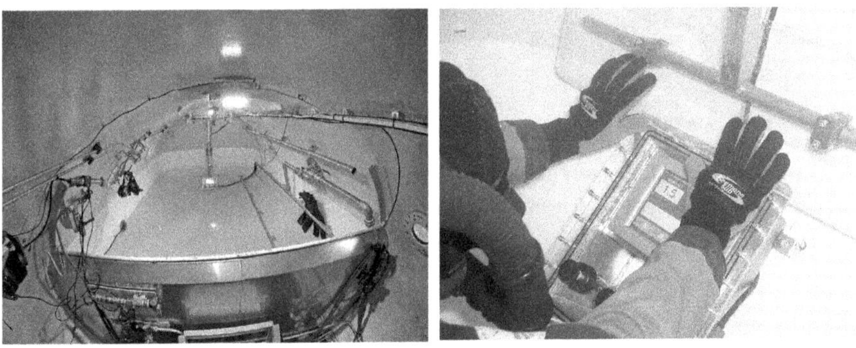

FIGURE 2.5 *Left*, at the back of the hyperbaric chamber, a push-on ergometer can be seen connected to a computer which displays the parameters of the diving process; *right*, a diver, during research on *UBA* in a wet pot inside a hyperbaric chamber, pushes on a barrier of the ergometer and uses a control dive computer.

Source: Author's own photos.

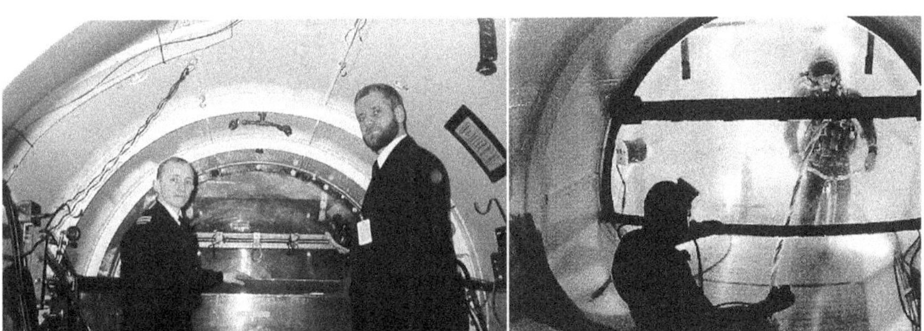

FIGURE 2.6 *Left*, full-diameter wet pot in *DR – DC Toronto*;[38] *right*, the author in front of the barriers on the underwater bicycle.

Source: Author's own photos, courtesy of DRDC Toronto, former DCIEM Canada,

FIGURE 2.7 The full-diameter wet pot of the Navy Experimental Diving Unit, Panama City: *right*, a vertical bicycle situated inside the chamber.

Source: Author's own photos, courtesy of US NEDU, Panama City.

However, when using other means than a swimming simulator, the possibility of calculating precisely the rational workload for the diver is smaller. Most often, measurements and investigations performed during the aforementioned tests were similar to those described earlier. At this stage of research, the measurements are supplemented with measurements of leak-tightness, like investigations on ergonomics, for example, evaluation of construction fit for the diver's body, loosening of clamps on the diver's body, visibility impairment caused by venting breathing gas to water, handling facility of the valves and other parts of the *UBA*, leak-tight connections. It is also possible to determine the frontal resistance of the *UBA*.

The swimming simulator can be employed to measure the breathing gas composition inhaled by the diver. In other cases, only the periodical sampling of the breathing gas is possible. However, it is also possible to assemble the gas analyzers inside the *UBA*. Indications of gas analyzers can be transmitted by means of telemetry and supplemented by monitoring of the physiological parameters transmitted by telemetry as well.

These types of research are rarely presented in the literature. Successful results of research on *UBA* and unmanned laboratory verification of the brief initial design are the bases for further manned investigations. Some experiments performed with the participation of men, and examples of the equipment used, will be described further.

Normobaric, or at a shallow depth, manned verification of some brief initial designs can be performed in two stages: without immersion in water and with immersion in shallow water.

At the first stage, measurements are often performed using a bicycle ergometer. The diver breathes using a *UBA* and exercises on a bicycle ergometer. This way, the workload can be controlled. However, it is completely different from the physical exercise that the diver should perform at depth. The work control is possible because the *UBA* is not carried by the diver, for example, the *UBA* is hung up or laid before the diver (Figure 2.2).

The health and competency of experimental divers employed in research on new *UBAs* are carefully examined. To protect the health of experimental divers, prophylactic examinations are performed. Aside from prophylactic examinations, the divers are frequently subjected to work tests during experiments. The tasks of prophylactic examinations are quite different from those of work tests. The aims of exercise stress tests are to establish a representative human sample and learn about processes occurring in the diver's organism during exertion. An experimental diver becomes a human etalon.[39] Exercise tests performed with the use of a bicycle ergometer enable determining the relationship between each diver's exertion and his lung ventilation, heart rate, oxygen consumption, and carbon dioxide emission to normal and hyperbaric conditions. Based on the diver's total characteristic and the experimental results obtained, approximate conclusions can be made concerning the processes occurring in the *UBA* tested. To have a sufficiently high degree of approximation, the diver's competency characteristic is obtained during exercise similar[40] to that of the *UBA* test.

Except medical parameters, most often, during experiments with the divers exercising on a bicycle ergometer, the following process parameters are investigated:

- *Ventilation*, for example, composition of the inspired and expired breathing medium, composition of the gas in the presumptive dead spaces, gas flow in the mouthpiece, etc.
- *Heat distribution*, for example, temperature inside the carbon dioxide absorber, temperature of the inspired and expired gas at the different places of the breathing loop, etc.
- *Performance of the breathing gas regeneration systems*, for example, determination of the carbon dioxide protection time, carbon dioxide ducting, effect of temperature and gas humidity on the canister packing, package resistance for ventilation changes, etc.

The aforementioned working parameters can be checked by means of *built – in – UBA* sensor systems with telemetry data transmit or periodically during immersion in water. The procedure for manned normobaric verification of the brief initial design for great-depth *UBAs* and respiratory systems may be troublesome. This is due to the fact that the *UBAs* often begin normally functioning from a specific depth. At shallow depths, the diver is supplied with a hypoxic breathing medium. If it is possible, the *UBAs* are tested after some changes in their functioning have been introduced. The changes should be as small as possible and should demonstrate certain similarity in ratio with respect to physical effects that occur in the *UBAs* in real conditions. Based on the model of phenomenon, mathematical methods of scale enlargement and reduction can be used (Johnstone and Thring, 1957; Zlokarnik, 1991).

During tests of the *UBAs* designed for deep diving, it is possible to use the breathing mixture of the increased oxygen percentage and perform normobaric tests of the *UBA* for the new conditions established.[41] The method can be used provided that the *UBA* ventilation model is known. When an analytical mathematical model is not available, semi-empirical methods, for example, dimension analysis or an empirical one, can be used (Siedow, 1968).

The author's own research performed at shallow depths for *UBAs* with regeneration of the breathing medium *SCR* enables proposing measuring the following parameters: dead space,[42] composition of the inspired breathing gas, cumulative contamination of the breathing gas, work of the gas regeneration system, ergonomics of the construction, leak-tightness of the construction, etc.

2.3.1 Conclusion

Normobaric manned laboratory verification of the brief initial design is an important supplementary research program. It is necessary to pursue the *UBAs* tests. The results of the research will be presented further.

2.4 HYPERBARIC MANNED LABORATORY TESTS

The aim of the hyperbaric manned laboratory verification of the brief initial design is, first of all, to verify the decompression assumptions. Except for physiological assumptions, the decompression assumptions include assumptions concerning the composition of the inspired gas and take into account dead space of the *UBA* breathing system. As far as it is possible, research should be performed for all predicted situations that can occur during operating the system.[43] The results should be elaborated in accordance with the statistical reference (e.g., Wienke, 2008, 2016). There are different methods used to evaluate decompression safety. However, they are rarely presented in the literature. Whenever published, usually the procedures applied are not described. Therefore, from the scientific and practical point of view, the methods can be of low value. The authors very seldom give any descriptions of the values determining the decompression safety. For example, aquaCorps claims that the typical number of decompression accidents to regard a decompression system as "*safe*" should be less than 1% (Huggins, 1992). Most often, the number of decompression diseases per the total number of person decompression is treated as an indicator of the quality/safety of a developed decompression system. However, there is no information concerning the procedure employed to collect the data. It is not stated whether the examples were gathered during the hyperbaric laboratory verification of the brief initial design, in field tests, or during the stage of the author's supervision of the implemented system, or simply during individual operation of the system by the user (e.g., Lippmann, 1990; Lippmann and Mitchell, 2009). It would seem that the stage of experiments when the data was obtained is not essential. However, in reality, this claim is not finally justified. Land-based hyperbaric complexes with wet pot are the most convenient facilities to perform hyperbaric manned laboratory verification of the brief initial design. There are some publications presenting reviews of hyperbaric complexes with wet pot systems (e.g., Haux, 1982). Hyperbaric research is rarely reported in the literature. The Polish system and the results of the author's own research will be reported in the next chapters of this book.

Basic investigations of the diving system and the *UBAs* employed should be performed during hyperbaric manned laboratory verification of the brief initial design.

The number of investigations should be high enough to constitute the basis for statistical inference whether the system meets the technical specifications at the assumed confidence level. As was mentioned earlier, the system should not be tested for this reason at the later stages of experiments.[44] Justification for adopting such a procedure will be provided later.

Firstly, from the humanitarian point of view, delivering a diving system to the user is highly irresponsible without prior laboratory verification of this system. It is often the case that the harmful effects of the systems are caused by a delay in time,[45] and the effects can be of cumulative nature.[46] One can risk the statement that potential harm caused by such a system may be avoided during its operation. This is the case when a diving company makes sure that members of a diver team are rotated, at sufficient frequency,[47] replaced by healthy, well-trained younger divers without the "bad diving history." Then the number of dives performed by one diver can be high, and the number of cases[48] low. However, a diving system can be dangerous, and it may cause divers to be handicapped persons later in their lives. It will be extremely difficult to prove that a diver's health problems have been caused by the poor diving system, as accident statistics[49] are very good. Of course, it should not be excluded that during the phase of laboratory verification, such a situation may occur, but undoubtedly, it is a little probable. This is because of better and better measuring equipment that should indicate all deviations from the assumptions adopted.

Secondly, companies operating a diving system are not interested in revealing cases of diving sickness, because of financial compensation they would have to incur. Thus, the majority of small problems developed by divers are handled without proper documentation; sometimes they are even disregarded. Professional divers are not interested in revealing small problems either because they are under the pressure of their employers and diving groups. It is also connected with the possible loss of unemployment and getting bad opinion.[50] Therefore, divers dissimulate diving sickness or do not admit problems or even belittle them (Risberg, Møllerløkken, and Eftedal, 2019). For this reason, statistics can be false, and it is after years that divers can feel the effects caused by the poor diving system or small cumulative decompression incidents.

Thirdly, there are a few substantiated cases of exceptional resistance of individual divers to poor diving conditions, for example, *Alexander Lambert*. Diving companies are interested in recruiting young, healthy, and strong men of exceptional individual resistance to diving conditions. Therefore, the statistics based on the results of tests of diving systems provided by commercial companies are not reliable due to failure to obtain representative samples. Part of the data is also not reliable because experienced divers and diving supervisors use their "own patents" in tests, that is, their own modification of recommended diving technologies. Divers are also not interested in applying the worst scenarios accounted for in a diving technology. As a result, the system is often not tested fur such scenarios. Therefore, in case of emergency, the system may fail and cause irreversible damage. Many a time, a diver falsifies his/her exposure time at a given depth in order to get more money and have decompression time extended by a careful diving supervisor.

The only way to partly prevent the falsifying of the results during tests of diving systems is to place trust in the conscience of inquiring scientists. Experienced researchers guided by high moral rules will guarantee obtaining the reliable results when they supervise diving experiments. For this reason, the sequence of experimental procedures proposed here to be applied to diving systems should not be rejected. Based on the author's own research, the same measurements and investigations as in normobaric air are recommended during hyperbaric laboratory verification of the brief initial design. Initially, the investigations should be performed without immersion. In order to program the diver's exercise rate, the cycloergometer placed inside the hyperbaric chamber can be used. The next stage of experiments should be performed in the wet pot of the hyperbaric complex. In order to improve the quality of research, it is recommended to equip the hyperbaric complex with a swimming simulator and the exercise rate measuring device.

At this stage of the experiment, the diving system should be investigated in a water pot under different conditions. The attendant divers who use the already-verified diving technologies should ensure the safety of the divers serving as the subjects. An exchange of the attendant divers is needed more than once during experiments.[51] The divers who are released from the experiments should be successively decompressed in separate hyperbaric chambers of the diving complex. In order to test new diving technologies and equipment, a hyperbaric facility is needed. The diving procedure requires access to a multichamber diving complex. During all manned laboratory diving experiments, except for the monitoring of the processes that occur in the diving system, versatile monitoring of all medical diver's parameters[52] is additionally recommended.

Experimental hyperbaric *systems* can serve a variety of many additional purposes:

- Scientific investigations
- Support for underwater works[53]
- Performing hyperbaric works[54]
- Lifeguard[55]
- Hyperbaric treatment[56]

2.4.1 Preparation

Turning on a *DGKN* – 120 hyperbaric *system* (Figure 1.13 *left*), which is a signal to check the *system* status – (Figure 2.8), will be presented as an example of a preparation *process*.

The *IT subsystem* should check the status of several *subsystems*, such as lighting, heating, internal *TV* and communication, internal and external regeneration, emergency,[57] automation,[58] fire protection, and so on after turning on the electric power of the hyperbaric *system* – process *N*º01. Turning on/off the control panel is shown in Figure 2.8. The preferred method for checking and visualization of the status of

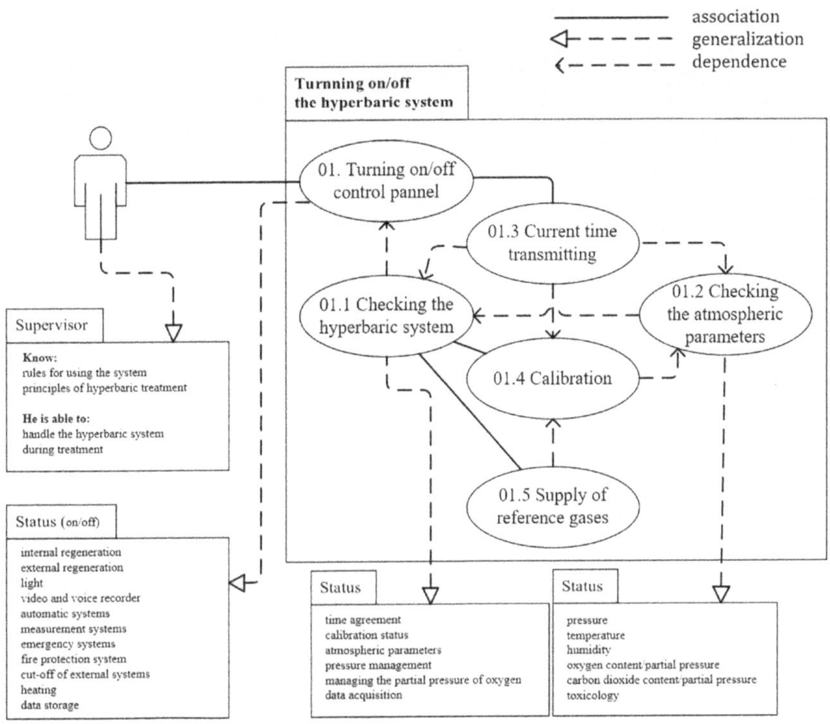

FIGURE 2.8 Use-case diagram for the hyperbaric system on/off *process*.

subsystems is to use *IT* support. An example of a synoptic board is presented in Figure 2.9, *left*.

This *IT system* emulates buttons on the computer touch screen for turning on/off *subsystems*. The *subsystem* on/off mode is indicated with a different color. An additional bar indicator is shown for each *subsystem*. It changes its color if the selected value of power consumption is too high or too low. The computer program can check all the systems and generate a report, but in most cases, it is a qualified personnel that checks the process manually (on/off). Then they make a report using a separate form outside the *IT* system. The *IT subsystem* used to turn on/off various *subsystems* should be duplicated by a traditional, manually operated *subsystem* (Figure 2.9, *right*).

For safety reasons, some mechatronic *systems* are controlled and operated manually outside the *IT system*, including cutoff of *system* (e.g., external *systems*; Figure 2.10), checking and activating of the fire *system* (Figure 2.11), and so on.

These *systems* require a reliable source of power, independent of the electricity supply, which is why they are most often powered pneumatically. The necessary bank of air under pressure is stored in pneumatic accumulators and fire extinguisher water in tanks (Figure 2.12).

In order to prevent them from being activated due to an error generated by the *IT system*, they are usually handled and checked separately.

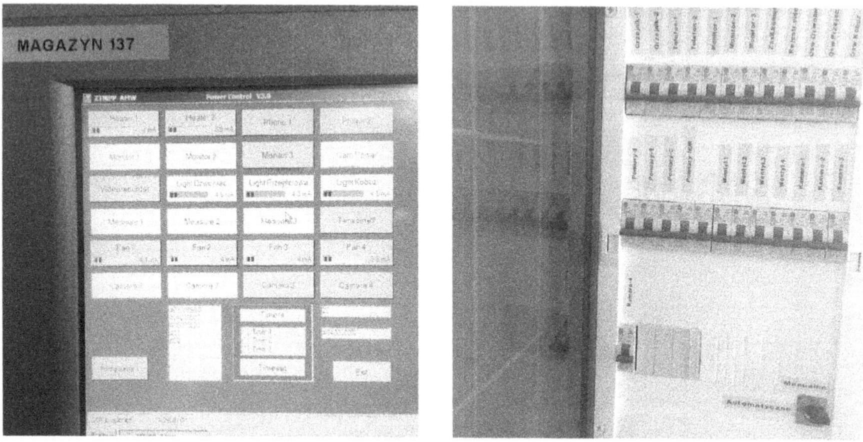

FIGURE 2.9 Turning on/off control panel at the Polish Naval Academy: *left*, computer touch screen display of the status of subsystems; *right*, manual on/off switches for various subsystems in the duplicated *IT* sub-system.

Source: Author's own photos.

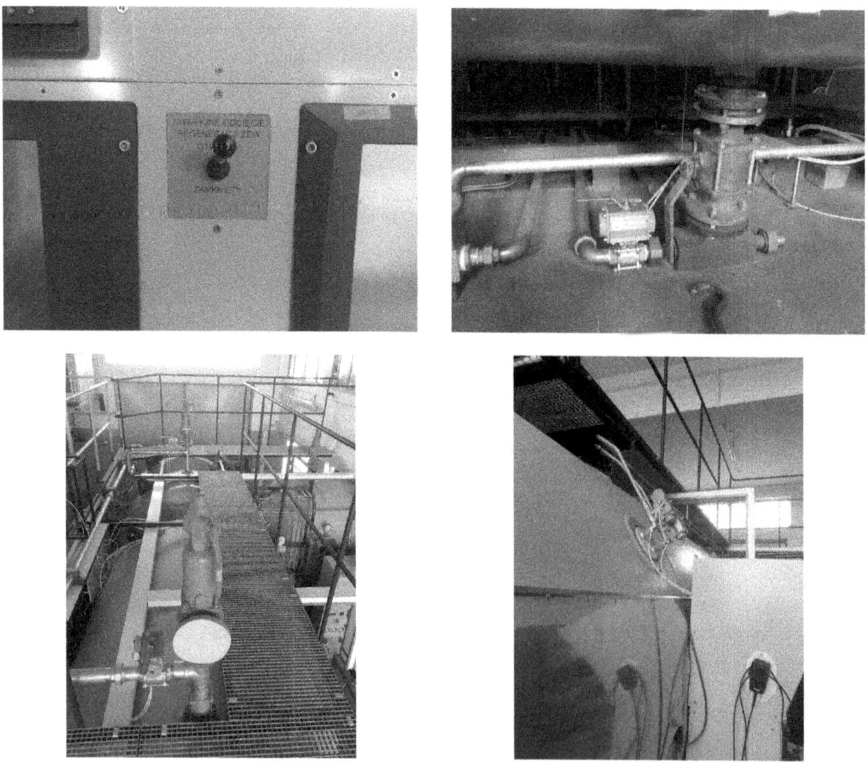

FIGURE 2.10 Cutoff *system*: *left*, open valve; *right and down*, different actuators.

Source: Author's own photos.

FIGURE 2.11 Fire system: *left*, trigger valve; *right*, water nozzle inside a hyperbaric chamber.

Source: Author's own photos.

FIGURE 2.12 Banks: *left*, pneumatic accumulator; *right*, water bank.

Source: Author's own photos.

The sequence of the checking *process* should be adapted to each hyperbaric *system*, because automation *subsystems* and emergency *subsystems*, in particular, are often different in almost each of the hyperbaric *systems* – *process* $N^{\underline{o}}$01.1 in Figure 2.8: checking the hyperbaric *system*. The same type of hyperbaric *system* may differ in some details because it may have been built and modernized at a different time or by a different contractor. Often, the same types of *systems* differ due to the owner's requirements or the local law. Thus, the checking procedure depends on *subsystems* the *system* is equipped with.

It is very important to keep the same real time in *IT processes*, such as data acquisition, video and voice recording, or recording the time of calibration of the measuring equipment – process $N^{\underline{o}}$01.3. Current time transmitting in Figure 2.8. The current time in important *subsystems* should be displayed, as is shown in Figure 2.13, obligatorily checked during $N^{\underline{o}}$01.1: checking the hyperbaric system process and periodically checking throughout the hyperbaric system use.[59]

Based on $N^{\underline{o}}$01.3, current time transmitting *process*, $N^{\underline{o}}$01.1, checking the hyperbaric system *process* can suggest a demand for $N^{\underline{o}}$01.4 calibration *process* (Figure 2.8). Based on $N^{\underline{o}}$01.5, supply of reference gases *process*, $N^{\underline{o}}$01.1, checking the hyperbaric *system process* can trigger $N^{\underline{o}}$01.4 calibration *process* to start (Figure 2.14).

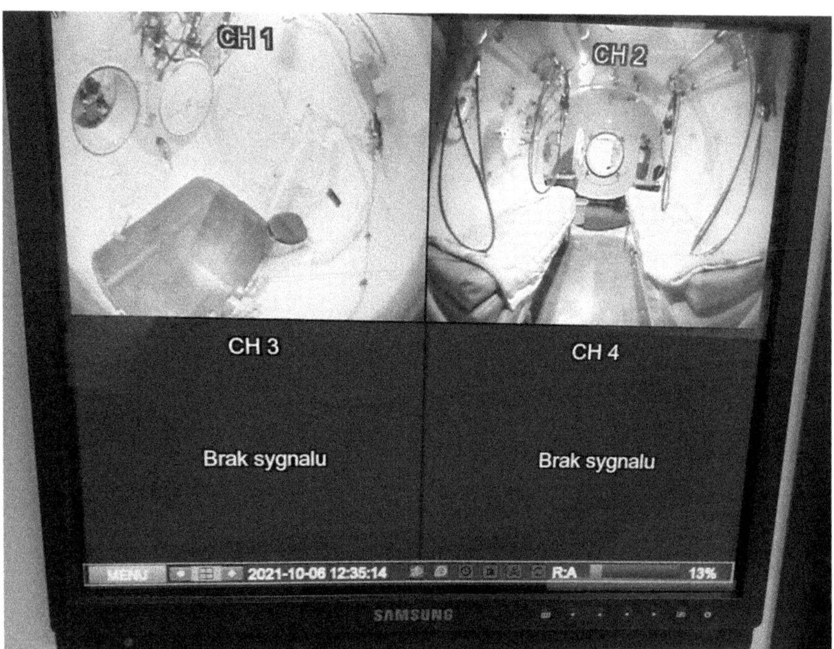

FIGURE 2.13 Current time in image recording *subsystem* displayed on *VCR* panel.

Source: Author's own photo.

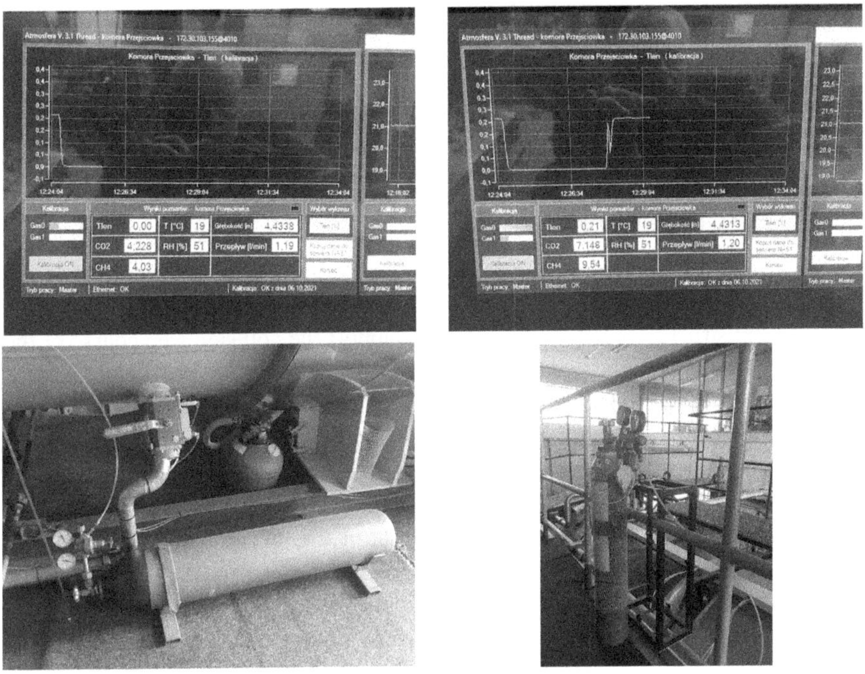

FIGURE 2.14 *Top*, analyzers calibration process; *bottom*, zero and proof gas release process, zero and proof gas mixture banks.

Source: Author's own photos.

During *N*º01, turning on/off the control panel *process*, the hyperbaric *system* should be adapted for conducting a collision-free diving operation by including/excluding elements and *subsystems*, depending on the planned hyperbaric exposure. Next, during *N*º01.1, checking the hyperbaric *system process*, the key elements of the structure needed to ensure a collision-free course of the planned diving *process* should be checked,[60] for example:

- Auxiliary power supply *system*[61]
- Lighting *systems*[62]
- Heating *systems*[63]
- Communication *systems*[64]
- Video and voice recording *process*[65]
- Data storage *system*
- Internal and/or regeneration *systems*[66]
- Fire protection *systems*[67]
- Cutoff of external *systems*[68]
- Data storage *process*[69]
- Automation *systems*, and so on

Process $N^{\underline{o}}$01.1, checking the hyperbaric *system*, can be programed for different diving scenarios or some tests, for example, pressure,[70] regeneration,[71] oxygen content maintenance,[72] and other tests.

The tests of some systems should be performed outside the chamber, for example, $CO / C_n H_m X_k$ conversion test.[73]

2.4.2 Hyperbaric Exposure

Hyperbaric exposure can be started if *process* $N^{\underline{o}}$01.1, marked as "checking the hyperbaric *system*" in Figure 2.8, is completed for selected elements of the hyperbaric *system* structure whose elements may support the collision-free course of the planned diving operation.

The $N^{\underline{o}}$01.2, checking the hyperbaric parameters *process*, plays the main role during the execution of hyperbaric exposure. It may be interrupted by *process* $N^{\underline{o}}$01.1, checking the hyperbaric *system* during a long hyperbaric exposure (Figure 2.15).

In many cases, *process* $N^{\underline{o}}$01.1, checking the hyperbaric *system* during execution of hyperbaric exposition, is different from that of the preparation phase. In many cases, during a hyperbaric exposure, the scope of *process* $N^{\underline{o}}$01.1, checking the hyperbaric *system*, differs from the scope of this *process* in the preparation phase of the hyperbaric system. Most often, it is focused only on part of *process* $N^{\underline{o}}$01.4, calibration, in which part of the process is related to checking gas analyzers. Often, saving measurement data, video, and talks is a very important part of the hyperbaric exposure execution, *process* $N^{\underline{o}}$01.2, checking the hyperbaric parameters

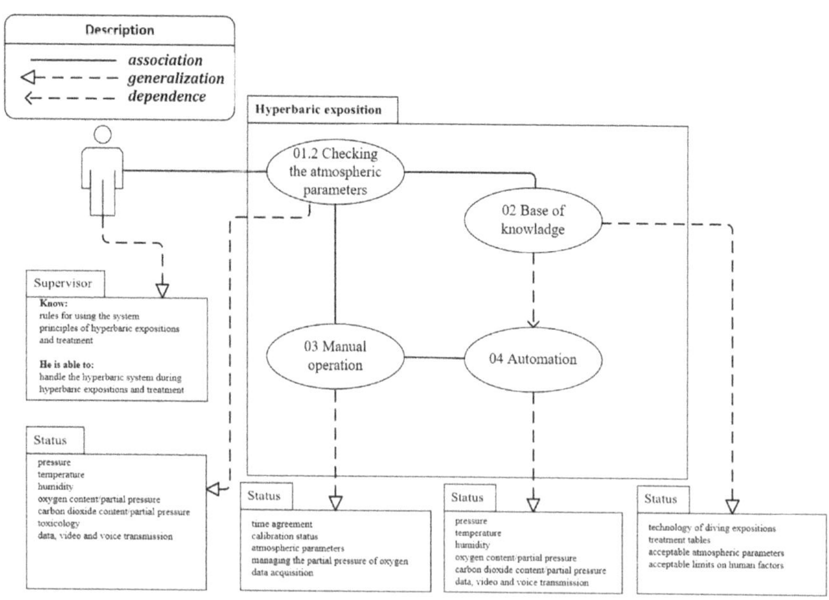

FIGURE 2.15 Use-case diagram for the hyperbaric exposition process.

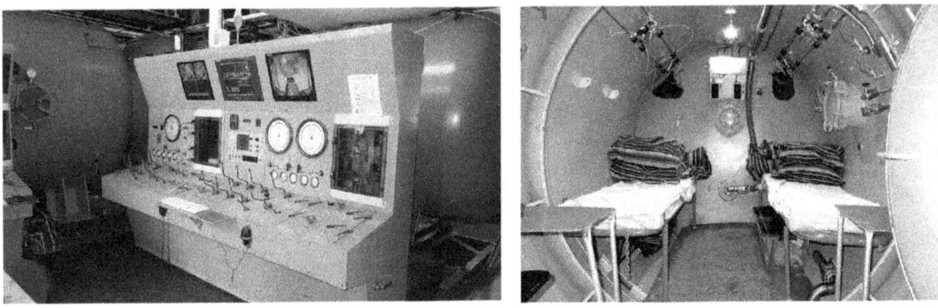

FIGURE 2.16 An experimental diving complex at the Naval Academy: *left*, one of the control/operating panels of the diving complex; *right*, the living chamber.

Source: Author's own photos.

(Figure 2.8). *Process* $N^{\underline{o}}$01.2 is also responsible for monitoring key divers' homeostasis chosen parameters during the execution phase.

The same process $N^{\underline{o}}$01.2 is also responsible for monitoring selected key parameters of divers' homeostasis during the implementation phase of hyperbaric exposure. Figure 1.13 and Figure 2.16 show the main view of a diving simulator.

2.4.3 Conclusion

It should be noted that in extensive investigations, the hyperbaric manned laboratory research is the most expensive and the longest stage of experiments. In order to avoid the high costs of the experiments, the statistical methods of experiment design should be applied. In any case, the reasons mentioned earlier should not exclude carrying out of the whole range of diving experiment. Examples of such research in which the author participated are presented in the next chapters of the book. The results obtained in the previous and current stages of the experiments have been positive and statically confirmed. Therefore, they can be the basis for undertaking the trials of the brief initial design in the field under real conditions.

2.5 VERIFICATION UNDER REAL CONDITIONS

Verification of the brief initial design of a new diving *system/UBA* at open sea requires training of the future users, which is the final stage of the introductory research. The issue is almost omitted in the literature or included in laboratory tests.

Based on the author's own research, it seems to be a well-grounded recommendation that future users of the system who are not experimental divers be allowed to do diving (Figure 2.17).

At this stage of the research, monitoring is costly and troublesome. Therefore, very accurate monitoring can be dropped. A better approach is to keep life functions of only less-experienced divers monitored. At this stage of the investigations, the divers perform only tasks listed in the research schedule. At this stage, commercial

FIGURE 2.17 Testing *SCR* at sea:[74] *left*, author is preparing to do an experimental dive; *right*, a diving bell.

Source: Author's own photos.

diving tasks should not be carried out; it is only simulated exertion that should be allowed. During research, the decompression process must be tested, and divers must be trained.

In addition, measurements should be made, and resistance of the operated diving equipment and systems to trying conditions[75] should be tested. The problems that should be investigated within the framework of the research are:

- *Correctness of the storage procedures.* Preparation and performance of the system, for example, resistance of the regeneration system to vessel vibration in the course of storing a carbon dioxide canister prepared for use, resistance of the diving equipment to different positions while stored, abrasion of an absorber package during vessel vibration and high environmental humidity, housing and packaging durability and their usefulness, etc.
- *Safety system.* For example, standard and emergency communication systems, emergency breathing systems, standard and emergency decompression procedures, emergency equipment and procedures to be applied in case of dangerous diving accident, etc.
- *Logistics and evacuation support.* For example, receiving and preparing breathing mixtures, packaging regeneration systems, organization of transporting divers under pressure in case of failure of or fire in the system, etc.
- *Organization of diving process.* For example, testing the system for mobilization time to reach operational readiness, duration of emergency and rescue procedures, etc.

2.5.1 Conclusion

The field verification stage of the brief initial design is the natural continuation of the manned laboratory research stage. Laboratory research followed by and supplemented with field verification has to be complete and well-performed. All the system elements are tested in real conditions that are significantly more trying and uncomfortable than the laboratory conditions. Field experiments should ensure the fluent transition from research to diver training. Common work done by the scientists and the working divers can contribute to better understanding of the diving system, which may lead to cause enhanced confidence of the divers in the system. In case of necessity, the divers will not hesitate to ask the scientists or the experimental divers questions, as they know each other. Common work will give definite advantages, as they will be able to share their observations and remarks concerning the system and bring attention to the problems. Due to the mutual recognition of their professional competence, the system will be more effectively supervised, and the observations and remarks provided the system operators will be considered reliable.

2.6 SUPERVISING

The supervision of the diving system's operation exercised by the author is an additional stage of research aimed at verifying the operation of the diving system. This problem is totally neglected in the literature or included laboratory tests. A dozen years ago, organizations of divers[76] or commercial associations of divers[77] started to collect voluntary reports concerning diving accidents. The reports are analyzed and promulgated among broad groups who are interested in this problem. However, the technical problems have not been reliably elaborated.

In light of the experience earned during the author's own research, it is not recommended that the whole staff of the scientific/research group be employed in the supervision phase due to the cost. Neither should the whole technical capabilities of the establishment involved in implementing the system be mobilized. As a rule, only a few representatives of the scientific/research group and the technical personnel are employed in this stage of research. Sometimes, the experimental divers who took part in the earlier stages of the implementation research are also employed. The presence of the scientific/research group should make sure that findings recorded and measurements performed are used for the further evaluation of the diving system; they will enable to increase the statistical sample without errors. Additionally, the presence of the implementation scientific/research group representatives will be comfortable for the users and will be helpful in an emergency. However, this generates additional costs of operating the system. For this reason, this stage of research should not be artificially prolonged.

2.6.1 Conclusion

The supervision of the diving system operation exercised by the author is an additional stage and can be treated as continuation of the implementation research. The time during which the author supervises the diving system operation should be based on the technical, sociological, and economic analysis.

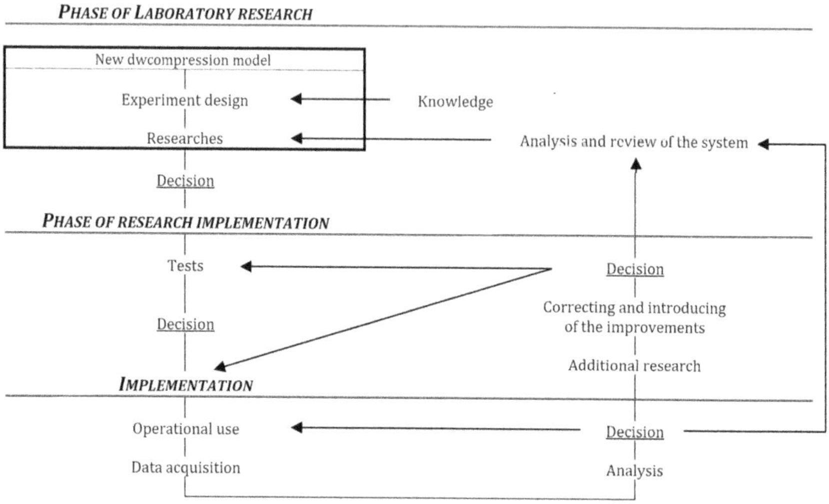

FIGURE 2.18 Flowchart showing the implementation of new methods of decompression in the Polish Navy.

Source: Kłos (2000).

2.7 COLLECTING DATA

The methodology recommended for decompression table validation and the implementation of new decompression schedules is shown in Figure 2.18.

In the laboratory research phase, the already-existing models are used to develop new mathematical models of decompression based on previously collected data, and/or new findings are modified. The tables thus obtained are tested. The experimental guidelines are established in accordance with the principles contained in the *Helsinki Declaration*. In most countries, human experimentation requires approval by the *EC*[78] or review board. Chamber tests, where pressure and time can be accurately controlled, are essential for the laboratory research phase, but pool tests and open water tests may also be performed. Such experiments should be continued until the results, based on medical and scientific considerations, justify a transition to the implementation phase. The laboratory phase may be stopped at any time. If changes in methods or a model are needed as a result of the experimental findings, they must be approved by the *EC*. The number of laboratory tests required to start the implementation phase may vary according to the model being tested. If an entirely new model is being tested, the number of trials must be large enough to secure an adequate confidence level for acceptance of the results. On the other hand, if the model has been previously tested and only minor changes are introduced, the number of tests needed may be smaller.

In the implementation phase, tests are performed under operational diving conditions in accordance with the standard diving rules with high concern to diver safety. It is the most experienced operational diving teams that are assigned the task of managing dives, with the assistance of well-equipped medical and scientific teams to monitor the medical and physiological status of the dive subjects and the safety of the diving procedures being tested. This is to ensure that a sufficiently large quantity of high-quality data is being obtained while maintaining a high level of safety for experimental divers. This will also form the basis for the decision to subsequently approve the decompression tables for operational use.

However, this is not the end of the research process. During operational use, the decompression procedures and the results of the dives carried out will be monitored by operational diving teams. If any problems arise, they should be directed back to the research phase. If decision about the withdrawal of a decompression procedure or about its modification should be made, the process will have to start anew (Figure 2.18).

2.7.1 Validation

The onset of *DCS* symptoms may be treated as a statistical phenomenon since it is seldom a case that divers subjected to the same decompression procedure all show the same reaction or symptoms. Because of the variability in the biological processes underlying reactions to decompression stress that are observed, it is most convenient to acknowledge a formal statistical correlation between the results obtained from experimental dives and the results expected to confirm the safety and adequacy of the mathematical model used to describe them. The use of statistical methods sometimes produces simple and relatively easy-to-interpret results. Still, the verification of decompression tables from a small number of experimental dives is statistically questionable because, in such a case, it is difficult to definitely prove their safety. As for decompression tables, the goal should be to statistically verify the decompression required for each depth and the bottom exposure time. A contradiction arises at this point as, on the one hand, the aim is to acquire the highest possible number of experimental results,[79] while on the other, this procedure is very costly and risky.

The systematic study of the phenomena accompanying decompression is a complex problem because there is a lack of precise measurement methods to monitor the processes taking place in body tissues. The mathematical models used to describe the decompression process reflect only a part of the total phenomena. They usually involve efforts to fit experimental data by using relatively simple mathematical functions. Such mathematical models, however, should be treated as calculation methods used for deriving decompression procedures, not as mathematical models of the physiological processes taking place during decompression.

Validation is performed to verify the assumptions made in the decompression model by establishing a decompression sickness *DCS* risk function based on experimental dive data (Kłos, Nishi, and Olszański, 2002).

Figure 2.18 shows a recommended methodology for validating decompression tables and implementing new decompression schedules. In the laboratory research phase, new mathematical models of decompression are developed based on already-existing models, using previously collected data and/or new findings. The tables thus produced are tested. The experimental guidelines are established in accordance with the principles contained in the Helsinki Declaration (Figure 2.18). In most countries, human experimentation requires approval of the Human Research Ethics Committee *EC* or a review board (Kłos, 2021). Chamber tests, where pressure, time, and other parameters can be accurately controlled, are essential for the laboratory research phase, but pool tests and open water tests may also be performed during laboratory tests. Such experiments should be continued until the results, based on medical and scientific considerations, justify transition to the implementation phase. The laboratory phase may be terminated at any time. If any changes in the methods or the model are needed as a result of the experimental findings, they must be approved by the *EC*. The number of laboratory tests required to start the implementation phase may vary according to the model being tested. If an entirely new model is being tested, then the number of trials must be large enough to achieve the confidence level $\overline{P}$ required for acceptance of the results. On the other hand, if the model has been previously tested and only minor changes have been introduced, then the number of tests needed may be smaller.

In the implementation phase, the tests are performed under operational diving conditions in accordance with the standard diving rules, with high concern given to diver safety. It is the most experienced operational diving teams that are tasked with managing dives, assisted by well-equipped medical and scientific teams responsible for monitoring the medical and physiological status of the dive subjects and the safety of the dive procedures being tested. This is to ensure that a sufficiently large quantity of high-quality data will be obtained while maintaining a high level of safety for the experimental divers. This will also form the basis for the decision to subsequently approve the decompression tables for operational use.

However, this does not end the research process. During operational use, the decompression procedures and the results of the dives carried out will be monitored by operational safety diving teams. If any problems arise, they should be directed back to the research phase.

If the decision to withdraw the decompression procedure or modify it should be made, the process shown in Figure 2.18 will have to start anew.

2.7.2 Binomial Distribution

Assuming that the results of the decompression performed may be described by a binomial distribution,[80] the probability of the occurrence $P(n,N)$ of n events of *DCS* symptoms in a random sample of general population N, where the underlying probability of *DCS* is ρ, is given by (Kłos, 2007b):

$$P(n,N)=\binom{N}{n}\rho^{n}(1-\rho)^{N-n}\Bigg|\,n=1,2..N;\binom{N}{n}=\frac{N!}{n!(N-n)!} \tag{2.1}$$

The discrete probability distribution is normalized to 1:

$$\sum_{n=1}^{N}P(n,N)=1 \tag{2.2}$$

Hence, the joint probability for more than n cases of *DCS* is given by:

$$P_{>}(n)=\sum_{i=n+1}^{N}P(i)=1-\sum_{i=0}^{n}P(i) \tag{2.3}$$

And the probability of onset of less than n cases of *DCS* is given by:

$$P_{<}(n)=\sum_{i=0}^{n-1}P(i)=1-\sum_{i=n}^{N}P(i) \tag{2.4}$$

Analogous to the earlier calculations, the probability of less than one *DCS* incident is:

$$P_{\geq}(1) = 1 - \sum_{i=0}^{0} P(i) = 1 - P(0) \rightarrow P_{<}(1) = 1 - P_{\geq}(1) = P(0) \qquad (2.5)$$

2.7.3 Uncertainty

A validation procedure for the decompression profiles to be tested requires making decisions about the uncertainty–confidence level that we want. If a given decompression profile is repeated on several individuals or on the same individual on several occasions, the results of such an experiment may differ widely. An acceptable confidence level should be selected, and an upper limiting value n_{max} for the occurrence of *DCS* should be defined. Thus, a decompression procedure that will keep the number of incidents below our limiting value, for example, below $n_{max} \leq 5\% N$ at a confidence level of 90%, where N represents number of experiments, can be accepted. The probability of rejecting a procedure by establishing a rejection criterion based on a set number n_{max} of *DCS* incidents in a selected number N of experimental dives may also be calculated. In this way a relatively simple statistical procedure may provide significant information, that is, on the number of dives needed, to validate a procedure at an accepted level of risk, or to determine a maximal risk associated with the onset of a set number of incidents after a selected number of dives.

If the estimation average error of the probability limits is α_0, the confidence level of this estimate is $\bar{P} = 1 - \alpha_0$. The confidence interval[81] for this confidence level, where the right and left limits for n events of *DCS* in N dives, with the exception of $n = 0$ and $n = N$, can be determined and may be expressed as:

$$\begin{cases} P(\rho \geq P_r) = \sum_{x=0}^{n} \binom{N}{x} P_r^x (1 - P_r)^{N-x} = \dfrac{\alpha_0}{2} \\ P(\rho \leq P_l) = \sum_{x=n}^{N} \binom{N}{x} P_l^x (1 - P_l)^{N-x} = \dfrac{\alpha_0}{2} \end{cases} \Bigg| \; 0 < n < N \qquad (2.6)$$

Example: For $N = 9$ dives, at $\rho = 0.05$, the probability values for events are:

$p(0) = 0.90^9 \cong 0{,}630$ $\qquad p_{>}(0) = 1 - p(0) \cong 0.370$

$p(1) = \dfrac{9!}{1!8!} \cong 0.05\, 0.95^8 \cong 0.298$ $\qquad p_{<}(9) = 1 - P(9) = 1 - 110^{-12}$

$p(2) = \dfrac{9!}{2!7!} 0.05^2\, 0.95^7 \cong 0.063$ $\qquad p_{\geq}(2) = 1 - p(0) - P(1) \cong 0.071$

$p(9) = 0.05^9 = 1.9510^{-12}$ $\qquad p_{>}(2) = 1 - p(0) - p(1) - p(2) \cong 0.008$

Using the preceding formulae (2.6), a table can be produced of the probability of selected joint events occurring in some number of dives N at a given probability ρ of *DCS* symptoms (Table 2.1). It shows the probability of the occurrence of zero, one, and two or more incidents of *DCS* $n = \{0, 1, 2+\}$ in N experimental dives if the actual probability ρ of *DCS* is $p = \{0.01, 0.05, 0.10\}$. The *DCS* probability ρ should be interpreted as an average value for the onset of *DCS*.

The confidence level $\overline{P}$ informs on how many of 100 conclusions may be wrong, on average, after performing S_N experiments consisting of N experimental dives. The confidence level $\overline{P}$, that is, specified by the probability of error α_0, is defined as the probability ρ of the DCS onset of each particular event:

$$\left.\begin{array}{r} \forall_{0<n<N} P\left(P_l<\rho<P_r \middle| S_N=n\right) \\ \forall_{n=0} P\left(0\leq\rho<P_r \middle| S_N=n\right) \\ \forall_{n=N} P\left(P_l<\rho\leq 1 \middle| S_N=n\right) \end{array}\right\} = 1-\alpha_0 \tag{2.7}$$

Using equations (2.6) and (2.7), a probability interval $P_l<\rho<P_r$ of DCS onset may be easily calculated for $0\leq\rho<P_r$, the absence of DCS or the onset of DCS incidents in all subjects $P_l<\rho\leq 1$:

$$\begin{array}{c} \left\{\begin{array}{l} P_l=0 \\ P_r=1-10^{\lg\frac{\alpha_0}{N}} \end{array}\right| n=0 \\ \left\{\begin{array}{l} P_l=10^{\lg\frac{\alpha_0}{N}} \\ P_r=1 \end{array}\right| n=N \end{array} \tag{2.8}$$

TABLE 2.1
Probability of $n=\{1,2,2+\}$ *DCS* Incidents for a Given Number of Dives and Probabilities of *DCS* Risk

Number of Dives	Number of *DCS* Incidents	Probability of Joint Events for Definite *DCS* Risk ρ		
N	n	$\rho=0.01$	$\rho=0.05$	$\rho=0.10$
5	0	0.95	0.77	0.59
	1	0.05	0.20	0.33
	2 and more	0.001	0.02	0.08
10	0	0.90	0.60	0.35
	1	0.09	0.32	0.39
	2 and more	0.004	0.09	0.26
20	0	0.82	0.36	0.12
	1	0.16	0.38	0.27
	2 and more	0.02	0.26	0.61
50	0	0.60	0.08	0.01
	1	0.31	0.20	0.03
	2 and more	0.09	0.72	0.97

Source: Wienke (2003, 2008, 2016).

The number N of experimental dives required without $n \equiv 0$ any DCS symptoms at the assumed values α_0 and P_r may be calculated by the transformation of the preceding formula (2.8) to the form:

$$N = \left.\frac{\lg \alpha_0}{\lg(1 - P_r)}\right| n = 0 \tag{2.9}$$

Example: From formula (2.9), we can calculate the number of experimental dives required at $n(DCS) = 0$ at the assumed values α_0 and p_r; for example, for $\alpha_0 = 0.05$ and $p_r = 0.01$, we obtain $N = 298$, and for $\alpha_0 = 0.01$ and $P_r = 0.01$, we obtain $N = 458$. The calculations for these cases when $0 < n < N$ are more complicated and are usually done numerically. The results of these calculations are to be found in the tables. For example, for three DCS incidents in nine dives, we have:

$n = 3\, N = 9$

$$P_{\geq} = \sum_{i=3}^{9} \binom{N}{i} P_l^i (1 - P_l)^{N-i} = \frac{\alpha_0}{2} \Rightarrow \frac{\alpha_0}{2} = 84\, P_l^3 (1 - P_l)^6$$

$$+126\, P_l^4 (1 - P_l)^5 + 126\, P_l^5 (1 - P_l)^4 + 84\, P_l^6 (1 - P_l)^3$$

$$+36 \cdot P_l^7 \cdot (1 - P_l)^2 + 9 \cdot P_l^8 \cdot (1 - P_l) + P_l^9$$

$$P_l \cong 0.075 \mid \alpha_0 = 0.05$$

$$P_l \cong 0.042 \big| \alpha_0 = 0.01$$

$$P_{\leq} = \sum_{i=0}^{3} \binom{N}{i} P_r^i (1 - P_r)^{N-i} = \frac{\alpha_0}{2} \Rightarrow \frac{\alpha_0}{2} = (1 - P_r)^9$$

$$+9\, P_r (1 - P_r)^8 + 36\, P_r^2 (1 - P_r)^7 + 84\, P_r^3 (1 - P_r)^6$$

$$P_r \cong 0.701 \big| \alpha_0 = 0.05$$

$$P_r \cong 0.781 \big| \alpha_0 = 0.01$$

The calculations are shown in Figure 2.19.

Testing and validating decompression procedures is a long-lasting process which comprises several phases. At each test phase, decompression tables are closely analyzed and corrected or modified, if necessary (Figure 2.18). Should a modification be necessary, the approach to the problem may have to be changed entirely, and the whole process restarted. The basic problem in developing decompression procedures is finding the number N of experimental dives needed to prove that the decompression profiles that have been tested are sufficiently safe. In order to answer this question, it is necessary to establish the definition of "safe" decompression. A 5% risk of

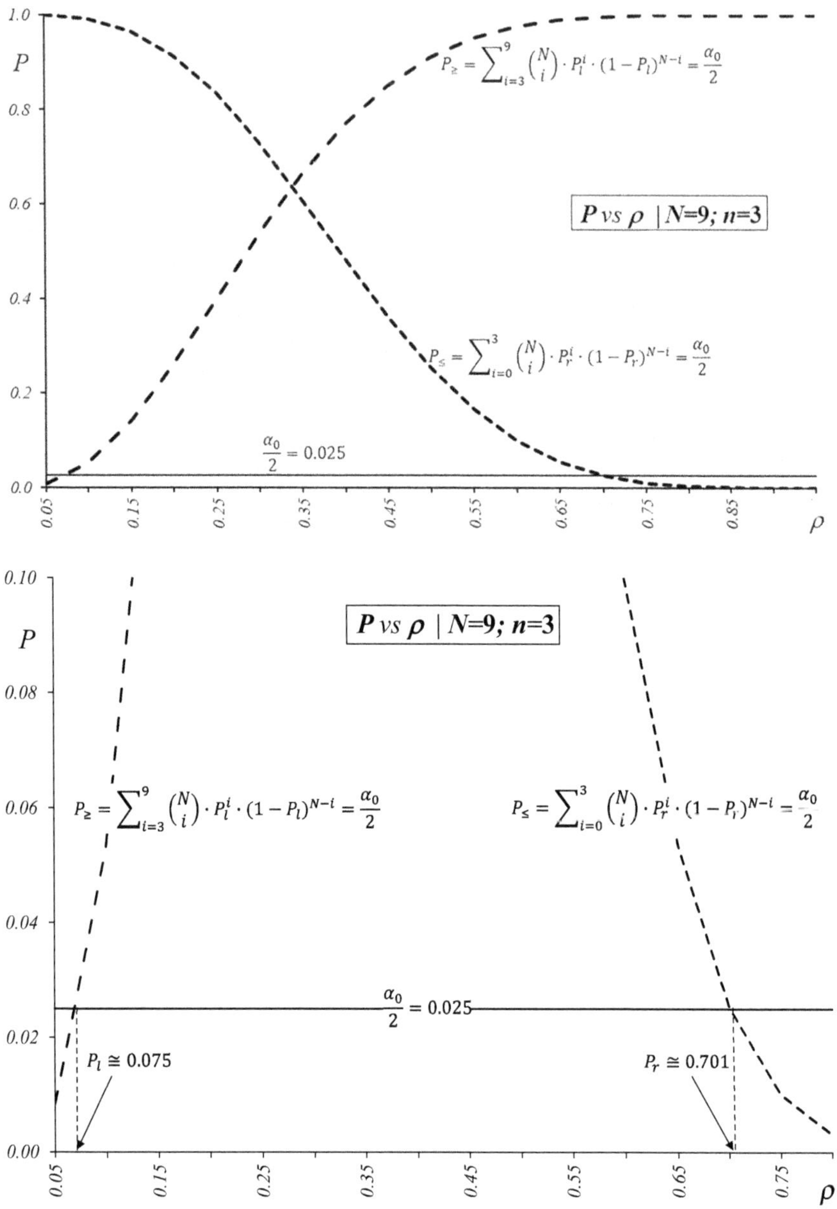

FIGURE 2.19 Graphic representation of confidence level with Bernoulli's trials P *vs.* ρ for $N = 9$ and $n = 3$ as an example.

DCS may be acceptable for some military diving purposes. On the other hand, for recreational diving, the risk should be less than 1% at a confidence level of $\overline{P} = 95\%$. The estimation error of the probability limits $P_l < \rho < P_r$ is α_0. Hence, the confidence level of this estimate is $\bar{P} = 1 - \alpha_0$. In order to find the number N of dives necessary to verify a decompression profile within a specified limit, a low probability of error α_0 must be assumed. Then we can calculate, from the binomial distribution, the number N of experimental dives and the incident rate n that will give us the confidence interval $\bar{P}$, which has the probability equal to or less than the limit value assumed α_0. The confidence intervals $\bar{P}$ of ρ value for $n = 0$ and $n = 1$ incident of DCS at $\bar{P} = \{95, 99\}\%$ confidence levels are shown in Table 2.2.

Figure 2.20 shows the maximal probability $P_r \equiv \rho_{max}$ for $n = \{1,2,3\}$ incidents of DCS as a function of the number of dives N.

Before we start testing decompression tables, we would like, ideally, the validation procedure to have the following characteristics:

- The number N of the experimental dives required to be small
- The experiments to involve a relatively small risk ρ of DCS
- The final result to provide a simple and clear answer about the safety of the decompression model tested

TABLE 2.2

Confidence Limits of DCS Risk ρ (%) for Zero $n = 0$ or with One $n = 1$ Single DCS Incident in Relation to the Number of Experimental Dives Performed at $\bar{P} = \{95,99\}\%$ Confidence Levels

Number N of Experimental Dives	Number n of DCS Incidents	Confidence Limits of DCS Risk ρ in $[\%\ DCS]$ at Specified Confidence Level $\bar{P} = \{95, 99\}\%$	
N	n	$\bar{P} = 95\%$	$\bar{P} = 99\%$
5	0	0.00–45.07	0.00–60.19
5	1	0.51–71.64	0.10–81.49
10	0	0.00–25.89	0.00–36.90
10	1	0.25–44.50	0.05–54.43
20	0	0.00–13.91	0.00–20.57
20	1	0.13–24.87	0.03–31.71
50	0	0.00–5.82	0.00–8.80
50	1	0.05–10.65	0.01–13.94
80	0	0.00–3.68	0.00–5.59
80	1	0.03–6.77	0.01–8.92
90	0	0.00–3.27	0.00–4.99
90	1	0.03–6.04	0.01–7.97
100	0	0.00–2.95	0.00–4.50
100	1	0.03–5.45	0.01–7.20
300	0	0.00–0.99	0.00–1.52
300	1	0.01–1.84	0.00–2.45
460	0	0.00–0.65	0.00–1.00

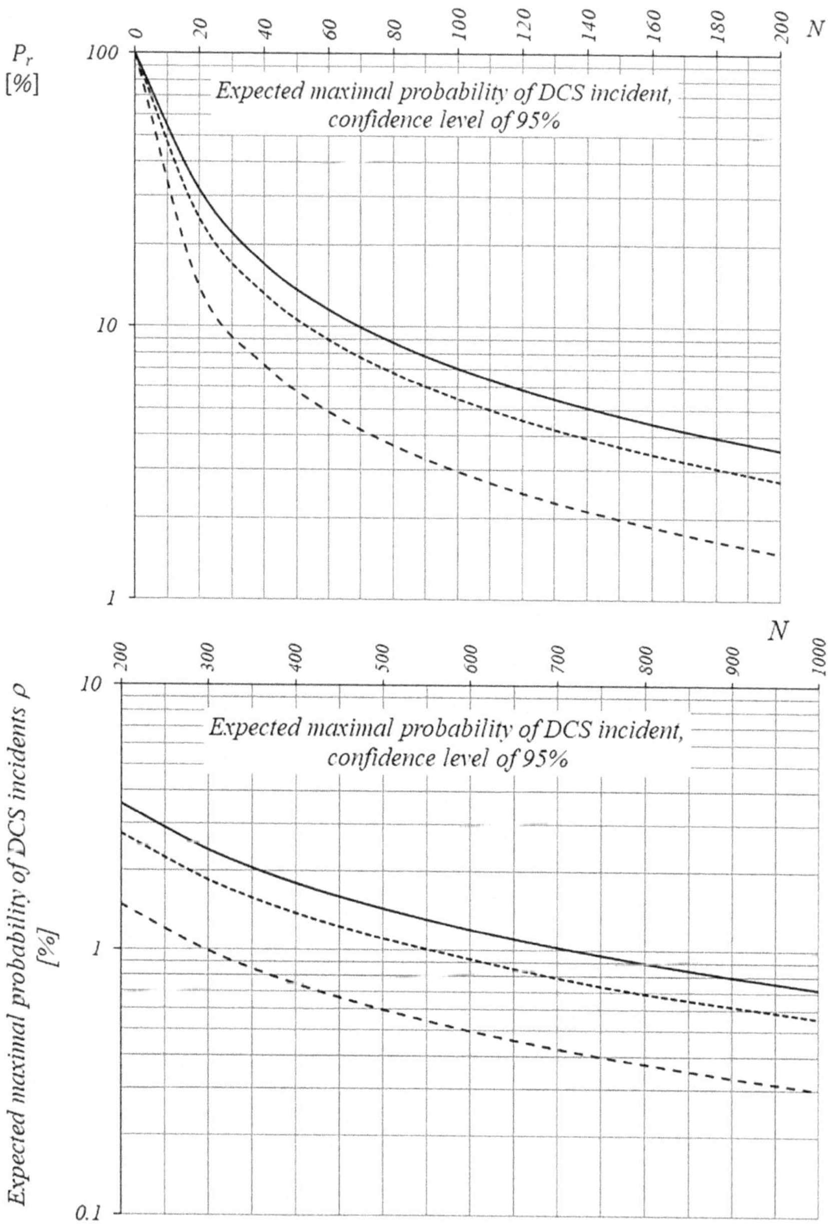

FIGURE 2.20 Expected maximal probability $P_r \equiv \rho_{max}$ of *DCS* incident for confidence level of $\bar{P} = 95\%$.

The aforementioned expectations are obviously contradictory. As shown in the preceding example, if we aim at reducing the number of dives N and at minimal *DCS* risk ρ, we cannot, as a rule, achieve a clear answer concerning the safety of the decompression profile tested. A simple and clear answer requires a relatively high number of tests N. This demands considerable financial means and a relatively long time to complete the study.

Example: After completing $N = 20$ experimental dives without $n = 0$ any incidents of DCS, it may be assumed that the probability ρ of DCS remains between $\rho \in (0,13.91)\%$ at a confidence level of $\bar{P} = 95\%$. In order to obtain a DCS risk below $\rho < 1\%$ at the same confidence level $\bar{P} = 95\%$, about $N \cong 300 \mid n = 0$ experimental dives must be done with any DCS. If it is necessary to assume a higher confidence level, such as $\bar{P} = 99\%$, and we would like to conduct experimental dives for each depth and time, the $\rho \in (0.00,1.00)\%$ confidence interval requires the completion of at least $N \cong 460$ experimental dives without $n = 0$ any DCS symptom for each entry in the decompression table.

Because the use of confidence intervals to validate the entire decompression table would require an extremely large number N of experimental dives, some researchers prefer to assume that the DCS risk is constant $\rho = const$ for the whole range of depth and bottom times generated by the mathematical model of decompression. In this case, the expected probability ρ of DCS is applied to a whole population of events, regardless of the decompression profile tested. This involves dividing the profiles tested into certain groups[82] and then running statistical tests for these groups. Of course, the representative quality of such tests depends on the quality and variety of the selected profiles. Such procedures are justified when the experimental dives ascribed to the particular groups are comparable to bottom times and depths. But when this selection is spread over a whole range of depths, for dive duration from several minutes to dive times leading to saturation of the body, such procedures could be wrong.

In order to develop more practical decompression validation procedures affordable under financial and time constraints, it will be necessary to investigate other strategies applicable to reducing the number of the required dives while still keeping in mind the statistical considerations presented in this book.

2.7.4 Sequential Analysis

In the context of significance testing in statistics, we can define two basic kinds of situations – most often, it is reject–support or accept–support. In reject–support testing, the null hypothesis H_0 is the logical opposite of what we actually believe, and rejecting it supports our theory by the following strategy:

- State as a statistical null H_0 and alternative H_1 hypothesis.
- Gather data.
- Using statistical theory, show from the data that it is likely H_0 is false and should be rejected for H_1.

2.7.5 Significance Tests

The statistical null hypothesis H_0 can be true or false. In this way, it can be rejected or not rejected. When H_0 has not been rejected, we can only postulate that there are not basis to reject H_0. It does not mean that we accept H_0. That is why we should perform the rejection[83] of H_0. Consequently, before undertaking the experiment, we can be certain that only four possible things can happen (2.10) (Table 2.3).

$$\forall_{i\in\{0;1\}} P\left(H_{1-i} | H_i\right) \equiv \alpha_i \tag{2.10}$$

Note that there are two kinds of errors represented in Table 2.3. Many statistics textbooks present the view that type I error rate must be kept at, or below, $\alpha_0 < 0.05$, and that type II error rate must be kept low also. Significance tests minimize type II error rate also.

Statistical power, which is equal to $\beta = 1 - \alpha_1$, must be kept correspondingly high. Ideally, power should be at least $\beta \geq 0.80$ to detect a reasonable departure from the null hypothesis H_0. In statistical power β calculation, binomial distribution plays a significant role.

Generally, there are two reasons for introducing sequential methods into statistical analysis. One is to solve more efficiently a problem which has a fixed sample solution. The other is to deal with problems for which no fixed sample solution exists. In this report, only the first is taken into consideration.

The principal subject of this report is sequential hypothesis testing and related estimation problems. Most of the problems studied in statistical inference exist as fixed sample solutions.

Although the probability of type II error is almost never known to an experimenter, these probabilities can be calculated for hypothetical values of the parameter of interest. If, for example, someone wishes to test the hypothesis:

- $H_0 : \rho_0 \leq \rho$, the decompression profile is dangerous – risk of DCS is equal or higher than ρ_0.
- $H_0 : \rho_0 > \rho$, the decompression profile is safe – risk of DCS is lower than ρ_0.

The function that is responsible for allocating probabilities, $P\left(X = x\right)$, is known as the probability density function L of random variable X. Figure 2.21 presents the probability density function for two DCS risks: $\rho_0 = 0$ and $= 0.2$.

Next, and critically important to the experimenter, is to assume that the null hypothesis H_0 is to be rejected if the true decompression risk is $= 0.2$, rather than the hypothesized value of $\rho_0 = 0$. Without this supposition, we cannot count type II error α_1. If the experimenter wishes to be confident that the null hypothesis H_0 will be rejected when the true decompression risk is $\rho = 0.2$, the statistical power

TABLE 2.3
Type of Errors vs. Type of Decisions into Statistical Inference

		State of the World	
		H_0	H_1
Decision	H_0	Correct H_0 non-rejection	Type II error α_1
	H_1	Type I error α_0	Correct H_0 rejection

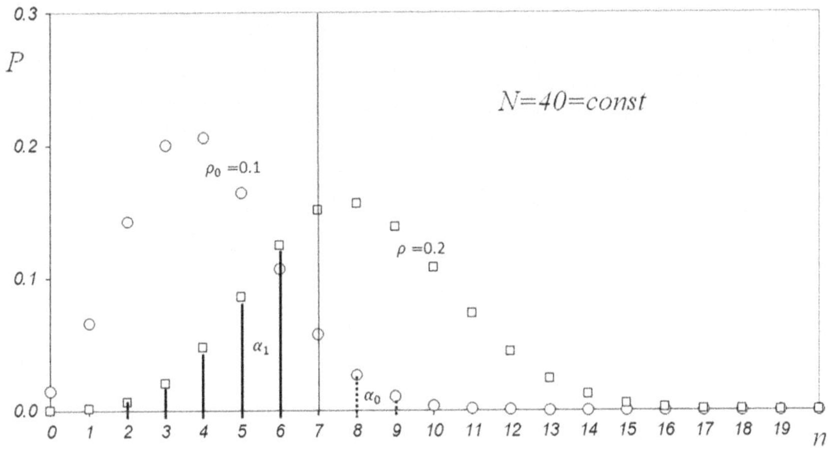

FIGURE 2.21 Example of probability density functions for ρ_0 and ρ *DCS* risk and error probabilities for $N = 40$ experimental dives.

$\bar{P} = (1 - \alpha_1)$ of the test must be acceptably large.[84] The significance level, the probability of type II error, and the power β of the test are indicated in Figure 2.21.

Figure 2.22 illustrates converse cumulative distribution function F of *DCS* occurrence for the same test as in Figure 2.21.

To ensure a statistical test has adequate power β, one usually must perform analyses prior to running the experiment, in order to calculate how large N is required. The statistical theory, of course, cannot tell us what will happen with any particular experimental dives. However, through the concept of a sampling distribution, it can tell us what will tend to happen in the long run S_N, over many experimental dives of a particular size N. A sampling S_N distribution is the distribution of a statistic over repeated samples N. *DCS* risk ρ, in typical error propagation, should have binomial distribution with the most probable value $\hat{\rho}$. Let A be an event comprising N multiplication of decompression profile tests which produce n onset of *DCS*. For A, it is assumed that $\hat{\rho} = \rho_0$, and probabilities density function $L_{\leq}(n, N) = \alpha_0$:

$$A : \left\{ \hat{\rho} = \rho_0, L_{\leq}(n, N) = \sum_{i=0}^{n} P(i, N) = \alpha_0 \right\} \tag{2.11}$$

In order to determine the *DCS* risk for a test profile, as a rule, a set of two complementary hypotheses should be built – null H_0 and an alternate H_1 hypothesis:

$$\begin{cases} H_0 : \rho \geq \rho_0 \\ H_1 : \rho < \rho_0 \end{cases} \tag{2.12}$$

Rejecting H_0 will support what the experimenter actually believes.[85] In statistics, this kind of testing is much more common in the context of significance testing. Sometimes this kind of strategy cannot be used, and it is only a test acceptance–support testing strategy that can be used.

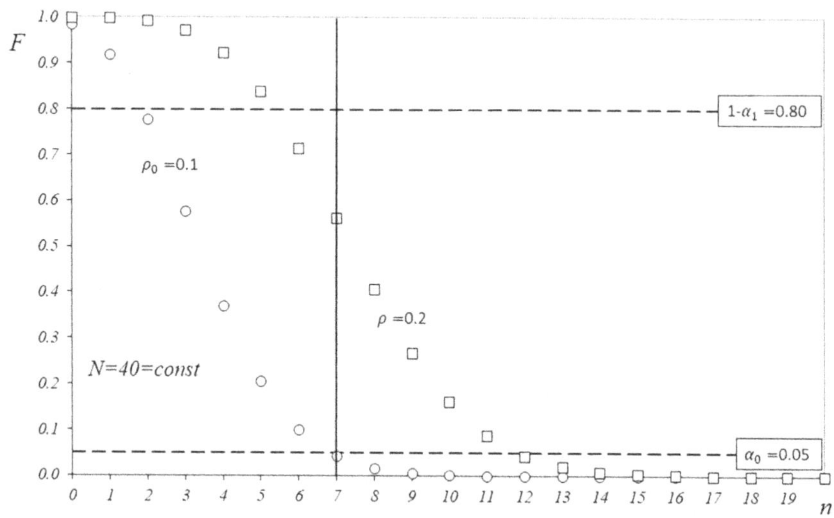

FIGURE 2.22 Example of converse cumulative distribution functions F of DCS occurrence for $\dot{V}_E$ and $\rho = 0.2$ for $N = 40$ experimental dives.

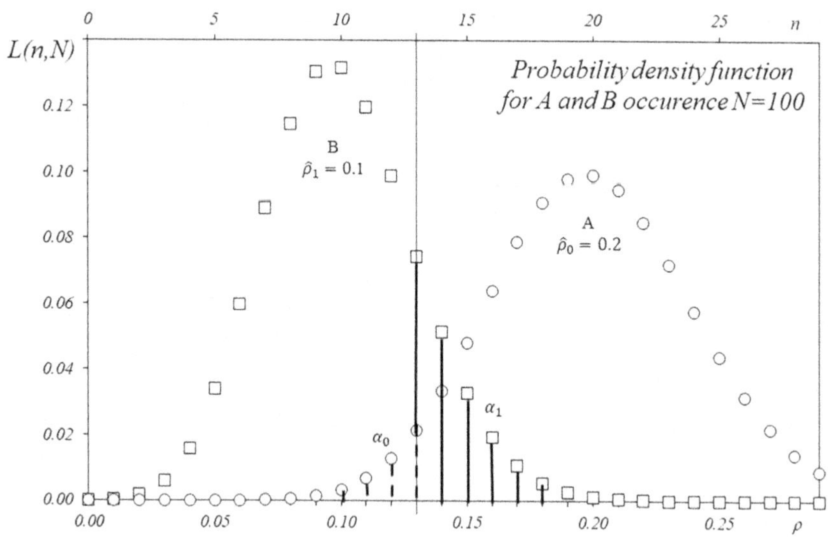

FIGURE 2.23 Example of probability density function L for occurrences A and B for $N = 40$ experimental dives, described in the text.

The formula $\rho = n / N$ defines experimental DCS risk. If the cumulative statistical function of whichever events for DCS occurrence included in set $(0; n)$ is small, smaller than critical value[86] α_k, then there is a low probability of an A event. However, if A has occurred, it is evident that it has other probability density distribution, for example, for B event with $\rho_1 < \rho_0$ (Figure 2.23).

With the precision[87] of this statistical inference, the probability of A occurrence can be accepted for an observed experimental frequency of DCS, or it can be estimated from critical probability by using statistical tables. The rejection of H_0 is excluded, with a probability of error on α_k level.[88] Acceptance of H_0 shows that, on experimental grounds, there is no basis for H_0 rejection. It is not proven that H_0 is true. It is only suggested that assumptions, on the basis of the statistical experiment, have to be changed,[89] and in this way, the statistical inference regarding the rejection of H_0 can be statistically justified. This choice, in many cases, enables us to minimize the number N of necessary experiments for taking a statistical decision. Of course, it still remains a type II risk error. Because of difficulties in taking into consideration this type of error, many scientists often disregard type II errors, but this can lead to wrong decisions during statistical inference. Additional assumptions need to be made to take into consideration type II errors. In this case occurrence, A and its probable density function are treated as a theoretical distribution,[90] and occurrence B, with $\hat{\rho} = \rho_1$ and probabilities density function $L_{\geq}(n,N) = 1 - \alpha_1$:

$$B : \left\{ \hat{\rho} = \rho_1,\ L_{\geq}(n,N) = \sum_{i=n}^{N} P(i,N) = 1 - \alpha_1 \right\} \tag{2.13}$$

Like an occurrence which is the result of a decompression profile test.

The distance Δ between densities function L of B and A is defined as $\Delta = \rho_1 - \rho_0$ (Figure 2.23). The difference Δ represents demanded resolution between probability density functions and the most probable values. The experimental value of DCS risk, for N experimental dives, was connected with DCS occurrences n, as mentioned. For A and B events, value $\rho = n / N$ cutoff values of cumulative distribution F functions for A and B, like α_0 and α_1, respectively (Figure 2.24).

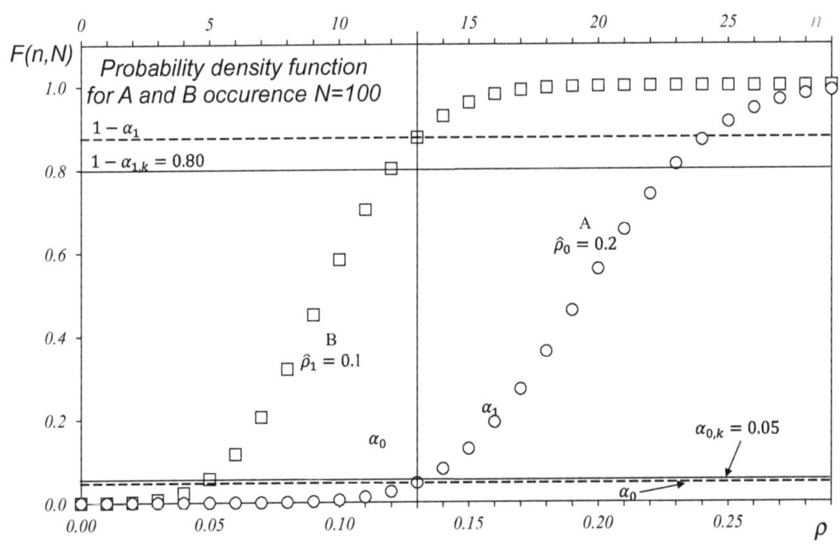

FIGURE 2.24 Probability distribution functions F for occurrences A and B, described in the text.

If these probabilities are lower for critical values $\alpha_{0,k}$ and $\alpha_{1,k}$, then the occurrence of A event is hardly probable $\alpha_0 < \alpha_{0,k}$, and it is a high probability of B event,[91] that is, rejection of H_0 is statistically justified. This is a much more precise method of interference than the previous one, because it gives the most probable value of DCS risk ρ.

Example: In the reference literature, various methods of statistical inference (validation) are found. One of the simplest methods is that developed at the Naval Medical Research Institute (*NMRI*): it limits the number of experimental dives N to a maximum of $N \leq 40$ (Figure 2.25).

Having completed 28 dives free from *DCS* incidents, it is assumed that the profile tested is safe. But if under the tests a case of *DCS* has appeared, the dives should be continued to a total of 40, on condition that no DCS incident has been recorded before. If it does, the procedure is arrested, and the profile is considered to be dangerous, and therefore rejected; but if, in the course of 40 expositions, only one case of *DCS* has taken place, the profile gets accepted as safe enough. It is not possible to use the inference dependent on the rejection H_0 in requiring a small N. That is why the hypothesis set is described as:

$$\begin{cases} H_0 : \rho_0 > \rho_1 \\ H_1 : \rho_0 \leq \rho_1 \end{cases} \tag{2.14}$$

For this set of hypotheses, it can be described by this equation:

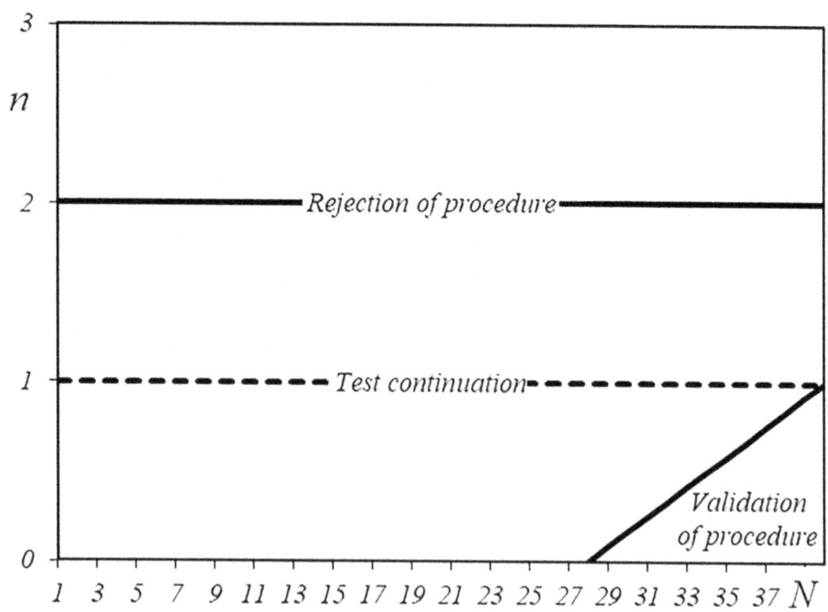

FIGURE 2.25 Chart of validation procedure used by the Naval Medical Research Institute for a single decompression profile.

Source: Huggins (1992).

$$\begin{cases} 1-\sum_{x=0}^{n}\left[\frac{N!}{x!(N-x)!}\rho_1^x(1-\rho_1)^{N-x}\right]=1-\alpha_1 \\ \sum_{x=0}^{n}\left[\frac{N!}{x!(N-x)!}\rho_0^x(1-\rho_0)^{N-x}\right]=\alpha_0 \end{cases} \tag{2.15}$$

which can be solved numerically. For $N = 28$, $n = 0$, $\alpha_0 = 0.05$, and $1-\alpha_1 = 0.80$, *DCS* risks are $\rho_0 \cong 0.101$ and $\rho_1 \cong 0.056$. Solving again by equation (2.15) for $N = 40$, $n = 1$, $\alpha_0 = 0.05$, and $1-\alpha_1 = 0.80$, $\rho_0 \cong 0.113$, and $\rho_1 \cong 0.073$. In other words, it is possible to distinguish *DCS* risk between higher than 10% ($\rho_0 > 0.10$) and lower than 7% ($\rho_1 < 0.07$), with 95% of significance level and statistical power of 80% (Figure 2.26). Thus, the *NMRI* procedure is a good tool only for the elimination of risky decompression profiles, in contrast to the method based on binomial distribution, where it is possible to estimate the maximum risk of *DCS* for a decompression profile.

An answer from sequential analysis is by no means more complete than that derived from the binomial approach, but it is obtained using a faster method. The previous results show that they are not a basis for rejection of H_0 with $\alpha_0 = 5\%$ of error for the *NMRI* procedure. Rejection of H_0 gives a basis for the acceptance of decompression profile that can generate no more than 10% *DCS* risk. Thus, the *NMRI* procedure is a good tool only for the elimination of risky decompression profiles, in contrast to the method based on binomial distribution, where it is possible to estimate the maximum risk of *DCS* for a decompression profile. This answer appears unsatisfactory, but determination of *DCS* risk $\rho = 0.01$ level by means of the binomial model needs:

$$N = \left.\frac{\lg \alpha_0}{\lg(1-\rho_1)}\right| n = 0 \rightarrow N \cong \frac{\log 0.05}{\log(1-0.01)} \cong 298 \tag{2.16}$$

$N = 298$ experimental dives without $n = 0$ any *DCS* for $\bar{P} = 95\%$ confidential level. In this way, the *NMRI* procedure appears to be an efficient screening method.

2.7.6 Monte Carlo Simulations

The *Monte Carlo* technique is an important numerical tool for studying the relationship of variables in complex systems of equations. The general idea behind the *Monte Carlo* techniques is to generate random variations in the variables, which are then inserted into the appropriate equations to arrive at some result. The starting point is the pseudo-randomization function supplied with most computers, or the use of special mathematical formulas. In most situations, the pseudo-random numbers generator generates every value between 0 and 1, which has an equal probability of being hit. For generating pseudo-random number L from an a to b interval, with

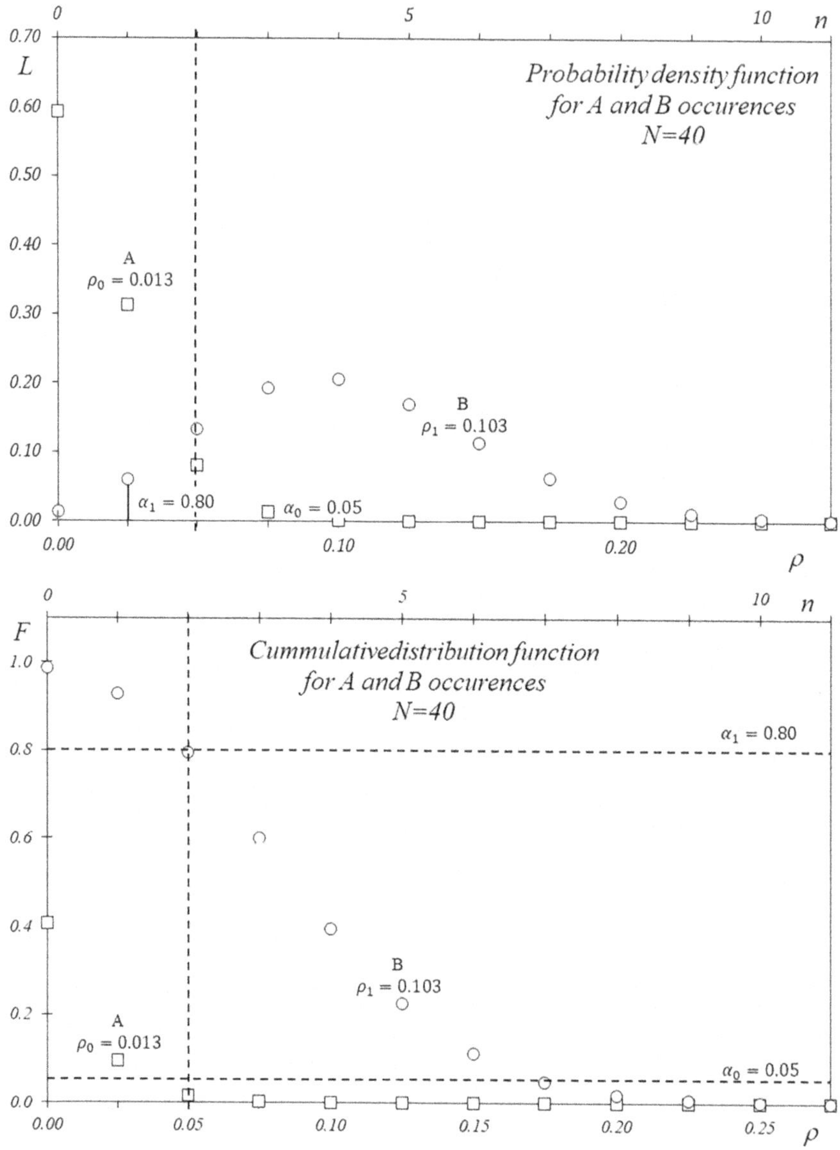

FIGURE 2.26 Probability density L and cumulative distribution functions F for the *NMRI* method.

the use of generator G, which generates pseudo-random numbers from 0 to 1, it is possible to use the formula:

$$L = G(b - a) + a \tag{2.17}$$

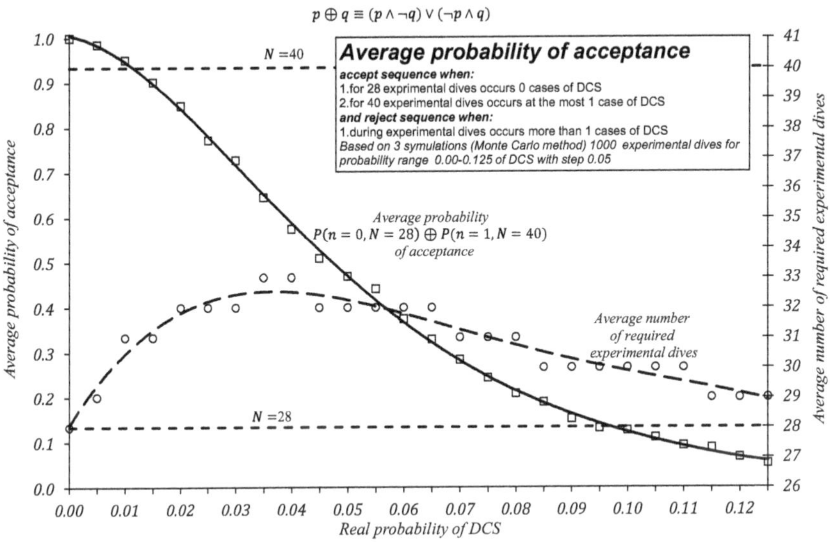

FIGURE 2.27 An average result of three times' repetition of 1000 *NMRI* procedure simulations, where *XOR* function is defined as $p \oplus q \equiv (p \wedge \neg q) \vee (\neg p \wedge q)$.

Example: For 1,000 repetitions of the *NMRI* procedure with *DCS* probability accuracy $\Delta = 0.001$, the result of one dive simulation can be written as:

$$\begin{cases} n = 0\, for\, int(1000G) < int[1000\rho] \\ n = 1\, for\, int(1000G) \geq int[1000\rho] \end{cases} \tag{2.18}$$

Figure 2.27 shows an average result of 1,000 *NMRI* procedure simulations repeated three times.

In Figure 2.27 and Figure 2.28, the points represent the result of simulations, and the lines are the result of smoothing the data. The solid line represents the average probability of the *NMRI* procedure acceptance, while the dashed line represents the average number of required dives. The horizontal lines show two thresholds of the *NMRI* procedure.

The Monte Carlo simulation is an effective tool for checking theoretical assessments. By using it during scientific investigations, gross and systematic errors can be avoided.

2.7.7 Conclusion

Statistical methods play many roles.

One of the primary roles of statistical methods is to better visualize data by collecting it in special charts, computing special agglomerated point measures, finding their distributions, and so on. This is the so-called descriptive statistics.

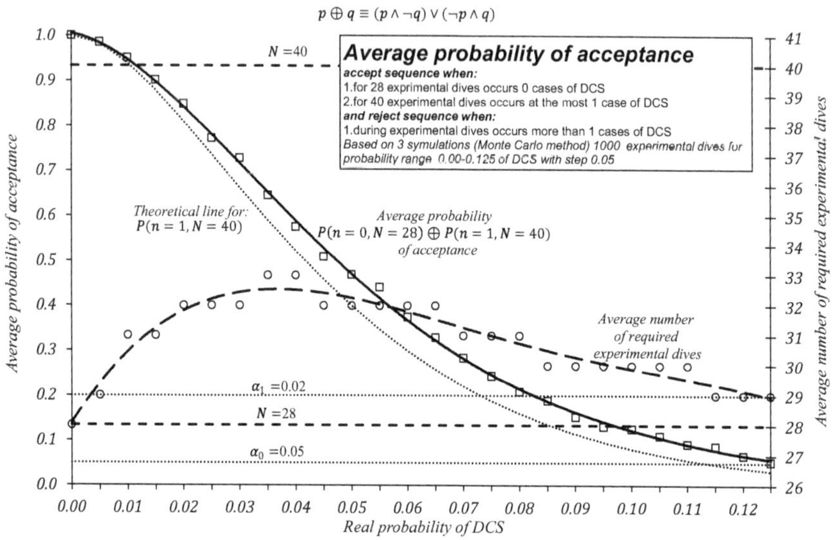

FIGURE 2.28 An average result of three times' repetition of 1,000 NMRI procedure simulations and theoretical lines,[92] where the *XOR* function is defined as $p \oplus q \equiv (p \wedge \neg q) \vee (\neg p \wedge q)$.

The second very important role of statistical methods is finding outliers from the typical data distribution of the phenomenon under study. These points may indicate hidden deterministic behavior in a series of data that looks like noise, which may reflect the mechanism of the phenomenon which is of interest for the researcher. In many cases, this role is neglected and outliers are removed from the dataset, and therefore, the deterministic mechanism is still not explored and remains undiscovered.

The validation process of diving decompression schedules is based on statistical methods. These methods are used to verify the assumptions made in the decompression model by establishing a decompression sickness *DCS* risk function based on experimental dive data. The presented analysis takes into account the first role of statistical methods, but whenever the proposed methods are used, one should bear in mind the second role of statistical methods.

2.8 CONTEXT

The necessity to employ these methods resulted from the specific requirements set for military technologies,[93] which, as broadly as possible, should be knowledge-based. The knowledge-based approach, by its very nature, allows continuous improvement of the adequacy of the predictions made, the estimation of the level of risk when diagnosing deviations from the repeatability or precision of the model, and the possibility of adapting the technology to the changing requirements of the user resulting from tactical considerations of its use.

2.8.1 Diving

In each project, there should be defined assumptions relating to its objective. Propositions of the *process* used to achieve it[94] comprise a specification[95] of the minimum and optimal structure for the proposed *system* that can ensure consistent, stable, and effective implementation of the proposed *process*, taking into account the impact of the *process* on the environment and the impact of the environment on the system and the occurring *process*.

Initially, only processes in which the diver's body was exposed to pressure were considered as dives, while dives using one-atmosphere diving suits were considered as use of an underwater vehicle. Now, due to the required training and the necessity to carry out tasks autonomously, the term *diver* is also extended to aquanauts using a one-atmosphere suit.

Different goals can be set for the diving process. Depending on the expected end goal of the diving process, an appropriate system should be established which is capable of ensuring the sustainability of the readiness to achieve the diving objectives. Different subdivisions can be applied according to the purpose of the diving process – for instance:

- Extreme
- Sports
- Recreational
- Technical
- Professional
- Rescue
- Military, etc.

The minimum *system* capable of ensuring continuity of the diving *process* conducted for a specified purpose is merely a theoretical construct. In extreme dives, which are conducted in order to set up a record, system minimization is most often applied. Under these conditions, the usual main assumption is to enhance the *system* with minimum security measures or ensure redundancy[96] of some of the elements of the system.[97] The elements that go beyond the minimum structure of the *system* to achieve the objective of the diving *process* are a result of a risk analysis and feasibility study.

Illustratively, the minimization of the diving system can be analyzed using the example of dives performed by sponge divers described in historical travel books.

The minimum *system* needed to carry out the diving *process* consists of only a trained diver. The element of the diving training system can be divided into two subsystems: swimming lessons and training in diving whilst holding one's breath. However, the feasibility study shows that a trained diver will work more effectively if he/she is equipped with a sponge cutter, which, according to the risk analysis, can also be used to assist in the protection of the diver. If a diver is well-trained and can withstand a longer period of time with his breath held, then the effectiveness analysis for the diving process in order to harvest sponge shows that he should be additionally equipped with a net for collecting sponges.

The feasibility study shows that sponges are unlikely to grow close to the shore, and that the effectiveness of swimming far from the shore is unlikely to be acceptable;

thus, it is necessary to ensure that the diver is able to travel by boat. When focusing only on the diving system, the boat should be considered as an element of the system environment, that is, the context[98] surrounding the diving system under consideration. It should be noted that in this context, there are also other elements that affect the diving system for sponge harvesting, which should be taken into account in the feasibility study, such as market demand or the threat of a shark attack.

Subsequent risk analyses and technical feasibility or economic viability studies may relate to the use of an air pump or air compressor, an air supply system with or without filtration, and an air supply regulator; the use of a mask and fins; and access to health care, social security, or pension systems. Defining the objective to be achieved by an established process can thus be used to establish the structure of the system that supports it, taking into account the environmental impact on the *system* and the impact of the *system* on the environment.

For the division proposed here, the distinguishing feature of these types of dives is precisely the definition of their different objectives. Extreme dives are usually carried out in order to break records, but they can be a part of a rescue operation or a quickly changing tactical scenario for a combat situation. Typical examples of extreme dives include free diving where a breath-holding diver is dragged to an extreme depth riding on a weighted sledge or, alternatively, where the diver reaches the desired depth on his/her own using fins either after being initially pushed to a certain depth by the second diver or reaching it by himself/herself.

Sport dives, in turn, are carried out for competitive purposes. For example, underwater spearfishing championships previously consisted of catching fish in a defined time, though nowadays photographic competitions are more common. Recreational dives are carried out for pleasure.

Technical diving originated from the desire to explore greater depths and for a greater duration than had previously been possible, the achievement of this goal requiring the use of special, complicated equipment. Technical dives are often implemented, as well as extreme dives, to break records of depth, time spent underwater, or to explore some areas, such as wrecks or caves.

Professional diving involves the use of a variety of technical equipment for conducting commercial underwater activities. A specific type of professional diving involves operations to save lives. Rescue of property is part of typical commercial professional diving. Lifesaving dives are characterized by different dynamics and determination from other types of diving. Moreover, they use other procedures and carry additional equipment for the victims. Technical dives have emerged from military dives, but their purpose is nonmilitary.

The objectives of military dives may vary, for example, in combat dives, the objectives are usually different from those of *MCM* dives.[99]

According to the aim of the diving process, different approaches are used to build a *system* ensuring that the process runs smoothly, efficiently, and safely.

2.8.2 Conservatism

During underwater works, both divers and diving supervisors prefer to carry out decompression according to low-risk regimes;[100] therefore, a *conservatism function* has been introduced into decompression programs, using, for example, the method of calculation for elevated inert gas content compared to actual, assuming an increased depth compared to the actual depth, assuming longer than actual time at depth, or using an asymmetric model for saturation/desaturation of theoretical tissues.[101] Some programs use more than one of these methods or a combination of them to increase decompression safety. These methods are called conservative methodologies or are referred to simply as *conservatism.*[102] *Conservatism* is often specified by indicating how much longer and deeper the decompression profile should be compared to the actual one. Sometimes you can find a numerical expression for *conservatism,* for example, in percentages related to the saturation gradient.[103] Diving technologies may be categorized according to their level of *conservatism.*

In the case of extreme dives, the technology of diving is chosen individually for the diver. Most often, the concept of *conservatism* for this type of dives does not occur, as the typical practice is to exceed the typical safety limits to which conservatism refers. An occurrence of a diving accident is considered to be part of extreme diving, and the security measures are related to the development of treatment systems as close as possible to the place where extreme diving is organized.

In technical diving, technology is often dedicated to the task and the current advancement of the divers' training. Often, technologies, and especially decompression schedules, are selected for the current course of diving by computer systems supporting the process of diving, taking into account not only changes in depth and time of stay but also variations in temperature, gas consumption, or changes in *conservatism* settings. Technical dives should not cross the threshold of zero *conservatism,* as is the case with extreme dives.

Part of technical diving is scientific diving conducted in order to get to know the water environment, both animate and inanimate. Due to the fact that the diving is performed by scientists, also advanced in age, *conservatism* is selected for them individually at an elevated level.

In sports diving, the level of conservatism exceeds the typical *conservatism* level of 30%, which is the limit for diving with safe, albeit strenuous, decompression. It is consensually agreed and is identical for all diving participants. The differences lie in the way of preparing for the sports diving process, which constitutes a hidden handicap of individual competitors.

In the case of recreational diving, the level of *conservatism* is selected in such a way that the risk[104] of diving-related illnesses *DCI*[105] is lower than 1% (Huggins, 1992).

Part of recreational diving is rehabilitation diving, less frequently practiced in the form of medical rehabilitation[106] and more commonly in the form of social rehabilitation.[107]

In professional diving, medical standards are usually more elaborate and rigorous than in recreational diving. Conservatism is often set at the same level as for recreational diving but no longer applies to the entire population that can dive but to a selected population that meets higher requirements than the average individual in the professional diving population. If, for professional dives, the risk of *DCI* is higher than 1% in the estimation as for an epidemiological risk, for a selected population meeting the special requirements set in medical standards, similarly as in normal procedures of work protection, divers must give informed consent to derogations, and the risk incurred should be compensated both in terms of health protection and financial compensation.

Rescue dives are characterized not only by increased risk of an occurrence of the *DCI* symptoms but also by the risk of a technical accident due to the failure of the equipment used.[108] An increased risk of *DCI* is only allowed when directly saving lives.[109] Hence, most often in rescue operations, conservatism is maintained at the level typical for professional diving. This applies, of course, to the group of carefully selected and trained divers provided with safety support by specialists in various fields needed in the rescue operation.

Military dives are characterized by different levels of risk of *DCI*. For *EOD*[110] dives, the same rules apply as for rescue dives. For *MCM* dives, depending on the importance of the mission, it is possible to approve the threat of the occurrence of *DCI* symptoms from the level of *EOD* dives to combat missions *SRT*,[111] for which the typical *DCI* threat is determined at the level[112] of 5% in terms of estimation, as for the epidemiological threat, for a selected population meeting the special requirements set out in medical standards.[113]

This shows that military diving technologies should be based, as broadly as possible, on knowledge. Other technologies can be bought. Even with technical dives, broader but available knowledge is sufficient to make decisions about the risks to be taken, which is enabled by the algorithm embedded in the computer memory to support decision-making concerned with the decompression process used.

The context of the use of diving technologies as elements of a warfare system seems to be particularly important. The aim should be to make the technology's *conservatism* as effective as possible and to create a risk R within the limits of $R \in [1,3]\%$ in terms of its estimation, as for an epidemiological threat, for a selected population meeting the special requirements set out in medical standards.

2.8.3 Physiology

The basic requirements for conducting diving operations using self-contained diving devices result from the risk of *decompression sickness DCS* and *central nervous system oxygen toxicity*[114] *CNSyn*.

Among the basic decompression parameters, the following assumptions for exposure can be distinguished:

- Maximum planned depth of hyperbaric exposure
- Exposure time at maximum depth
- Breathing mix used
- Diver's predisposition to undergo a hyperbaric exposure

Usually, the depth H is a complicated functional dependence on the time of exposure t: $H = f(t)$, which can be approximated by a pair of parameters: maximum depth of exposure H_{max} and time t of stay at maximum depth $t = f(H_{max})$. An extremely important parameter for planning effective, safe exposure and decompression is the oxygen partial pressure p_{O_2}, which is typically a function of the exposure and decompression times t: $p_{O_2} = f(t)$. During descent and presence at the bottom, it is most common to use a single operational mixture to supply the diving apparatus, but depending on its design, it can maintain the partial pressure of oxygen p_{O_2} within very narrow or very wide ranges.

Planning to perform long-term intensive work in diving operations outside the saturation zone is troublesome due to the lack of possibility to design adequate decompression (Brubakk and Neuman, 2003). Strenuous exertion induces the need for an adequate response from the diver's body. Most often, the response is to include anaerobic changes, causing the formation of lactic acid in the muscles, which, by acidifying the blood, causes a decrease in the efficiency of oxygen transportation through hemoglobin (Bühlmann, 1984, 1995).

These effects differ with individuals θ and, for one and the same individual, strongly depend on the current training $\theta_i = f(t)$, preparation,[115] or even attitude to diving. It is difficult to take some individual parameters into account when planning adequate decompression for the population. During saturated dives, time is given for stabilization prior to the decompression process. This time depends on the type of work performed earlier and can amount to up to two days.

As mentioned earlier, in a simplified approach to the decompression planning process, two parameters are used: the maximum depth of exposure reached H_{max} and the exposure time at maximum depth $t(H_{max})$, calculated from the beginning of the immersion to the initiation of the decompression process. The maximum exposure depth H_{max} may be reached just for a moment[116] with the decompression process starting from the moment when the pressure can only decrease, except for the required stops at decompression stations. If the stops at the decompression stations become longer, the decompression schedule may also change. This way of planning the decompression process is based on the worst-case scenario selection procedure. This increases the *conservatism* of the adopted decompression procedure in a variable way, depending on the diving conditions and the diving supervisor's approach to decompression planning. It should be noted that an unjustified increase in *conservatism* may result in inefficiency in the diving process due to the significant oversizing of the decompression process. In accordance with the tradition of the Polish Navy, the worst-case scenario method was used to select the decompression process for the project.

The decompression and compression processes may also occur at constant pressure when the composition of the breathing mix changes (Brubakk and Neuman, 2003). Isobaric decompression processes are accompanied by oxygen decompression. Replacement of the gas mixture with pure oxygen results in the washing out of inert gases by oxygen from the tissues. Oxygen is removed from the body, mainly through metabolic consumption. Isobaric decompression occurs at oxygen decompression stations. Due to its toxicity, oxygen can only be used at relatively shallow decompression stations. It is accepted around the world that oxygen decompression can be started from a depth of $12\,mH_2O$. The Polish Navy has a long tradition of using oxygen decompression starting from $15\,mH_2O$, but in order to maintain interoperability in diving, oxygen decompression was planned from $12\,mH_2O$. The breathing space of the diving apparatus could be ventilated with oxygen as early as $15\,mH_2O$.

Decompression can also be accelerated by using oxygen-enriched gas mixtures. The mechanism is similar, and the use of such mixtures can begin before the transition to oxygen decompression. However, it is possible to use a mixture other than one enriched with oxygen. For example, by substituting travel mixture as *Trimix Tx* with, for example, *Nitrox Nx*0.3, in such a case, even at constant external pressure, *Nx* will elute helium from tissues and replace it with nitrogen, that is, the replacement of the breathing mix will cause the phenomena of isobaric decompression. It should be noted, however, that nitrogen is much better soluble in tissues containing fat than helium is, so planning a long period of isobaric decompression if the breathing mix is changed from *Tx* to *Nx* may prolong the decompression process instead of accelerating it. If this is not taken into account, symptoms of decompression sickness *DCS* may occur. Such a mechanism of decompression sickness is known as *DCS* induced by the counterdiffusion process (Predictive Studies IV, 1978; Lambertsen, 1986).

The diver's predisposition to hyperbaric exposure constitutes a complex of issues that may include the following:

- Physical build
- Physical and mental state prior to diving
- Mental build
- The current state of training
- The scope of the hygiene rules applied, etc.

In the past, diver's physical build was a basic parameter allowing him/her to perform military service as a diver or to work as a professional diver. This was related to the types of diving suits. Too tall a diver did not fit in a suit, while for a short diver, a reefed suit posed a threat. For example, a crack in the crotch belt in a classic suit caused the escape of the helmet and the inability to release air from the suit, leading to a diver being thrown up to the surface or hanging in the depths on a diving hose jammed with underwater obstacles. These days, it is possible to produce a suit adapted to the diver. In the past, a strong body build was required for a classic diver, because the equipment worn on the surface weighed just under 100 *kg* and required sufficient physical strength to put it on and enter the water. Nowadays, the equipment and devices for professional divers are much lighter but still require considerable physical strength to put on and move around with outside the aquatic environment.[117] Also, performing underwater works, although carried out with the use of tools and various types of equipment, often requires considerable physical effort, so that some works have to be performed only under saturation conditions.

The use of semi–closed circuit breathing apparatus requires certain predispositions. The physiological parameters *human factors HF*, which influence the level of diving safety, are indicated here with the symbol θ. For example, for *SCR CRABE SCUBA*, this is primarily the breathing module: *ventilatory equivalent for oxygen* ε.

The causes of certain phenomena[118] are not always known and should therefore be the starting point for further scientific research. It is commonly believed that older divers require more conservative decompression procedures, but quite a large percentage of divers over 45 years of age exhibit a significant increase in individual immunity to strenuous decompression. This phenomenon was observed in the course of monitoring the formation of the free gaseous phase in veins during experimental dives, although earlier cases of individual immunity to difficult decompression had been noted.[119]

This phenomenon does not occur only as an individual property θ_i but can also change with the current psychophysical state[120] of the diver. The influence of attitude or determination is not only observed when diving. There are many reliable attitude-related events, such as the occurrence of the phenomenon of imaginary pregnancy in humans and animals. It is believed that the diver is built up in 80% of his/her mental structure, especially the ability to operate effectively in stress and resistance to panic.[121] There is also a conviction that attitude is correlated with immunity to central nervous system oxygen toxicity *CNSyn*. It was also attempted to promote the view that deep knowledge-based consciousness is a contribution to facilitating the development of anxiety leading to panic in stressful situations. Indeed, it is observed that less-educated divers show greater resistance, but once the fear barrier is broken, the development of stressful situations leading to the emergence of panic is extremely violent.

Experienced dive supervisors try to gain the trust of divers to become their trustees, which makes it easier for them to assess the current psychophysical condition of the diver. When assessing this state, it is important not to underestimate any symptoms, not even minor ones. It is good practice for the supervisor to have eye contact with the diver[122] during the preparation for diving, with simultaneous examinations of the state of muscle tension and body tremor[123] or the evaluation of the heart rate,[124] etc. It is important to check the coordination of movement once the equipment is put on.[125]

It is important to detect signs of fatigue, early signs of illness, stress, etc., as they have a direct effect on the safety of decompression.

When repeating a dive, it is important to note the saturation of tissue with inert gases[126] remaining after the previous decompression. Lowering the pressure value, as is the case during flights on airplanes or transport helicopters, may cause a serious imbalance in the pressure of gas in tissues resulting in the occurrence of symptoms of *DCS*. The project did not include any studies on the possibility of cyclic dives,[127] repeated[128] dives, or achieving a state of gas balance between tissues and breathing atmosphere[129] on the surface,[130] but during the progress of works, such studies may be conducted.

In addition to strength training, ensuring good physical fitness, and flexibility training, ensuring good muscle stretching,[131] it is also important to carry out adaptive training. Adaptive training is aimed at preparing the diver for hyperbaric conditions and decompression. Making the body adapted to these conditions by clearing the path of gas exchange between tissues and respiratory atmosphere contributes to the safety of diving. In the design of the project, it will be important to define the scope of such training and guidelines for strength or flexibility training.

Hygiene of life and work of a diver is a broad concept, as are the aforementioned complications resulting from strenuous exercise done prior to diving. An important element is to undergo scheduled strength, endurance, stretching, or adaptive training. One of the important elements is the aforementioned care for proper rest and antistress training. During the training, a well-balanced diet is crucial, limiting the possibility of accumulation of adipose tissue, but ensuring an adequate energy load. It is important to provide appropriate micronutrients and vitamins in their natural form, rather than in the form of dietary supplements. Dietary supplements or medicines may cause the presence of condensation nuclei in the blood and thus increase susceptibility to easier formation of the free gaseous phase during decompression. However, solutions to this problem will not constitute partial objectives of the project and need to be developed within the framework of other activities.

The basic variants of decompression assumptions may include:

- Effort
- Thermal comfort

It is always necessary to take into account the possibility of additional, unforeseen work, for example, related to changes of the hydrological situation necessitating work against the water current. In addition, the effort expended can lead to an imbalance in thermal comfort. For example, if divers have prepared themselves to expend light effort in cold water by wearing appropriately selected thermals, they may overheat in conditions of intense effort. Thermal comfort can also be lost by crossing the thermocline[132] or as a result of prolonging the stay underwater. Previously performed effort, lack of sleep, slight infections, overcooling, or overheating are often the cause of faster hypothermia in the diver. Overheating usually occurs before diving,[133] which later results in a more rapid cooling underwater. This is especially relevant to a standby diver prepared for immersion and waiting at the waiting position; therefore, it is necessary to pay attention to his/her protection against overheating, as during the rescue operation the diver may become so exhausted that he or she will only increase the risk.

The effect of excessive, long exertion in the course of diving or before diving on decompression has already been highlighted. Thermal comfort also has a significant impact on the safety of decompression. During decompression, the state of saturation with gases is maintained in a metastable equilibrium. Each change in temperature causes the disturbance of this equilibrium. Both cooling during diving and surface decompression in a warm decompression chamber, as well as overheating during diving and decompression under cold conditions, cause disturbances of this balance, which may lead to the occurrence of symptoms of decompression sickness *DCS*. The main variants of conducting decompression include:

- Response to contamination of the breathing mix
- Reaction to deviations in the composition of the breathing mix
- Acceleration of the decompression process
- Interruption of the decompression process
- Execution of the surface and compensatory decompression
- Accumulation of decompression time at deeper stations
- Extension of the decompression process

Contamination of the breathing mixture is always undesirable. The problems related to such contamination have been previously described and will not be analyzed in this study (Kłos, 2017). In regeneration devices, carbon dioxide CO_2 retention and dust emitted by the filling of the absorbent are frequent hazards. Exhaled carbon dioxide CO_2 may accumulate in the respiratory space as a result of an absorbent either being of a poor quality, not changing the medium frequently enough, or the effect of "channeling," where the breathing medium slips through the filter bed without coming into contact with adequate amounts of the absorbent, etc. It is advisable to seek cooperation with a supplier who guarantees that the required quality of the sorbent will be maintained.

Furthermore, adequate laboratory control over the stored sorbent should be maintained. These issues will not be part of the project, but the established *SOP*[134] procedures resulting from previous work will be maintained and revised as necessary (Kłos, 2009). Sodium lime dusting is associated with its form. Sometimes it is tableted and highly resistant to abrasion and breakage; other times it is loosely formed in various extrusion shapes (Kłos and Kłos, 2004).

A nondusting form has also been introduced, where the sorbent is stabilized in a substrate made of plastic. However, the most common form available on the market is crushed lime which has been sorted in terms of size of granules on sieves. Such a form is susceptible to dusting the particles due to mechanical shocks, for example, during transport or packing of scrubbers. For this reason, the sorbent should be dedusted initially on the screens before packing into scrubbers, and then on the screens the scrubber should be air-washed. During transport of the prepared apparatuses, the lime in the scrubber may be subject to shock and crushing. This phenomenon is accompanied not only by dusting but also by the formation of passages enabling the regenerated breathing mix to pass through the scrubber without sufficient purification. It is therefore necessary to ensure careful transportation of the rebreathers. Soaking the internal layers of the breathing tubes can be a protection against dust by forming a water film on the internal surface of the tube to which the dust emitted by the filter can adhere. Carbon dioxide CO_2 sorbents are usually a mixture of highly caustic hydroxides and can cause micro burns in the respiratory tract. Due to the possibility of lumping, the sorbent is stored in a dried form, which has a lower CO_2 chemisorption efficiency. That is why before use, it should be moistened by performing several exhalations into the scrubber. Soaking the tubes also creates an environment to activate the filling of the scrubber. Partial depletion of the sorbent can occur when it is stored in partly unsealed containers or a filter. Additionally, insufficient water tightness of the device may cause the formation of a highly corrosive hydroxide solution during diving, which, in turn, may cause severe burns to the respiratory tract.

Carbon dioxide CO_2 accumulated in the circulating breathing mix causes a number of adverse effects, such as toxic effects, increased respiratory action, constriction of blood vessels[135] in the periphery of the body, and relaxation of blood vessels in the brain, causing an increase in cerebral blood flow, which leads to a greater sensitivity to the central form of central oxygen toxicity *CNSyn* (Kłos, 2012).

The impact of a general change in the breathing mix composition on decompression was discussed in the context of isobaric decompression and counterdiffusion phenomena. When designing hyperbaric exposures, especially decompression schedules, it is extremely important to take into account the existing fluctuations in the composition of the breathing mix inhaled by the diver. This is particularly important for semi–closed circuit rebreathers, where fresh premix is dynamically mixed with regenerated breathing mix.

Even with oxygen decompression, it is never assumed that the diver will breathe pure oxygen. Nowadays, high-purity oxygen[136] can be used because it is available at a reasonable price, but such a gas allows for decompression in open circuit apparatuses.[137] In diving devices with a semi-closed circuit, there are problems with effective ventilation of the breathing space with oxygen and washing out from the body the inert gases accumulated in the circuit of the breathing mix inhaled by the diver. Therefore, the issues of ventilation must take into account these phenomena, and the assumptions related to the effectiveness of the ventilation process must be validated during experimental dives (Kłos, 2007a, 2023). The issues of rebreather respiratory ventilation with both premix and oxygen are key issues and are thus the main focus of research. In the project, it was necessary to take the following into consideration:

- The time of breathing from the rebreather on the surface before the start of the diving process
- The effectiveness of ventilation of the breathing space with fresh premix during the process of descent
- The influence of body cleansing from inert gases[138] on the composition of the breathing mix inhaled by a diver
- The effectiveness of washing the respiratory space with a fresh premix during the stay on the bottom[139]
- The effectiveness of washing the respiratory space with fresh premix before starting the ascent process
- The necessity to wash the respiratory space with fresh premix at decompression stations
- the effectiveness of the process of exchanging the breathing mix with oxygen before commencing oxygen decompression
- The necessity of washing the breathing space with fresh oxygen at oxygen decompression stations

The ventilation model of the breathing space of the rebreathers is the main basis for planning safe hyperbaric exposures and subsequent decompression. The evaluation of the effectiveness of the ventilation process should be continued during training on a hyperbaric simulator, where reliable data from the diving process can be collected. The database of conducted dives provides the basis for progress building through knowledge and an important element of diving safety monitoring, on which the periodic safety analysis of diving operations should be based. The effectiveness of flushing the breathing space of a diving apparatus is an important element in assessing the training of divers. The ability to use the procedures for flushing the breathing space of the diving apparatus should be periodically checked during mandatory training on a hyperbaric simulator.

The most common procedure used to accelerate the decompression process is oxygen decompression. It is commonly believed that such an acceleration constitutes a 2.5-fold reduction of decompression time using an operating mixture (Bennett and Elliott, 1993; Brubakk and Neuman, 2003). For safety reasons, this parameter should be determined for a specific case of its use or rounded up to simplify the system.[140] However, rounding leads to less-effective[141] decompression schedules; hence, there may be a need to generate two or more tables, for example, school, training, combat used for exercises, combat used during warfare, extreme use in antiterrorist operations, etc., differing in the risk of an occurrence of symptoms of decompression sickness DCS or central oxygen toxicity $CNSyn$.

Decompression acceleration may occur in emergency situations, such as the need to bypass some decompression stations in case of injury or loss of breathing mix. This must be combined with an appropriate compensatory action. Compensatory decompression may involve oxygen exposure in normobaric or hyperbaric conditions, as in the case of the use of surface decompression procedure.[142] Further development of this issue and of the accumulation of decompression time at deeper stations will be discussed in the context of implementation of the decompression process.

In order to increase safety, the Polish Navy traditionally adopts several scenarios in which the decompression process is extended,[143] that is:

- A diver's actions require a lot of physical effort underwater
- A diver feels coldness in the course of diving and decompression
- A dive is one in a series of dives
- A diver is untrained or has an individual predisposition to develop decompression sickness DCS
- A diver is obese or his/her body weight exceeds $80\,kg$
- A diver is over 40 years of age

If one of the aforementioned problems is diagnosed, the conservatism of the approach to decompression is increased by using extended decompression. Extended decompression is based on adopting the next, longer diving time from the table for the same depth.[144] If two of the aforementioned circumstances occur, the diving time is increased by moving two lines down. In the case of a greater number of impediments, these are not taken into account and are treated as if there were only two of them. Therefore, the last two lines of the allowed stay times at the bottom, for each depth appearing in the decompression table, play the role of emergency times used only in the extended decompression procedure. Attention is also paid to other factors influencing the increase in the risk of occurrence of the symptoms of decompression sickness DCS, discussed in the next paragraphs.

This approach requires both the development of diving technology with planned effort,[145] posing a potential risk of developing symptoms of decompression sickness DCS, and technology taking into account the lack of any additional work by the diver, where the main problem is the risk of developing symptoms of central nervous system oxygen toxicity $CNSyn$.

2.8.4 Organization

In the implementation of independent dives in the marine environment conducted by Polish Navy personnel, for the classic decompression designed with decompression stations every $3\,mH_2O$, the diving technology should ensure the emergency decompression process, with the possibility of having the last decompression station at a depth of $6\,mH_2O$. Such a procedure may be useful in the case of changes in wave height in the diving area or in the event of diving in locations with high average wave heights. In such a situation, the possibility of a diver remaining at the station depth of $3\,mH_2O$ is problematic. For organizational reasons, it is best to assume that the duration of the emergency decompression process should be the same as for the basic decompression, and the stay at the $6\,mH_2O$ station should be the sum of the stay at the $3\,mH_2O$ and $6\,mH_2O$ stations.

The Polish Navy has traditionally required that the decompression system be able to compensate for difficult decompression transition conditions, be capable of surface decompression, provide for compensatory decompression, etc. Such a system should consist of hyperbaric procedures, including therapeutic procedures, and provide the required technical and medical equipment to carry them out. The system should be supplemented with a subsystem for the evacuation of injured divers to a specialist hyperbaric center or hospital. The elements of the system securing the realization of therapeutic processes can be borrowed or developed anew and dedicated especially to the system of conducting diving works. Due to the use of autonomous *self-contained underwater breathing apparatus SCUBA* equipment, the system of therapeutic recompression tables *TT*[146] should provide for therapeutic procedures in case of a diver being "thrown" from considerable depths, bypassing decompression stations and, as a result, explosive decompression.

The use of *semi–closed circuit rebreathers SCR underwater breathing apparatus UBA* and mixtures with increased oxygen content requires the development of a recruitment process for divers and maintaining those divers in good physical condition. The system must take into account the health parameters of the divers. The assumption here is that they will be compatible with the requirements for military divers. Additional criteria can also be added to the qualification process.[147] The adopted standard of physiological parameters of a diver is a guideline for the development of a procedure for the implementation of an effective decompression process.

A higher level of risk of developing symptoms of decompression sickness *DCS* than considered acceptable for recreational divers was assumed for this decompression process.[148] The system may also include procedures for decompression as for recreational divers, which will be used during training.

Due to the fact of the development of technology for the diving process using semi–closed circuit rebreather *SCR*, it should include developed and validated effective procedures for the process of breathing space flushing. The technology should assume that there will be no need for the implementation of the breathing space flushing process while at the bottom. However, an effective flushing procedure can always be used in the case of a suspected loss of control over the composition of the breathing mix that the diver is breathing. These procedures increase the oxygen content in the circuit, and their effect on the possibility of occurrence of central nervous system oxygen toxicity *CNSyn* should be investigated.

An important parameter of a diving operation is the ability to load a diver with work. The value of workload is related to the process of ventilating the breathing space of the rebreather, as well as to the process of decompression and increasing risk of *CNSyn*. In this case, medium-heavy[149] workload and rest[150] have been assumed. The influence of thermal protection of a diver on the decompression process must also not be overlooked.

Risk analysis shows a need to use emergency breathing subsystems. Emergency breathing systems may be based on the capability to transition to breathing from an integral cylinder set using a redundant[151] breathing system. This process prevents the loss of control over the main respiratory system, not the loss of breathing mix reserve. From the perspective of planning the exposure process and subsequent decompression, the use of an emergency respiratory system with a different mode of action from that of the main system should take into account variations in the risk of *CNSyn*. If it is considered necessary to develop procedures for the use of respiratory systems independent of the rebreather used,[152] an account should also be taken of the phase of diving for which transition to emergency decompression will be possible.

In order to secure the described basic and emergency processes, it is required to develop a minimal system[153] consisting of the following:

- Procedures for qualification for diving with the intended rebreather and breathing mixtures
- A specially designed and validated procedure for basic decompression, together with the rules for the workload of the diver
- Specially designed and validated emergency decompression procedures
- Procedures for flushing the breathing space of the diving apparatus, taking into account the increased risk of developing symptoms of central nervous syndrome caused by oxygen toxicity *CNSyn*
- Borrowed or newly developed therapeutic procedures in the event of an occurrence of diving sicknesses *DCI*
- Hyperbaric and medical equipment to secure the process of treatment of *DCI* and other possible cases
- Transport subsystem to a hyperbaric facility
- Procedures for estimating the risk of *CNSyn*
- An integral emergency breathing subsystem of the rebreather
- Emergency rebreathers
- Diver thermal protection subsystem

If the requirement for acceleration[154] of the basic decompression process by using oxygen at the last decompression stations is introduced, the system must be further extended not only by a validated procedure for the oxygen decompression process but also by the elements resulting from the risk analysis carried out for this process.

The decompression procedure will need to determine at what stage it is safe to transition to oxygen decompression due to past exposure and increased oxygen partial pressure, taking into account the resulting increase in the risk of central nervous syndrome caused by oxygen toxicity *CNSyn*. Regardless of the time and depth of the transition to oxygen decompression, an effective procedure should be sought for the process of ventilating the breathing space of the rebreather with oxygen.

The decompression acceleration algorithm will depend directly on its effectiveness. It is necessary to consider the procedure of switching from oxygen decompression to decompression using a gas mixture at any time.[155] It is best if the decompression time between the decompression process using a respiratory mixture and oxygen is calculated according to a simple rule. In this case, it is sufficient to specify the duration of the stay at the stations for the basic decompression schedule using oxygen decompression. Emergency decompression using a breathing mixture will be the calculation of the time spent at stations breathing oxygen. The change of the breathing mix during decompression will require the development of rules for flushing the breathing space of the device. In case of a failure of the oxygen supply from the rebreather, it is also possible to consider extending the system with external oxygen breathing systems.[156]

The use of an oxygen decompression system and emergency decompression when breathing the circulating breathing mixture may become complicated if the requirement for marine systems, with the last decompression station at $6\,mH_2O$, is maintained. Such a requirement for all decompression schedules may make the primary decompression process too conservative, and thus ineffective, from the point of view of military diving operations. In this case, it is possible to dispense with this requirement for emergency decompression and to develop two decompression systems depending on whether or not the dive will be protected by other oxygen sources in the event of a malfunction in the supply from the integral cylinder set of the rebreather.

For marine systems, it seems attractive to use the surface decompression process. For some decompression stations, the process of surface decompression enables interruption of the decompression process in water, rapid ascent, and entering the decompression chamber. Then the compression and completion of the rest of the decompression is performed on the surface in the decompression chamber. Such a procedure reduces the diver's risk associated with being underwater but exposes the diver to stress associated with interrupting the decompression process. Evidence shows that interruption of decompression can last up to $7\,min$.[157] It is a relatively short time for ascent, reaching the decompression chamber, partial removal of the diver's gear, and recompression. Surface decompression systems, as the basic procedures for the decompression process, are used for tethered dives carried out from a surface platform equipped with a decompression chamber.[158] For independent dives, such procedures only serve emergency functions.

The system for conducting underwater works may not provide for repetitive dives but should normalize the rest time that should elapse before the next dive.

2.8.5 Equipment

The most often used mixture for supplying open circuit apparatuses is the premix, which fills the cylinders supplying the demand regulator, although it is possible to have designs that produce the breathing mix during the diving process from gas components.[159] One of the options to consider is an open circuit apparatus supplied with air and oxygen which adjusts the oxygen content x_i in the dynamically composed air–oxygen mixture, depending on the diving depth H, similarly as for semi–closed circuit rebreathers, for example, the Polish *SCR KSZYK SCUBA* (Kłos, 2000, 2008, 2023). In such a case, it is necessary to know the function realizing the algorithm of mixing gases from components $x_i = f(H)$ and to assess on the basis of oxygen content x_i its partial pressure p_i, as well as the partial pressures of other gases $(p_1..p_\bullet)$. The partial pressure of oxygen p_i and other gases $(p_1..p_\bullet)$ is the basis for estimating decompression, exposure to various forms of oxygen poisoning, and toxic effects of gases.[160]

The situation will become more complicated if the types of breathing mixes are changed during the diving process.[161] In this case, the possibility of counterdiffusion, including isobaric, should be taken into account. Both the composition of the breathing mix $(x_1..x_\bullet, x_i)$ and the partial pressures[162] of its components $(p_1..p_\bullet, p_i)$ for the different types of rebreathers can represent complex functions of depth H: $(x_1..x_\bullet, x_i) = f(H), (p_1..p_\bullet, p_i) = f(H)$.

Open circuit apparatuses adapt to the breathing action of the diver; hence, the limitations of workload W are related to the training of the diver and the constraints of the breathing system. However, it is not advisable to plan for heavy effort[163] during independent and tethered dives outside the saturation zone, for reasons given earlier.

In closed circuit rebreathers *CCR*, in addition to maintaining the composition of the breathing mix $(x_1..x_\bullet, x_i) = f(H,t)$ function of depth H and time t, the reliability of the algorithm for maintaining the partial pressures of the gas components of the breathing mix $(p_1..p_\bullet, p_i) \cong idem$ in response to forced actions is important. For example, to changes in the consumption of oxygen $\dot{v}$ as a result of a change in workload W necessitating, most often,[164] changes in the oxygen stream $\dot{V}_i$ dosed into the respiratory circuit, $\dot{V}_i = f(\dot{v})$.

Devices of this type are designed in such a way as to economize on the supply of the breathing mix; hence, they have to be able to react moderately quickly to forced action. Otherwise, minor fluctuations in the maintenance of a certain depth $\uparrow H \downarrow$ could result in an unjustified loss of the breathing mix. Hence, deviations from the stable oxygen content state $x_i \neq idem$, for example, with rapid changes in the workload[165] W, may cause instability in the algorithm for maintaining the composition of the inhaled breathing mix $\uparrow (x_1..x_\bullet, x_i) \downarrow$, thus posing an additional risk. The algorithm used in closed circuit rebreathers is ideal for sustained light effort. Therefore, even with the use of combat oxygen devices, it assumes planning rather low speed of movement during a combat mission. The planned effort should be relatively constant, both based on purely mechanical and electronically controlled mechatronic actuators,[166] so that the system of supplementing the deficiencies in oxygen $x_i = f(\dot{v}, H)$ or the mix $(x_1..x_\bullet) = f(H)$ was able to correctly interpret the situation and respond with an adequate action.

In a semi–closed circuit rebreather *SCR*, fresh breathing mix is dynamically mixed with the regenerated one. Determining the function $(x_1..x_\bullet, x_i) = f(\dot{v}, H, t)$ of the composition $(x_1..x_\bullet, x_i)$ of the breathing mix inhaled by the diver depending on the consumption of oxygen $\dot{v}$, depth H, and exposure time t is a fundamental problem when planning safe hyperbaric exposure and subsequent decompression. Most often, such devices perform a predefined function to obtain the composition of the $(x_1..x_\bullet, x_i) = f(\dot{v}, H)$ mechanically.[167] Predefining an algorithm to maintain a constant, assumed composition of the breathing mix inhaled by a diver $x_1..x_\bullet, x_i) = f(\dot{v}, H) = const$ requires the adoption of assumptions concerning the individual parameters of a diver θ and the permitted tolerances for deviations from these values $\Delta\theta$. Subsequent selection of divers requires checking the fulfilment of the assumptions underlying the design of the algorithm for maintaining the assumed composition of the breathing mix inhaled by the diver $x_1..x_j, x_i) = f(\dot{v}, H)$.

The assumed individual parameters θ do not always have to be met by a diver, especially when accompanied by such phenomena as sports vagotonia,[168] bradycardia,[169] significant muscle tissue development, being overweight or underweight, etc. Hence, divers should be initially classified to diving using semi–closed circuit rebreathers with regard to the fulfilment of the assumed individual parameters θ when designing the device and diving technology.

Individual parameters of divers θ change with the age and the degree of diver preparation/training. Annual medical screening does not focus on these changes. Its task is to check the general state of health and not the level of training or fulfilment of individual parameters θ being the basis for the safe use of diving apparatuses. Therefore, onetime certification to dive with special diving devices may no longer be justified as diver's age, or if their training is not kept up-to-date.[170] It seems good practice to check the essential for safety individual parameters of a diver θ during each training course to obtain higher diving qualifications.

In the process of diving safety analysis in the Polish Navy, it is crucial to properly prepare a dive in terms of the diver's condition, efficiency, and decompression, because the individual parameters of a diver θ sometimes undergo significant fluctuations as a result of adaptation training or its discontinuation. Control of individual parameters of a diver θ is possible during mandatory training using hyperbaric exercise simulators, preferably hyperbaric swimming simulators,[171] as part of the prevention of potentially dangerous situations.

The level of diving safety is closely related to the risks posed by the equipment and technology used.[172]

The presented project focuses on the elements of the diver's equipment, that is, diving apparatuses, and the risk of decompression sickness *DCS* and central nervous system oxygen toxicity *CNSyn* in the tactical context of military operations. Important elements of the diving equipment include *XBS*,[173] diving suits,[174] gas analyzers,[175] decompression meters, etc.

2.8.6 Tactics

In military diving, the use of diving equipment as structural components of a system ensuring the implementation of processes resulting from the adoption of specific tactics[176] plays a dominant role. In the understanding of the tasks of the described project, dives constitute an element of the system aimed at ensuring the protection of our own forces and of public safety, which can be divided into three subsystems: *mine countermeasure MCM*, *explosive ordnance disposal EOD*, and *improvised explosive device disposal IEDD*.

Mine countermeasure MCM activities are processes implemented in a mine warfare system. The elements of the mine warfare system are specialized vessels, traditionally divided into two groups – mine sweepers and mine hunters – but nowadays, this division has become blurred, and it can be assumed that modern navies operate a single class of vessels, that is, *mine countermeasures vessel* MCMV. The first of these are the traditional minesweepers, which, when passing over the minefield, drag the minesweepers that simulate the signature of ships, detonating mines or undercutting an anchor mine and neutralizing it with gunfire on surface. The second type of vessel has developed into an *MCMV* and is a comprehensively equipped vessel for searches for mines and neutralizing them, in particular intelligent mines. Typically, *MCMV*s are equipped with autonomous *AUV*s[177]/*UUV*s[178] used to detect, locate, and then mark on a digital chart potential targets that can be neutralized by other autonomous *AUV*s or *ROUV*s.[179] Such vehicles can carry special explosives for the disposing of mine-like objects, placing near them dedicated explosives to be detonated remotely after the vehicle has withdrawn to a safe area, or with a self-propelled explosive charge.

Most commonly, a ship is also equipped with an *ROUV* fitted with various sensors to scan the water ahead of the ship,[180] and a towed system equipped with similar sensors for scanning the seabed. Such a vessel may be equipped with an intelligent sweeping system simulating signatures of different classes of ships. It is most often towed by an *MCMV*-based unmanned *autonomous surface vehicle ASV*. The ship is a part of the network-centric[181] system of larger mine countermeasure forces.[182] Such units usually have divers prepared to carry out rescue operations[183] and serve as one of the elements of dealing with the threat posed by sea mines.

It is possible for *MCM* divers to perform dives from small vessels operating from the shore or carried by ships.[184] Their tasks may include reconnaissance, less often disposal or partition of the defense infrastructure of hydrotechnical objects or coastal defenses.[185]

As for *MCM* activities, the presented project focuses on rescue operations by *MCM* equipment, neutralizing intelligent mines and jamming maritime electronic warfare assets.

The *EOD* activities focus on eliminating dangerous explosives, such as unexploded military explosives *UXO*,[186] being remains from wars or military exercises, etc. Due to the risk involved, these activities are treated as combat activities carried out in conditions of peace.[187] An important variation of these activities is counter-*IED* operations carried out to clear an area of *UXO* not recognized as typical military ordnance. At least theoretically, counter- *EOD* operations pose a lower risk, because the measures applied result from catalogued algorithms of handling *UOX* of military origin, which can typically be deactivated.

Unfortunately, even relatively primitive *IED*s can only be removed by controlled detonation, and the only measure to reduce the impact of detonation is to rapidly divert the shock wave in a different direction[188] or to use protective screens.[189] In some cases, it is possible to use jamming[190] against electronic devices employed; however, they are often relatively primitive electrical devices which can be deactivated only mechanically, for example, by shooting at them with various kinds of ammunition[191] or with a cumulative *explosively formed penetrator EFP* charge, burning them with a thermite charge, melting an explosive material, etc. The project, by its nature, is connected with *MCM* and *EOD* activities. *IED* activities are excluded, as *IEDs* are closely related to the parameters of physical fields generated by diving equipment and are not directly related to the diving technology, although the processes of flushing the breathing space of the device are extremely important as regards noise emitted into water.

A risk analysis conducted from a tactical point of view is done to identify levels of threat other than levels of threat identified through an analysis done from a point of view of diving process. This is obvious, as these two analyses relate to different objectives. However, tactical planning of diving operations must also take into account the conditions in which diving technology is to be used.

From a tactical point of view, in a hypothetical scenario of an operation to be carried out by a land force at a specific location, the goal of analysis of risk involved in carrying out a clandestine reconnaissance operation and possible preparation of separation of underwater minefields by divers is to identify the probability of effective execution of the task at the assumed level of confidence.[192]

The protection of navigable waterways should be attained effectively and quickly. It is not always possible for *MCMV* units to be called in to deal with drifting mines spotted by vessels which do not have dedicated equipment to deal with such a task. In this case, typical drifting mines may be disposed of by *VSW* divers/swimmers,[193] who, operating from helicopters, may be dropped into the water and place a remotely activated charge in proximity of a drifting mine[194] or on it. Intelligent mine countermeasures on a navigable route cannot always be performed by underwater vehicles or self-propelled charges due to a region or identification of the mine.[195] Since vehicles are more easily and more likely to be detected by sensors installed in such mines, divers are used, as they have a high capability of masking themselves against the background of normal physical fields occurring in water. A diver can precisely set a small non-electrically ignited cumulative charge, place a mine to be detonated by a self-propelled charge, or prepare it for towing. *IED* activities are practically always carried out by swimmers or divers deployed from a *VSW*.

To counteract terrorist attacks or actions in an asymmetric conflict, where suicide divers are used, employed are unmanned equipment, manned equipment, *special response team* SRT diving operations, etc. Knowledge of the capabilities of divers in such cases is extremely important, to estimate not only the risk to their own forces but also the potential of terrorist/suicide divers.

During the implementation of the project, data relevant to risk of decompression sickness *DCS* or central nervous syndrome caused by oxygen toxicity *CNSyn* will be compiled for use in tactical risk analysis of typical scenarios of underwater operations in relation to.

2.8.7 Research

At the very beginning of work on decompression systems applicable to diving devices in Poland, it was noted that domestic progress is possible only through seeking deterministic models.[196] Consistent adherence to this assumption by three generations of researchers has led to the development of reliable and effective, though arduous, methodology.[197]

In the applied research approach, the methods for predicting performance of an ergonomic diver-rebreather system were divided into a number of interrelated models[198] (Figure 2.29).

On the basis of the validation studies conducted to date, it seems that the analytical breakdown into partial models does not generate any significant errors in their synthesis into a holistic model and their use according to the holistic approach.

Ventilation modelling by means of differential equations has been known for centuries. These are relatively simple equations that can be solved with mathematical analysis methods.

However, their usefulness, until recently, has been questioned on the grounds that some parameters of these models cannot be determined with sufficient accuracy.[199] Therefore, ventilation engineering is based on semi-empirical models derived from the approximate solution of the differential equation by means of dimensional analysis methods.

The barriers related to the use of the deterministic approach to the ventilation process were first overcome about 15 years ago. This made it possible to master the modelling of the deterministic ventilation of a submarine, a hyperbaric chamber, or a mining pit (Kłos, 2007a, 2023). The use of these achievements in industrial and sanitary engineering could lead to a change in the approach to ventilation problems.

Earlier studies on ventilation of semi–closed circuit rebreathers with continuous dosing of the breathing mix were successfully completed about 20 years ago (Kłos,

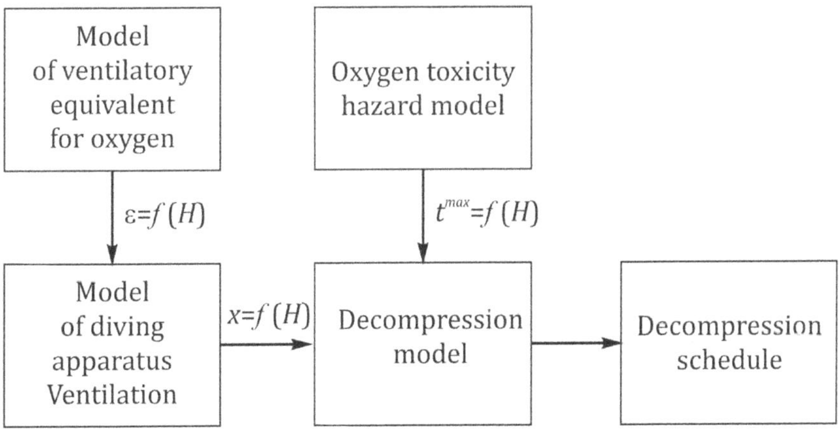

FIGURE 2.29 Model sequence.

2000). However, such devices have been withdrawn from national use. In the current approach to modelling a diving system construction with a metering bellows dispenser, the initially holistic approach has been replaced by an analytical one (Kłos, 2023). The basis for planning decompression schedules is based on several models, shown in Figure 2.29:

- Modifications in the *breathing module*[200] ε in the function of depth H: $\varepsilon = f(H)$
- Modifications in the composition of the breathing mix inhaled by the diver $(x_1..x_j, x_i) = f(H, \varepsilon, \varepsilon_k, x_w, t)$, depending on depth H, breathing module ε, design parameters for a semi-closed breathing space of a rebreather ε_k, premix composition x_w, and time t
- Changes in hazard of central nervous syndrome caused by oxygen toxicity *CNSyn*, depending on oxygen partial pressure p_{O_2} and exposure time t
- Designing decompression[201] as a function of hyperbaric exposure time t and pressure H, taking into account the influence of interfering factors such as temperature T, the influence of previous exposures on decompression, increased risk of central nervous syndrome caused by oxygen toxicity *CNSyn*, additional workload W, immersion and ascent sequence, transport to the place of immersion, stress load before and after diving, etc.

The function $\varepsilon = f(H)$ of changes in the breathing module[202] ε, depending on the depth H, is not crucial for the reliability[203] of determining the values of stable oxygen content x_s from the ventilation model. It has become significant because, unexpectedly, the other models shown in Figure 2.29 are determined with much greater reliability. When searching for solutions for the differential ventilation model of a rebreather containing a system of bags placed one in the other $\frac{\partial}{\partial t}(x_1..x_j, x_i) = f(H, \varepsilon, \varepsilon_k, x_w, t)$, the breathing module ε began to play a key role.[204] This made it necessary to conduct research on the reliable determination of the dependence of the breathing module $\varepsilon = f(H)$ on depth H.

The studies of the breathing module ε are described in the available literature with respect to atmospheric pressure, as they constitute an important indicator of athletes' training and form the basis for many studies concerning effort medicine or work standardization (PN-EN-14143, 2005). The hitherto-uncritical transposition of values for normobaric conditions into hyperbaric studies was a barrier to an adequate development of diving systems. The results of the studies concerning the breathing module ε in specific hyperbaric situations can be found in few specialist publications (Predictive Studies IV, 1978). However, the application of these values is still associated with a decrease in the required accuracy and precision of ventilation modelling. The determination of the relationship $\varepsilon = f(H)$ regarding changes in the respiratory module ε as a function of depth H for predefined conditions was performed by conducting experimental exposures in simulated conditions on a representative group of experimental divers.[205]

The ventilation model can be used to calculate the stable oxygen content as a function of depth H: $x_s = f(H)\big|\ x_s(H, W) \equiv x_i = idem$, environmental parameters, and effort W. After considering the fluctuations encountered Δx_s in a stable oxygen content x_s, for a predefined depth H and workload W, the model constitutes a contribution to decompression modelling (Figure 2.29).

In semi–closed circuit rebreathers *SCR* supplied with *premix*, there is a relative decrease in oxygen content[206] from the concentration in the premix x_w to a stable value[207] x_s. This phenomenon is caused by obtaining the equilibrium condition when the regenerated mixture is mixed with the fresh one. The dynamics of this process can be modelled similarly to other ventilation processes (Kłos, 2023). Based on previous long-term studies, it can be concluded that the ventilation process of the *SCR*s can be described in sufficient detail based on the ideal gas law. The differential equation built on the basis of the molar balance can be accurately calculated. Such an approach is not widely used, owing to the difficulties encountered in establishing certain parameters for this equation. Through many years of research on the process of ventilation of submarines, hyperbaric chambers, mining pits, etc., it was possible to propose a reliable approach to this issue; hence, the modelling was based on a fully deterministic, algebraic analytical model (Kłos, 2007a, 2023).

Research into the process of rebreather ventilation without human participation is carried out using a station of accumulated simulators of metabolic oxygen consumption $\dot{v}$, respiratory action $\dot{V}_E$, and depth H. The simulator is, in many respects, an original solution on a world scale created as a result of many years of own research. It is distinguished by the fact that a liquid medium is used to simulate oxygen consumption $\dot{v}$ and not a gaseous one, as in other similar technical solutions in the world (Frånberg, 2015). Because of this, adding it negligibly disturbs the volume of gases in the ventilated space. The simulator setup features an extremely short pathway between the pump simulating lung ventilation $\dot{V}_E$ and the tested respiratory system, which does not cause compression and expansion of the breathing mix in the line, inducing respiratory action.

Concurrently, research is conducted in simulated hyperbaric conditions with participation of experimental divers. Based on the results of these studies, the reliability of the applied modelling of the ventilation and gas exchange process in the ergonomic diver-rebreather system is determined. Studies conducted without the participation of experimental divers give an answer on the credibility of the applied mathematical model in relation to the processes taking place in the real object, that is, the rebreather. Research with experimental divers allows us to determine the decrease in the reliability of the mathematical model in the course of rebreather ventilation when accompanied by a disturbance associated with the biometric diversity[208] of divers *HF*,[209] marked here as θ.

While oxygen is needed to maintain hemostasis in humans, it is also a carcinogenic factor. Theories exist that postulate that its effects are the main cause of ageing of aerobic organisms.[210] In hyperbaric conditions, it is more dangerous, revealing some forms of toxicity not found in everyday life. Under hyperbaric conditions, oxygen exhibits pulmonary,[211] central,[212] and somatic toxicity[213] (Kłos, 2014b, 2014c, 2014d, 2014e, 2023). The most serious form of oxygen toxicity in short-term military dives is central toxicity *CNSyn*.

Experiments with central nervous syndrome caused by oxygen toxicity *CNSyn* often revealed less-specific symptoms, such as anxiety, paleness of the face, tremor of the lips and eyelids, nausea, cramps, confusion, lack of coordination, visual and auditory hallucinations, narrowed vision, or speech disorders. These symptoms rarely precede the seizure. The onset of generalized convulsions is sudden. The seizure starts with a tonic phase, usually lasting $30 s$, during which the diver loses consciousness and the respiratory function stops. This is followed by a clonic phase, with uncoordinated movements of the whole body. The duration of the whole attack is usually about $2 min$. If the poisoning has occurred in the water, the safety diver should try to wait until at least part of the consciousness and respiratory action have returned; a period of apnea of up to $3 min$ can be allowed.

An ascent of a diver without restoring respiratory action may cause pulmonary barotrauma, as oxygen toxicity and loss of respiratory action are usually accompanied by laryngeal clenching. Severe forms of central nervous syndrome caused by oxygen toxicity *CNSyn* are fatal. There are some empirical models in the world to predict the risk of this form of *CNSyn* (Kłos, 2023). The presented approach is based on the broadest published systematic research on modelling the risk of central nervous syndrome caused by oxygen toxicity *CNSyn*, relying on mathematical survival analysis[214] (Gerth, 2002; Kłos, 2007b, 2014d, 2023). The model was adjusted on the basis of the results of studies conducted by the British during World War II (Donald, 1947, 1992) and their continuation by the Americans until the late 1970s (Harabin, 1993). Our own research on the adequacy of this approach to *CNSyn* modelling was completed with the validation of the model established by the US Navy (Kłos, 2012, 2023).

The estimation of the level of relaxation[215] following oxygen exposures is most commonly performed using the empirical model[216] proposed by the Oceanic and Atmospheric Administration (NOAA, 2017).

The estimation of safe decompression schedules was based on the abandoned[217] $ZH - L_{12}$ model developed by Bühlmann for the Swiss army (Bühlmann, 1984, 1995). The model takes into account military conditions, such as the special selection of a group of divers, keeping divers in the condition to undergo strenuous decompression, increased level of approved threat, etc. It can be assumed to be extremely safe. The model was validated during previous studies (Kłos, 2011, 2016).

Decompression planning must be based on predicted scenarios for diving operations, safety measures applied,[218] pre- and post-diving rest conditions, transport conditions, etc. The assessment of these parameters can be carried out through experimental dives under simulated conditions, which allow the determination of the permitted ranges of selected exposure parameters. Validation of the approach, as in previous stages of the study, can be achieved by statistical means using sequential analysis (Wald, 1947; Kłos, 2007b).

2.8.8 Conclusions

The deterministic models of ventilation of the semi-closed respiratory space of the semi–closed circuit rebreather *SCR* and decompression presented in Figure 2.27, as well as the statistical model of the risk of *CNSyn*, were established with such reliability that the empirical model $\varepsilon = f(H)_\theta$ of a change in the respiratory module ε, along with depth H for a selected population of experimental divers θ as a parameter, became the modelling element with the greatest effect on the safety of conducting a hyperbaric exposure. A selected population of divers θ was a standard population admitted to military diving in the Polish Navy. Additionally, population θ should undergo the test for deviations from the function $\varepsilon = f(H)_\theta$ describing changes[219] in the respiratory module ε with depth H, as well as oxygen tolerance test *OTT* (Kłos, 2013, 2023).

A negative result of the oxygen tolerance test is a premise to plan diving operations only with a lower risk of central nervous syndrome caused by oxygen toxicity *CNSyn* and to take precautions when planning the treatment process using oxygen treatment tables $oxy - TT$ procedures (Kłos, 2013, 2023). In turn, deviations from the established function $\varepsilon = f(H)_\theta$ mean that, for safety reasons, the diver should have a differently designed decompression. Negative results of these tests tend to disqualify the diver from diving training with the use of *SCR*s, for formal organizational reasons.[220]

The ability to continuously improve the precision of decompression planning involves constant improvement of the empirical model $\varepsilon = f(H)_\theta$ for a change in the respiratory module ε together with depth H. Each dive providing reliable data[221] on its course enhances the database to be used in the inference on the precision of determining the function $\varepsilon = f\left(H\right)$ and can be used as a guideline to change the decompression schedule. Operational dives cannot provide reliable data for such inference, but training and simulation exercises give such a possibility.

The inference regarding increasing the risk is connected with the diagnosis of deviations from the empirical model $\varepsilon = f(H)_\theta$ of changes in the breathing module ε with depth H. Data collected during training and practice in the diving simulator help identify divers for whom the decompression procedure may be potentially dangerous. Diver testing on a diving simulator may be used to eliminate *DCS* hazards. If the results of the research show no deviations from the model $\varepsilon = f(H)_\theta$ and an increased risk of symptoms of decompression sickness *DCS* is observed, other causes should be sought, for example, deviations from the technology[222] of conducting underwater works or casuistic reasons.[223] Monitoring deviations from the empirical model $\varepsilon = f(H)_\theta$ of a change in the respiratory module ε with depth H is a potential tool of work for an officer responsible for the safety of diving in the Polish Navy.

2.9 STRATEGY

Decompression consists in causing a controlled disturbance of the state of equilibrium leading to the formation of a gradient of partial pressures of gases between the breathing atmosphere and tissues. This gradient is the driving force behind the decompression process. The aim of such a procedure is to safely bring the body's gas balance to a state of equilibrium with the reference atmosphere which the decompression process is aiming at.

Decompression and other models, such as oxygen toxicity hazard models, were discussed earlier (Kłos, 2023). Decompression models have been discussed in other publications (e.g., Bühlmann, 1984; Kłos, 2011). The principles of safe decompression strategies will be discussed in the next paragraphs, while the principles of treatment strategies will be discussed in Chapter 4.

The effectiveness of the decompression process increases with the increase in the partial pressure of oxygen p_{O_2} in the breathing media inhaled by the diver. The toxic effect on the human body is revealed at elevated oxygen partial pressures p_{O_2}. The possibility of increasing the partial pressure of oxygen is limited in the breathing species inhaled by the diver. Therefore, planning the decompression process requires a compromise between the efficiency of the decompression *process* and the toxic effects of oxygen.

In most situations, the symptoms of oxygen toxicity[224] *OSyn* are more dangerous than the symptoms of decompression sickness *DCS*; therefore, in many strategies, the hazard of *OSyn* is assumed to be at the much lower level than the acceptable level of hazard of *DCS* (Figure 2.30).

The hazard function *H* and the risk function *R* are functions of the survival analysis (Kłos, 2023).

The most frequently used ranges of oxygen partial pressures p_{O_2} due to the hazard of *DCS* and *OSyn* are presented in Figure 2.31.

Table 2.4 shows a broader representation of the ranges of oxygen partial pressures used in the dive technique. This matter will not be presented here in more detail, because it has been described earlier (Kłos, 2012).

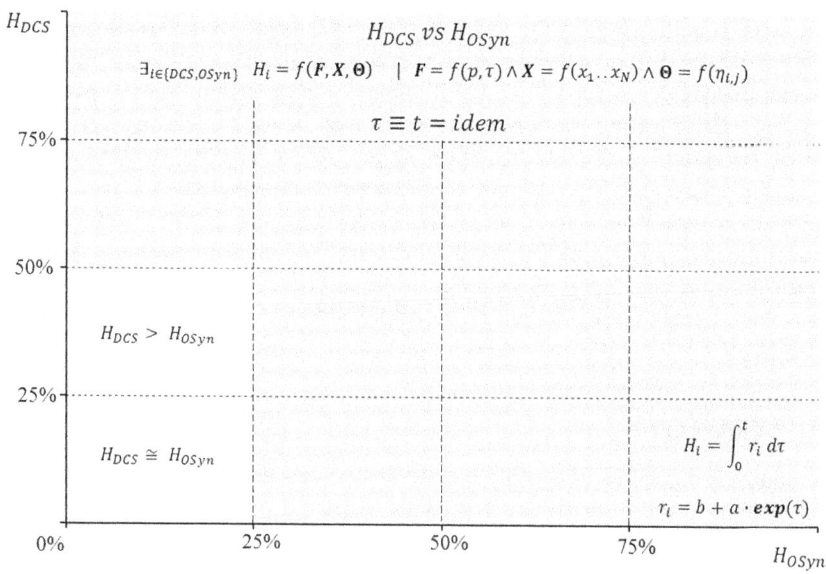

FIGURE 2.30 Comparison between the hazard of decompression sickness and oxygen toxicity, where *H* and *R* represent the hazard function and the risk function, respectively; *p* is the maximum pressure during hyperbaric exposition; $x_\bullet$ represents the fraction of gases in the breathing mixture; τ is the time and *t* is the exposition time; $\eta_\bullet$ human and environmental factors; but *a* and *b* any other model factors.

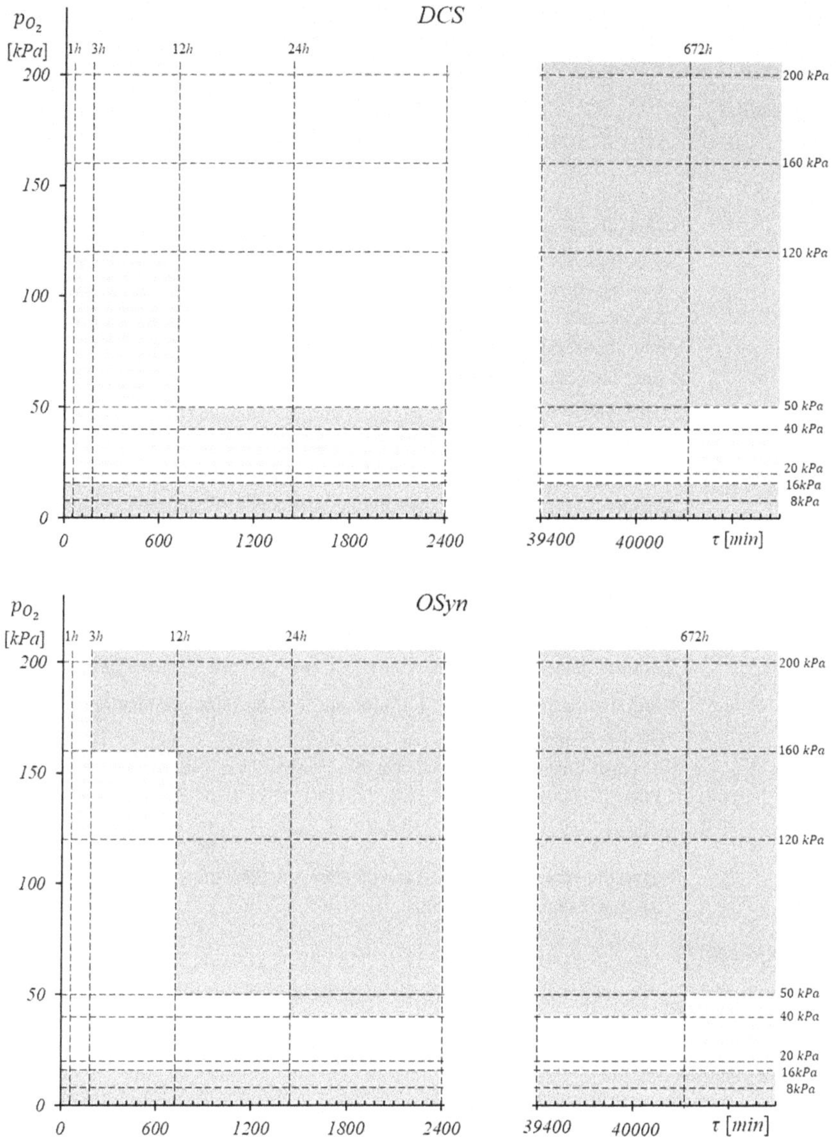

FIGURE 2.31 The most frequently used ranges of oxygen partial pressures due to the hazard of, *top*, decompression sickness and, *bottom*, oxygen toxicity syndrome.

Figure 2.32 presents only one of the central nervous system oxygen toxicity models for the central nervous syndrome *CNSyn* hazard *H* for short exposures to high oxygen partial pressure levels p_{O_2}. Other oxygen toxicity models have been discussed in other publications (e.g., Arieli, Yalov and Goldenshluger, 2002; Shykoff, 2007; Kłos, 2012).

TABLE 2.4
Frequently Used Oxygen Partial Pressure Ranges

Oxygen Partial Pressure (*MPa*)	Specification
0.010	Extinguishing gases type *INERGEN* – it is safe to breathe in limited time. A mixture of nitrogen, argon, carbon dioxide, and oxygen in which oxygen partial pressure falls down to the level of 0.008 *MPa* under the condition that partial pressure of CO_2 will be 0.005 *MPa* (Fire Research, Test, Development, and Education Centre, 1993).
0.012	Lower limit of safety due to hypoxia.
0.016	First symptoms of hypoxia.
0.021	Normal oxygen partial pressure in atmospheric air.
0.035 – 0.040	Typical saturation exposures.
0.050	Maximum oxygen partial pressure during saturation dives and the beginning of lung oxygen toxicity.[225]
0.10	Pure oxygen breathing on surface.
0.16	The most commonly accepted upper limit of safety for nitrox dives out of the saturation zone[226] (Hamilton, 1989; NOAA, 2001).
0.20	Treatment table *Cx*30[226] worked out by *COMEX* in 1986 (Comex Marseille, 1986).
0.24	Suggestion to use nitrox *Nx*0.4 at a pressure of 0.6 *MPa* in the treatment of diving diseases (Rutkowski, 1990).
0.25	The upper limit of permitted combat dives with oxygen[226] (Butler and Thalmann, 1986).
0.28	20 *min* test of oxygen tolerance (NOAA, 2001) and oxygen treatment tables.[227]
0.30	Suggestion to use nitrox *Nx*0.5 at a pressure of 0.6 *MPa* in the treatment of diving diseases[228] (Rutkowski, 1990).

Source: Kłos (2012).

2.10 SELECTED DIVING RISK FACTORS

A safe transition from a higher-pressure atmosphere to a lower-pressure atmosphere is achieved by means of a planned decompression process, usually through changes in pressure and/or composition of the breathing mix as a function of time. However, the decompression process is influenced by a greater number of inherent[229] factors than merely changes in pressure and composition of the breathing mixture, the values of which should be maintained within certain ranges. There are instances where control over them cannot be exercised, leaving elements of residual risk[230] to the decompression process. The safety of decompression should be assessed, inter alia, by analyzing this risk for each implementation of the decompression process.

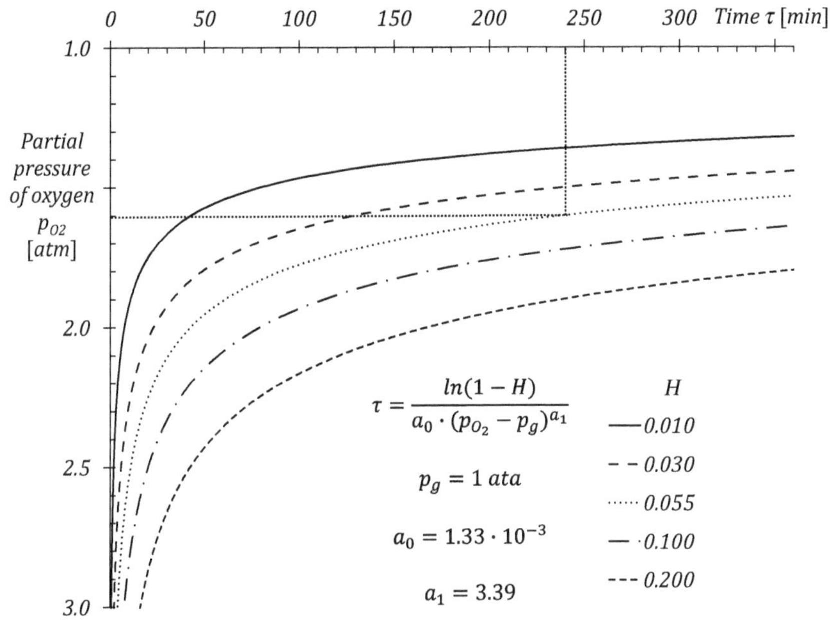

FIGURE 2.32 Oxygen partial pressure and exposure time limits in parametric relationship to hazard function H.

Source: Kłos (2023).

Every decompression system in use is subject to observation and assessment by users, and in some cases, also to diagnosis,[231] by the monitoring bodies set up for this purpose. The analysis of the observations collected leads to developing certain practices, which may begin to form part of the so-called *good practices GP*. Although conclusions derived from such collected observations usually refer to current theories, they cannot constitute precise rules of conduct, as they have not been verified in the scientific process and validated.[232] Hence, these observations are reluctantly used to devise some sets of recommendations,[233] although they are important prerequisites for analyzing the residual risk of decompression. If the conclusions do not yet constitute confirmed knowledge,[234] they may be passed on to other users only within a specific training system.

2.10.1 Decompression

In decompression, a controlled disturbance of equilibrium is caused, leading to the formation of a gradient of partial pressures of gases between the breathing atmosphere and the tissues. This gradient is the driving force behind the decompression process. The aim of such a procedure is to safely bring the body's gas balance to the state of equilibrium with the reference atmosphere which the decompression process is intended to achieve. The reference atmosphere to which the decompression process leads is usually the sea level. It can be the atmosphere surrounding a body of water elevated above sea level, where the pressure is reduced in relation to the normal pressure prevailing at sea level. It can also be the pressure in an aircraft, where the pressure is usually 1/4 lower than normal, or in a helicopter,[235] where the difference can be even greater. The reference atmosphere can be an underground cave or an underground corridor where the pressure can be much higher than the atmospheric pressure.[236] It may also be the habitat to which the diver returns after a trip from a saturation plateau.

The rate of decompression and the composition of the breathing mix are so selected that the partial pressure gradient guarantees a safe decompression process on the chosen path toward the reference atmosphere, but also that the process is as effective as possible. Decompression may also include decompression stops used to lower the current value of the partial pressure gradient before continuing toward the reference atmosphere.

To estimate the value of a controlled disturbance of equilibrium leading to a safe and effective partial pressure gradient of gases between the breathing atmosphere and the tissues, the initial saturation level must be determined. The initial state is estimated on the basis of the profile of the course of the dive immediately preceding the moment when the decompression process begins. Sometimes, previous dives and their decompression profile need to be taken into account. Most often, it is defined roughly using a nomogram in the form of a table, which takes into account the maximum depth achieved during the dive and the interval between the start of the dive process and the start of decompression. The same thing is done as the decompression progresses. It is tempting to make estimates for each, even the smallest change in depth, as permitted by an electronic decompression meter, which can be both its greatest advantage and its main drawback.[237]

2.10.2 Prevention

Special decompression procedures are used to prevent the occurrence of *DCS*. Most often, their descriptions are rather brief. It is required that the knowledge necessary for the proper use of the decompression system be acquired and tested during the obligatory courses and training undertaken at particular levels of a diver's career advancement.

There are exhaustive comments in the studies on decompression systems, but access to them is not universal. Rarely is the necessary knowledge provided in a comprehensive manner, as in the case of *Bühlmann's* decompression system (Bühlmann, 1984). At times, publications are produced that generalize the approach to decompression (Lippmann, 1990; Lippmann and Mitchell, 2009). Knowledge transferred during training courses is usually accompanied by some kind of printed course materials (Lewbel, 1984; Betts, 1994). However, such handbooks are not fully useful for acquiring knowledge applicable outside of the training system they refer to. Sometimes they are provided to a student with a note that only the person whose name is put on the cover can legally use the information contained therein (Comex Marseille, 1986; Imbert and Bontoux, 1987). This is due to the desire not only to protect the *know-how* but, above all, also to protect against misinterpretation of the information contained therein by an untrained user.

Most often, developing a decompression system involves access to substantial knowledge of the future user, which justifies the inclusion of only a sparing commentary. It is typically constrained to the exclusion of responsibility[238] of the entity developing the decompression system.

This particular way of introducing specialists to use the decompression system is justified by certain barriers in perception. The diver's skills should be considered in terms of highly qualified craftsmanship skills[239] which are acquired through an appropriate training system. This does concern the ability to use not only diving equipment but also decompression systems. It is possible to describe many of the conditions for a correct approach to underwater planning (Brubakk and Neuman, 2003). The number of special situations is so large that teaching them in a single cycle would be extremely long and boring, making applicability of such training questionable. This can be compared to learning all the options and capabilities offered by a multifunctional electronic device, such as a smartphone. As a rule, such comprehensive knowledge is not passed on to the user, as the functions not used on a daily basis will be quickly forgotten.

One can imagine divers to be selected and trained like astronauts are. That they are people with higher education in several fields of knowledge, with considerable practice in aviation and especially trained to achieve extraordinary fitness.[240] Going further, we could assume, for a moment, that a car driver should be an educated mechanic, trained in a manner similar to that of professional passenger aircraft pilots. But to postulate such an approach would, at best, be met with ostracism or, in a worse scenario, with an accusation of lacking an adequate reference to reality.

Modern cars are no longer supplied with technical instructions for their servicing.[241] Nor are they supplied with precise technical descriptions. The same applies to decompression systems. Sometimes, decompression systems are classified and their implementation requires involvement of the first user or the developer of the decompression system. Sometimes, information on the assumptions for decompression systems is patented, but even then, the scope of the disclosed know-how is limited (Kenyon and Butler, 2014).

In military systems and high-tech commercial institutions, efforts are made to transfer knowledge concerning the decompression systems used as accurately as possible. Even in such cases, however, there are not enough comments to the tables provided. They are replaced by mandatory operational risk analysis (US Naval Sea Systems Command, 2016). This forces the user to undertake their own studies or participate in additional courses in order to acquire the skills required to conduct such an analysis. In thc event of an accident or a potentially dangerous situation, the absence of such a document, or its perfunctory preparation, constitutes an aggravating factor for those responsible for the planned diving operation.

In most cases, the user would like the manual to be as comprehensive and informative as possible so as to cover all the possible situations. These requirements are mutually contradictory,[242] and for this reason, such an instruction manual probably does not exist in any field.[243]

2.10.3 Conservatism

As regards factors that increase possible occurrence symptoms of decompression sickness DCS, to reduce this risk, extended decompression can be applied. This procedure is called conservative, or simply referred to as *conservatism* J. Sometimes, *conservatism* is expressed as a vector of parameters x and called the J –factor:[244] $J(x_1..x_n)$. *Conservatism J* can be implemented in various forms. An example is the selection of a greater depth than the depth of exposure from the table. This recommendation also applies if the exposure depth is provided in the table. Most often, the level of conservatism for a selected decompression schedule is increased by artificially adopting a longer exposure at a selected depth than is the case in the air diving system used in the Polish Navy.

In the research on the acceptable values of partial pressure gradients Δ for the decompression process, the so-called gray zone concept is introduced. This gray zone contains gradients Δ which should not lead to the occurrence of symptoms of decompression sickness DCS, provided that there are no additional factors overlapping the decompression process. The upper limits of the gray area are the maximum permissible gradient values[245] Δ_{max}. The permissible lower limit for the gray area, which is often adopted in the range of $\Delta \in [0.75\ _{max}, 0.80\Delta_{max}]$ of the upper limit Δ_{max}, is rarely described. At this point, the so-called zero conservatism $J = 0\%$ is usually established. The maximum conservatism $J = 100\%$ is related to the zero partial pressure gradient $\Delta = 0$. For $J = 100\%$, decompression does not occur. If, therefore, decompression is planned well below the gray area, that is to say, with considerable *conservatism*, it is to be expected that the impact of the materialization of some additional aggravating factors can be compensated for. However, if decompression is planned close to or within the gray area, additional burdens[246] arising in a particular diving process may lead to the occurrence of symptoms of decompression sickness, despite the correct choice of decompression profile based on its parameters.

The responsibility of the dive planner and the person who is in charge of the dive is to assess the level of conservatism applied in relation to the estimated residual risk R of decompression. This can be done by conducting a thorough analysis of this risk R based on expert knowledge gained through experience.

2.10.3.1 Worst-Case Scenario Method

The worst-case scenario is commonly used for estimations. In diving technology, limitations concerning the scenario are sometimes needed as an attempt to take into account how all the possible circumstances that increase the risk of DCS may prevent working out a good solution. For example, using *Table* $3MW$, a maximum of two factors are considered (DMW, 1982). The materialization of more disturbing factors does not entail the need to increase conservatism when implementing decompression.

2.10.4 Variability

The decompression process is most often controlled by changes in the ambient pressure affecting the diver's body during exposure[247] and changes in the concentration of components of the breathing mix. Most often, any other factor influencing the balance between the pressure of gases in the tissues and its partial pressure in the breathing mix is treated as a factor disturbing the decompression process.

A change in the pressure is achieved through a controlled change in the depth in the water or a reduction in the pressure in the habitat. This decreases the pressure of the breathing medium administered to the diver, for example, from the breathing apparatus or the habitat atmosphere. With the change in pressure, the partial pressure of the breathing mix components decreases to a level lower than its tissue pressure. The difference between the partial pressure of the breathing mix component and its tissue pressure creates the necessary gradient for diffusion processes, allowing gaseous exchange. However, such a change can also be caused by the changed composition of the breathing mixture. A change in its composition will have the same effect of creating the required partial pressure gradient even at the same pressure value. The method of decompression without a change in pressure is called isobaric decompression, which can be carried out either with or without a modification in the composition of the breathing mix. For example, supplying a man who has never dived before with oxygen to breathe at sea level produces a partial pressure gradient of nitrogen of about $78\,kPa$, causing a specific "washout" of nitrogen from the body. Similarly, after the end of a dive, with the use of air as a breathing medium, a certain nitrogen charge remains in the body. This means that the pressure of nitrogen in tissues will be higher than the partial pressure of nitrogen in the atmospheric air. The resulting partial pressure gradient will be the driving force behind the process, which will lead to its elimination during surface isobaric decompression without any changes in the composition of the breathing mix.

Both of the described methods used to produce partial pressure gradients are most often used during dives with artificial breathing mixes. For example, when diving with helium-based breathing mixtures, in addition to lowering the pressure, decompression breathing mixes, such as various types of nitrogen–oxygen mixtures or air, can be used at the same time, while for additional acceleration of the decompression process, oxygen can be used in its final phase.

However, the partial pressure gradient produced is also influenced by a number of other factors, of which the following are very important: temperature, physical exercise, respiratory resistance, compression, immersion, etc. It is rare that these other factors are used to control the generated gradient. It is most often assumed that they should be kept constant. In some cases, this is not possible, and thus, these factors are treated as additional decompression loads.

2.10.5 Risk Factors

The Polish Navy traditionally adopted certain factors that increase the risk of decompression sickness *DCS* when a diver does the following:

- Performs hard work
- Is overcooled/overheated
- Is untrained or has an individual predisposition to suffer from *DCS*
- Is obese or has over $80\,kg$ in weight
- Is over 40 years old

Or a dive is an element of a diving cycle. The list is not complete, and it could be extended to include the following, for example:

- Specific hazards associated with the diving equipment, ancillary equipment, diving technology used, etc.
- Breathing resistance
- Decompression to less than atmospheric pressure or air transport following diving
- Local compressions caused by suit and equipment
- Forced position during work or decompression
- Flooding of the suit
- Diet
- Use of food supplements and medicines
- Dehydration, etc.

Some of the factors interfering with the decompression process are summarized in Table 2.5.

No one is likely to question the impact of the aforementioned factors on the increase in the risk R of decompression sickness DCS. However, the interpretation of these factors is usually controversial and therefore demands that they be precisely defined. It is difficult, sometimes impossible, or even unreasonable, to define many of the factors that increase the risk R of decompression sickness DCS in general.

2.10.6 Research

Most often, work on decompression is modelled on classical research in physics. In the initial research process, efforts are made to establish as many parameters as possible at the assumed level, modifying only those that can be controlled and reproduced with great accuracy and precision. The impact of changes in other parameters on the decompression process is then gradually examined – first with the set values of part of the parameters, then with the simultaneous changes of an increasing number of them. During the research, various strategies are adopted in order to determine as precisely as possible the functional relationships which define the influence of individual parameters on the decompression process.

Most often, research on decompression focuses only on determining the function of the influence of two variables on the decompression process: the composition of the breathing medium and the decompression profile, understood as the influence of ambient pressure as a function of time. Most often, only simple trapezoidal profiles[248] are tested in the laboratory environment.

In order to determine the level of risk caused by the omission of known and unknown interfering factors, research is conducted by building statistical models on the basis of the results of conducted experimental dives under simulated or real conditions. If such work is carried out in a cycle of scientific research under laboratory conditions, it is referred to as validation studies.

In addition to scientific research, monitoring of the hazards occurring during the operation of the decompression system in real conditions should be carried out. On this basis, risk models are built similarly to epidemiological monitoring. Monitoring

TABLE 2.5
Possible Factors Causing Decompression Accidents

1. Reduction in blood circulation due to obesity, hypothermia, physiological deterioration of circulatory efficiency (e.g., in elderly people), previous cardiovascular or diving illnesses, mechanical compression (e.g., of diving suits, no changes in position during decompression), etc.
2. Increase in the carbon dioxide content of the inhaled breathing mix or its accumulation in the body (hypercapnia) caused, inter alia, by poor diver's condition, effort, presence of dead spaces in diving equipment, high density of the breathing mix, etc.
3. Considerable effort prior to diving and muscle acidification occurring afterward. It is recommended to take $3\,h-6\,h$ of rest before diving (depending on the conditions of the future dive), and it is absolutely prohibited for a tired diver to dive.
4. Alcohol consumption prior to diving and prior to the end of decompression (also during compulsory surface rest, after diving, when isobaric surface decompression occurs). Alcohol reduces the surface tension of blood, which facilitates the formation of gas bubbles. It is recommended that $12\,h$ before and $12\,h$ after the dive, no beverages containing alcohol should be consumed.
5. Injuries such as tearing of tissue. In the areas of injury, a free gas phase may form, and not only in such obvious cases as in pulmonary barotrauma, but also in minor skin injuries.
6. Dehydration is often disregarded as a factor that can lead to decompression problems. Ignoring dehydration, however, may lead to decompression problems, despite the observance of a number of other recommendations. Most often, dehydration occurs in the following cases:
   - For long dives in salt water in a wet suit. In this case, water loss occurs through osmosis (especially important when using an open-system water-heated suit).
   - Hydrostatic pressure can cause increased diuresis.
   - Overcooling can cause increased diuresis.
   - Weightlessness causes increased diuresis.
   - During breathing with a dry breathing medium, dehydration occurs due to evaporation.
   - Caffeine, theine, and alcohol cause increased diuresis.
   - Some groups of medicines cause dehydration and many increased diuresis. Divers may only take medication under the supervision of a doctor.
   - Anxiety and nervous tension can cause increased diuresis.

   Increasing the level of fluids in the body (rehydration) is recommended to increase diving safety. It is best to administer mineral water or a limited amount of fruit juice for this purpose. A diver should drink a lot a few hours before the dive. Hydration before the dive should take into account the diving conditions. It is not always possible to urinate freely during diving or decompression. If the diver is decompressed in a dry chamber with the possibility to pass urine, fluids should also be provided during decompression.
7. Diving during symptoms of poor physical and/or mental aptitude, for example, hangover symptoms, sleep deprivation, fatigue, menstrual symptoms, headaches, disease symptoms, etc. These factors reduce the concentration of attention and thus lead to the possibility of confusion, for example, when checking the decompression profile, measuring the depth or length of stay, etc.
8. The use of dietary supplements, headache remedies, oral contraceptives, stimulants or antistress agents, and active plant-based products, such as herbal teas, may not be recommended before diving. One of the reasons for the harmful effects of such products may be "contamination" due to persistently high concentrations of chemicals in the blood. This may lead to an easier formation of a free gas phase. Some chemicals, such as alcohol (alcohol is a component of many liquid drugs, such as cough syrups), change the surface tension of the blood, thus facilitating the formation of the free gas phase in the blood.
9. It is important to consider various types of implants, even such typical materials as fillings of treated teeth. There are known cases of induction of tooth pain with poorly placed or improperly selected fillings. In these cases, it is necessary to consult one's dentist about the possibility of diving, and when treating teeth, it should be noted that the patient practices diving.

Source: Kłos (2000).

is carried out in order to establish whether the level of conservatism taken into account is high enough to mitigate the impact of disturbing factors occurring during the actual operation of the decompression system, other than those which were examined during the laboratory tests. Monitoring must take into account the impact of elements of the system, other than the decompression system, but having an indirect impact on the decompression process.

For example, general development and pressure training program; cycles of applied biological regeneration; health-care and physiological tests; control of hygiene, diet, and impact of underwater tools used; specific tasks; relaxation and psychological support; impact of water, underwater, land, and air transport; etc.

The level of *conservatism* should be increased if it is found that the adopted levels do not adequately reduce the level of risk to the decompression system, even if it is necessary to return to the laboratory testing stage for this purpose.

If the system seems too conservative, the level of conservatism can be gradually reduced. A reduction in the level of *conservatism* can be achieved not by manipulating decompression parameters but, rather, by studying the impact of the factors in the diving system on the safety of decompression. For example, in the Polish Navy, like in other countries, a system of periodic medical checks is in place as part of the admission to the diving profession. Such a system does not offer any support for diver-related medical services intended to strengthen safety and health protection of divers. It is merely a bureaucratic barrier required to be cleared to continue working as a diver. Such a system can be experimentally replaced by a doctor dedicated to a specific diving group. The responsibility of such a doctor would be to keep a group of divers in a good condition to perform combat operations. The scope of his or her responsibility would be to collect the obligatory results needed for the assessment and to transfer them in due time to the existing central register. Only accredited centers could commission the tests. Thus, he/she would replace the central selection committee by making a decision on his/her own. However, in this case, he/she could apply health-care treatments through additional tests, supervision, training recommendations, biological regeneration treatments, hygiene control, nutrition, referral to rehabilitation treatments or performing them independently, conducting antistress classes, organizing leisure activities, etc., just as it operates in the case of athletes' health protection. Such a system would also provide an opportunity to build a bond resulting in a relationship of trust. Therefore, many of the valuable comments would not become available for medical staff. The system could be introduced on a pilot basis over a three-year period in one particular military unit, after which it could be compared with the traditional solution.

The following example describes a typical element of a test administered to check a system's immunity to interference. An element is removed from the system, a new or redundant element is introduced or replaced, and the impact of these changes in the system structure on the processes that the system is designed to efficiently support is examined. Making such improvements in the system should be a permanent activity, as each system tends to degenerate rather than self-improve. The system is most often degraded by the introduction of bureaucratic structural elements, the effects of which are not supervised and evaluated but introduced on the basis of arbitrary decisions.

The lack of a systemic approach sometimes results in informal grassroots changes. For example, in the Polish Navy, the supervision of the safety of diving operations is not based on a scientific method of assessment or the application of the process approach described earlier. Hence, over decades, the diving practice of permanently growing conservatism concerned with air expositions has persisted. In the statistical staff assessment, these result in the opinion that the air tables used in the Polish Navy are very conservative (DMW, 1982). This is not confirmed by laboratory tests carried out when properly planning the exposure. Under controlled laboratory conditions, an excessive amount of free gas phase is observed in venous vessels (Kłos, 2020). These results were obtained in connection with research into new technologies, where experimental divers were secured by divers who breathed air and used standard air decompression tables. Although these were not systematic studies on the safety of the air tables used, it can be concluded from the measurements carried out that rather than being conservative, the tables were actually developed based on the assumption of a low level of *conservatism*. A brief analysis of the conservatism management procedures recommended with these tables, together with the occurrence of a moving gap between exposure times in the tables, shows that the low *conservatism* was probably at the heart of the concept behind their development.

2.10.7 Oxygen Toxicity

Under saturated diving conditions, the most important type of the toxic effect of oxygen is pulmonary oxygen toxicity. When diving outside the saturation zone and from the saturation plateau, the threat of central oxygen toxicity is called *central nervous syndrome*[249] *CNSyn*. *Central nervous syndrome* is a neurological condition caused by damage to or dysfunction of the *central nervous system*, which includes the brain, brainstem, and spinal cord.

For special procedures,[250] for example, oxygen diving, an important type is also somatic oxygen toxicity (Kłos, 2012).

When studying decompression, it is important to point out that its planning is a kind of competition between the risk of various forms of oxygen toxicity and the risk associated with decompression sickness *DCS* for the most effective decompression profile possible. This can be seen in some decompression systems, for example, in *US Navy tables* for operational *heliox Hx* dives (US Naval Sea Systems Command, 2016). These tables are not planned for a specific[251] *Hx* but for a specific range of *Hx* composition. When planning a dive, the ranges of maximum allowable oxygen partial pressures and the duration of exposure to them are considered first. The diving process may then be considered in terms of limitations related to the risk of decompression sickness *DCS* based on the partial pressure of helium in the inhaled breathing mix.

The efficiency of decompression, understood as minimizing the risk R, is often equated with minimizing the time spent in the aquatic environment. Reducing the time spent in the aquatic environment minimizes the risk posed by it: possible malfunctions of diving gear and equipment, overcooling, stress, loss of buoyancy control, inconvenience associated with relatively low ergonomics of the diving gear and equipment, weightlessness, distorted field of vision, etc. Minimization of the time spent underwater can be achieved by special technical equipment, such as diving from a bell or underwater vehicle, with possibly swift transport under pressure to a hyperbaric habitat, where decompression can be continued without risk of drowning. However, this equipment is expensive both in terms of investment and operation, hence not available to a wide range of divers. In combat dives, it is usually not possible to use such equipment. Therefore, it is possible to increase the effectiveness of decompression by accelerating it. Diving outside the saturation zone requires oxygen to be operated under relatively high pressure, which poses a threat of the occurrence of central oxygen toxicity symptoms $CNSyn$. For this reason, when analyzing the residual risk R of symptoms of decompression sickness DCS, the risk of symptoms of $CNSyn$ should also be considered.

2.10.8 Procedures

The specificity of some diving procedures may further increase the risk R of developing not only symptoms of DCS but also other diving diseases DCI.

For example, such an accompanying threat can be hydrostatic imbalance. With the use of special diving apparatuses with semi-closed respiratory circuit, a suction effect produced by negative pressure of the respiratory system may occur (Castagna, de Maistre, Schmid, Caudal, and Regnard, 2018). The phenomenon occurs in some positions of the diving apparatus relative to lung centroid (Kłos, 2000). This effect causes the narrowing of the bronchioli and hinders gas exchange. With prolonged exposure, it can lead to the appearance of fluid in the lungs.[252] This effect is well-known among long-distance, high-performance swimmers exposed to intense cooling and considerable effort (Alonso, Chowdhury, Borakati, and Gankande, 2017). At greater depths, it may be caused by an increase in respiratory resistance. Such an effect is not only limited to special devices but also applies to open circuit equipment, for which the respiratory support has not been properly set up, in the case of cooling as well as high-intensity physical exercise (Boussuges, Ayme, Chaumet, Albier, Borgnetta, and Gavarry, 2017).

Another example is carbon dioxide CO_2 retention.[253] Regenerative apparatuses pose an additional risk of a malfunctioning carbon dioxide scrubber. A moderately increased partial pressure of carbon dioxide may counteract a decrease in respiratory action when diving with high partial pressure of oxygen. Therefore, originally, the diving apparatus was designed as a system with large dead spaces.[254] However, it has been found that such a solution, by eliminating one threat, introduces another. Carbon dioxide accumulates in the body because the affinity of CO_2 to hemoglobin is significantly lower than that of oxygen, hence its retention in the body, which may lead to hypercapnia. However, significantly below the threshold of hypercapnia, a considerable effect of CO_2 retention appears upon increasing the risk of central oxygen toxicity $CNSyn$. It is associated with the effect of limiting the peripheral flow by CO_2 and increasing brain blood flow. The influence of CO_2 retention on the possibility of developing DCS is caused by the same phenomenon.

Passive heat protection with low-conductivity gases may affect the increased risk of symptoms of DCS. The idea of using such thermal protection was promoted several years ago and is still occasionally discussed nowadays. The use of *argon*

Ar for passive protection against heat loss in dry-type suits was intended to extend the period of time it was possible to remain in the water. However, *Ar* migrates through the skin, then through the bloodstream and lungs, to the respiratory circuit of a regenerative diving apparatus. If the breathing space of the device is not sufficiently ventilated, the *Ar* retention may reach a level that disrupts decompression processes. Argon was used in diving experiments to simulate changes in breathing mix density; hence, the decompression parameters for this gas are known. Argon also accompanies air diving, as its contribution to atmospheric air is important; thus, it is taken into account when developing air decompression tables. However, even with a slight increase of its content in the breathing mix, especially during long periods of stay, the decompression for argon mixtures must be significantly prolonged. Therefore, the use of *Ar* for thermal protection is an important problem for planning the correct decompression schedules.

An important element of the operational diving scenario is a change in the breathing mix and the resulting counterdiffusion phenomena having a significant impact on the safety of decompression. These problems have been described earlier and will not be discussed in more detail here (Kłos, 2019).

Some diving scenarios pose a greater risk of *DCS* than others. Among the dives that pose a potentially greater risk of *DCS* are saw-shaped dives, yo-yo dives, multilevel dives, and frequent repeated exposures. The selection of an unsuitable diving sequence, with repeated or repetitive diving procedures, may also pose an additional risk of *DCS*. The potential increase in the risk concerned with these procedures is associated with the disruption of the decompression process. The main route of decompression is gas exchange between the tissues and lungs via blood.[255] Ideally, this exchange should take place in the bound[256] or dissolved phase.[257] However, part of this exchange also takes place in the gas phase, constituting the nuclei of the gas phase in the blood. The lungs are not always able to filter out the resulting gas phase, which can pass through an *intrapulmonary arteriovenous anastomosis IPAV* in the lungs (Schipke and Tetzlaff, 2016). Such undisturbed physiological function of the lungs is popularly called a lung filter. The enlargement of the nuclei of the free gas phase to a certain size triggers the activation of the intravascular coagulation cascade, causing blood clotting, similar to the initiation of wound healing. This process is called complement activation (Kłos, Konarski, and Olszański, 2004). In these types of dives, at certain stages of execution of their scenario, there may occur cycles of compression and decompression of the free gas phase that disable the pulmonary filter and the free gas phase through the lungs and heart to the arteries. It is assumed that during decompression, a free gas phase may occur in venous vessels in sizes which do not cause complement activation. And for the same conditions, they should not lead to the secretion of a free gas phase in arterial vessels due to the pressure difference between venous and arterial circulation. However, rapid manipulation of changes in depth may cause a free gas phase to pass to the arteries and accumulate there and agglomerate to the extent harmful to the organism. An example is the observation of a surprising increase in the incomprehensible incidence of symptoms of decompression sickness *DCS* in divers working on fish farms (Cole, 1993; Lepawsky and Wong, 2001). This was surprising because they performed only short dives at shallow depths

(Kłos, 2000). The decompression tables most often refer to the depth[258] expressed in mH_2O for fresh water,[259] but diving in seawater or suspensions[260] exposes the diver to higher pressure than even an accurate depth measurement would suggest. Most often, the depth is determined by means of a pneumatometer[261] measuring directly the pressure exerted on the diver at the dive depth. As with any measurement, this one is inherently flawed. In addition, there exists a discrepancy in the interpretation of the exposure pressure measurement point.[262] Therefore, even if the dive depth measured with the pneumatometer appears in the decompression tables, it is recommended the next greater depth be taken to determine the decompression profile.[263]

2.10.9 Elevation

Often, decompression systems are dedicated only to marine applications where the reference pressure for isobaric surface decompression is the normal pressure. If the reference pressure is different, decompression schedules may be inadequate for a system developed for marine conditions. This process is encountered when diving in mountain lakes. Also, it is a problem that occurs when air transport is planned immediately after the dive.

Some tables take into account the elevation of the body of water above sea level, giving different decompression schedules for the adopted elevation ranges. For tables without such options, different forms of compensation may be proposed (Wienke, 1993).

Decompression systems for military use should specify the possibility of immediate air transport after the dive, as this enables the planning of diver recovery by air (Natural Environment Research Council, 1974).

Decreasing the pressure value, as in the case of flights by airliners or transport helicopters, can cause so serious an imbalance in the gas pressure in tissues as to trigger symptoms of decompression sickness *DCS*. Changes in the atmospheric pressure during a flight in an airplane[264] or helicopter may affect the equilibrium conditions after decompression, owing to a drop in pressure at the flight altitude.[265] This can cause an imbalance and a free gas phase, leading to symptoms of *DCS*. The procedures for estimating decompression should specify the threshold at which surface isobaric decompression will take place.

2.10.10 Work

In some cases, carrying out thorough scientific research does not translate into the possibility of using its results in practice. An example of this is the limitation of the amount of work expended. Even if it is possible to establish the percentage value of the increase in the risk of *DCS* as a function of workload in a unit of time, diving supervisors or divers are not able to measure the physical activity performed during the diving process. In addition, the safe level of physical activity performed strongly depends on individual predispositions, showing at the same time fluctuations for the same individual, depending on his or her mental state or current physical fitness.[266]

The effect of exertion on an increase in risk of *DCS* is usually caused by the transition to anaerobic metabolism and the secretion of lactic acid in the muscles, as a biochemical metabolite of anaerobic metabolism. Lactic acid from the muscles[267] penetrates the blood, causing a decrease in the exponent of hydrogen ions *pH* in the blood, causing significant changes in the effectiveness of oxygen transmission through the blood (Kłos, 2012). A decrease in efficiency of oxygen transmission results in abnormal gas exchange between tissues and blood. In this way, a biochemical barrier is created to eliminate excess gases dissolved in tissues during the decompression process. In such a case, the process of safe acceleration of decompression with oxygen or decompression mixtures is not in accordance with the model adopted in the research conducted to develop decompression schedules for safe decompression, so it may be inadequate. The impact of a decrease in efficiency of oxygen transport through hemoglobin on the risk of *DCS* would be difficult to reliably model; hence, this effect is not taken into account quantitatively in decompression studies.

Heavy work can only be conducted during saturation dives, because after the work is finished, divers can rest whilst still at the saturation plateau, thus restoring their regular homeostasis. Then it is possible to proceed with standard decompression. In dives outside the saturation zone, such a possibility does not exist, which is why the physical activity for such dives is limited.

For decompression disturbances, it does not matter how lactic acid has developed in the muscles. Distant effects may come from the previous dive, the current dive, or other pre-dive activities. Even normalizing the pre-dive rest time will not do much good if excess activity is performed beyond this limit, with insufficient physical training, leading to extensive lactic acid secretion in muscles. This is because the effects of physical activity on anaerobic metabolism can last for several days. It is unrealistic and unnecessary to demand that the rest time after a dive should be several days. The state of readiness of a diver must be assessed by the diving manager, who can rely on lactic acid measurements in the blood.

It should be remembered that after a dive, during the stay on the surface, isobaric decompression continues to occur, which has a direct impact on the amount of effort it is safe to exert following the dive. Therefore, the decompression system should include reliable limits of recommended rest time in the phase of isobaric decompression after the dive. However, the indications from the validation of the system should only be considered as an averaged value, as it may need to be extended depending on diving conditions or training status. Also, the individual capacity to compensate for post-dive changes is of great importance. It should be noted, however, that they may fluctuate depending on the circumstances and the current fitness levels of the diver.

It is generally considered that physical exercise during decompression has a negative impact on the decompression process (the effect of exercise during decompression, 1982). There are also credible indications suggesting the very opposite (potential fifty percent reduction in saturation diving decompression time using a combination of intermittent recompression and exercise, June 14–16, 2007). Undoubtedly, movement during decompression causes additional emission of a free gas phase into venous vessels. With an instrument for the detection of free phase gas or a typical ultrasound scanner, it is easy to ascertain. When conducting free gas phase detection in venous vessels, using the Doppler ultrasound technique, two types of measurements are carried out: in the rest phase and in defined motion (Kłos, 2020). The movement is to specifically force an increase in the proportion of free phase gas in venous vessels.

As mentioned earlier, when estimating the acceptable values of partial pressure gradients of gases during the decompression process, the existence of the so-called gray area is assumed. This includes partial pressure gradients, which should not lead to the occurrence of symptoms of *DCS*, provided that there is no overlap in the course of decompression with excessive, additional factors that burden the diver's body. If the decompression is planned below the gray area, the working of the muscles during decompression causes increases and decreases in pressure in the muscle tissue, resulting in a mechanical "pumping" of an additional portion of gas from the tissues to the blood, intensifying the decompression process. However, if the decompression is planned near or within the gray area, the additional "pumping" of gas into the blood may contribute to the occurrence of *DCS* symptoms.

2.10.11 Temperature

In general, changes in ambient temperature have a significant impact on the homeostasis of the human body, and therefore, they also have a considerable effect on the gas balance in the diver's body. Every temperature variation in the diving process has an impact on the process of gas exchange. For example, during research on decompression, a fast hot shower after the dive is used to provoke skin symptoms of decompression sickness. This effect is sometimes not easy to predict, for example, for air diving, the best effects were achieved when diving took place in a relatively cold environment and decompression in a warm environment (Gerth, Ruterbusch, and Long, 2007). However, for other dives, this relationship may be different.

The influence of temperature can be explained by physics. The most common phenomenon is that solubility of gases in liquids decreases with increasing temperature. However, the mechanism of inducing symptoms of the *DCS* is more complicated, for example, warming of the skin while taking a bath is accompanied by dilatation of blood vessels of the dermis and superficial layers of subcutaneous tissue and a decrease in overall blood pressure, causing an imbalance of gas in the tissues and blood and creating favorable conditions for the formation of a free gas phase. The Polish Navy uses the technique of transporting a diver in a dry bell following deep diving. After evacuation of water from the bell and its sealing, it is transported to a hyperbaric complex, which the diver enters under the current pressure. The first skin symptoms of the *DCS* are already observed whilst in transit to the complex, particularly on the hands and the face.[268] These symptoms are compounded during a dive protected by helium-based artificial breathing mixes, when the diver's comfort has been protected only by passive heat protectors. These symptoms intensify when the diver moves into the warm environment of the hyperbaric chamber of the complex. Then they gradually disappear. This effect is not observed during saturation dives, when comfort of the diver is ensured by active thermal protection measures.[269]

Changes in the diver's thermal comfort are usually accompanied by different phases of the dive, such as the process of preparing to dive, the time spent in the water environment, and the time after the dive. Thermal comfort is often subject to quite unexpected changes. For instance, when preparing for dives in cold waters without active thermal protection measures, it becomes important to choose appropriate protective clothing. When selecting thermal protection, it is necessary to take into account the level of planned effort and the time of stay. When planning a relatively long stay, it is not possible to assume a long-term effort; therefore, a relatively effective outfit for a diving suit is chosen. However, when a considerable effort is required, for example, when a diver is operating in a strong underwater current, he/she will need to work hard to swim against the current, which results in the diver becoming unexpectedly overheated. On the surface, a person can partly undress[270] and then dress again, whereas underwater, this is usually difficult.[271] Overheating increases the physical exhaustion of the diver during the diving process, which directly translates into safe decompression.

A further unexpected change in thermal comfort may occur as a result of flooding of a dry suit with water. If flooding occurs during the initial submerging in a dive or earlier, the dive can usually be interrupted or shortened in order to avoid escalation of the hindering conditions. However, if the suit floods late in the dive or during the decompression process, it is usually not easy to counteract the resulting problem situation. If the decompression tables in use are designed to use surface decompression, there is a way to shorten decompression in water. Surface decompression requires having the decompression chamber in close proximity, as surface decompression procedures guarantee time not longer than $t \in [5,7]\,min$ for a rapid ascent from depth, transport to the chamber, partial removal of diving gear, and return to pressure in the decompression chamber (DCIEM, 1995). This time is extremely short, so when a surface decompression procedure is planned to be used, it is important to practice all the operations to carry it out, because the resulting extension of this time may cause serious symptoms of decompression sickness (NOAA, 1976).

Dangerous effects of overheating are observed in safety divers. A safety diver cannot completely remove the diving gear, which means that he/she may overheat on the surface. Once in the water, he/she experiences a thermal shock. During an intervention dive, he/she may be exposed to further overheating or, on the contrary, to relatively rapid cooling, depending on the action to be taken during the intervention.

2.10.12 Successiveness

When undertaking repeated dives, account should be taken of the remaining saturation from the previous dive. The decompression schedule should include planned stops aimed at reducing the diver's saturation during isobaric decompression whilst resting on the surface after the dive. However, the surface decompression process is not the only factor setting limitations on the next dive. An important limiting factor is the mechanism of transition to anaerobic metabolism, which was described earlier. Often, the effect of saturation disappears after just $t \geq 3h$,[272] but the effects of intensive effort can persist for $t \in [2,3]\,days$. Therefore, if a diver had worked hard during the previous dive, the typically prescribed rest period on the surface may be inadequate.

Blood pH measurements[273] may only eliminate the diver during the initial diving test, as there is no reliable model taking this parameter into account when planning safe diving operations.

Consideration of the impact of expended effort, especially in a series of dives, is an important premise for the process of planning diving operations. Ignoring this fact is an error. The decision as to the proper course of action in such a case may only be entrusted to the knowledge, experience, and reasonableness of the decision-maker, which constitutes his/her sole responsibility in this regard.[274]

2.10.13 Predispositions

In the course of diver training, it is important to collect anthropometric and physiological data related to decompression safety. The basic datasets include nutritional status, total body weight, muscle mass, and fat mass. Also relevant are the levels of care given to the teeth, skin, eyesight, throat, ears, etc. What is important is the current, general condition of the body's fitness, with particular emphasis on flexibility, reaction to the long-term effort or work in a forced position, biological age, etc. Respiratory parameters, such as vital lung capacity, maximal oxygen uptake,[275] ratio of oxygen consumption to lung ventilation, etc., play an essential role. Also, other related parameters, such as oxygen pulse, oxygen breath, or bradycardia associated with descent, all play a major role. Sometimes, owing to the specificity of the performed diving service, not all parameters are equally important.

In addition to the fitness tests, it may be useful to test for resistance to rapid compression and tolerance to oxygen exposure. Nowadays, the need for testing to exclude heart defects, such as the presence of a *patent foramen ovale* in the atrial septum,[276] is gaining in importance.

The role of mindset and mental training is, relatively speaking, often disregarded. There is much evidence that mindset and the state of mental training play an important, if not fundamental, role in extreme sports or dangerous work. Quite often, a meeting of outstanding people who have proven themselves in extreme conditions is accompanied by the surprise that they are often people whose body structure, facial features, or behavior do not meet a common archetype. This shows that their distinguishing feature is mental rather than physical structure. Pre-diving predisposition is also important. This is not confined to the state of rest/lack of compliance with the prohibition to work prior to diving, etc., but also the mindset resulting from the analysis of the situation, the current condition, and the psychological burden caused by events that lie outside the realm of diving.[277]

2.10.14 Weight

There is a general belief that an obese diver is at greater risk of developing symptoms of *DCS* compared to a slim diver. This conviction is associated, for example, with greater solubility of inert gases in adipose tissue than in muscle tissue. There is also a conviction that a perfectly built athlete is more resistant to severe decompression than a person with an average muscle buildup, which is probably anchored in the general archetype rooted in society. In some cases, such an understanding of reality may be inadequate.

All diving mammals have a significant reserve of body fat. They can dive to considerable depths – for example, seals can dive down to 500 mH_2O, and some cetaceans down to 3,000 mH_2O. Seals can remain underwater for about 20 min, and cetaceans, on average, up to 50 min. Their anatomical structure, especially the ribbing of their lungs, is different from that of humans, which is why they are capable of diving to such considerable depths.[278] Nonetheless, the physical phenomena accompanying the diving process are the same. These animals likely suffer from DCS, just as it was probably the case with diving dinosaurs. These theories are based on studies of the skeletons of these animals and the finding of sterile bone necrosis. If this premise is accepted, it should also be assumed that such observations can only be made in older, if not aged, animals. However, the observation of marine mammals cannot be transferred directly to humans, due to the completely different structure and function of the peripheral part of the circulatory system. A thick layer of adipose tissue is the basic thermal insulation of these mammals. Through evolution, they have developed the ability to cut off the flow of blood to the subcutaneous tissue, so it is not saturated during immersion.

It appears that greater solubility of inert gases in fats than in water is essential in DCS processes induced by counterdiffusion (Kłos, 2019). The difference in the solubility of inert gases in the presence of significant blood supply to adipose tissue seems to be less significant. However, in contrast to animals hibernating in winter, the blood supply of human body fat is poor and not particularly variable.

For reasons of thermal protection, it is sometimes inexpedient to require the diver to lose weight in order to significantly eliminate body fat.

A strong body build can be an additional burden when considering the risk of DCS. The developed muscle tissue consumes more oxygen during physical effort. The parameter of oxygen consumption per unit of body mass is one of the typical parameters taken into consideration when assessing overall fitness. High values of this parameter are an important premise taken into account when assessing athletes' training condition. In diving, good general training and high muscle mass may be a premise of increased risk of DCS. This may be the case if the muscle tissue has developed as a result of workout other than swimming or diving. Strong muscle development requires efficient gas exchange. If the muscle buildup was caused by surface exercise training, it is possible that an efficient surface respiratory and circulatory system may not be efficient enough under external pressure. Those who exercise according to strength training, especially isometric training, can relatively easily switch to anaerobic metabolism. The harmful effects of this phenomenon on the decompression process were described in the workload analysis.

Muscles are relatively easily capable of pushing through significant amounts of gas dissolved in them during contraction. Thus, during the assessment of decompression risk, the gas emitted from the muscle and forced into the blood is measured by observing the amount of free gas in venous vessels (Kłos, 2020). Expanded muscle tissue can release significant amounts of inert gas into the blood as a result of effort, increasing the risk of decompression sickness DCS.

The principle of operation of semi–closed circuit apparatuses is based on ventilation models, whereas safe decompression is planned based on the maximum reduction of oxygen content in the breathing mix inhaled by the diver. Therefore, the model is based on the assumption of maximum oxygen consumption by the diver. If a diver is allowed to dive with such equipment with an overconsumption of oxygen, the decompression schedule assumed for the diver may not be adequate.

2.10.15 Loads

The load on the diver consists not only of the useful work he/she performs but also of the load associated with it:

- Respiratory resistance
- Ergonomic loads, such as suit resistance, tools, diving gear dimensions, etc.
- Parameters of the aquatic environment: lift, buoyancy, etc.
- Operation and other loads connected with the necessity of continuous concentration
- Psychological burdens, etc.

Respiratory resistance is a serious contribution to the diver's workload. It is unavoidable when diving, for instance, because of the increased density of the breathing mix.[279] Its impact is taken into account during the process of developing and validating decompression schedules for the diving equipment concerned.[280] Inspiratory resistance is particularly harmful, but moderate expiratory resistance can have a positive impact on a diving and decompression process.

The bronchial tree from the trachea to the bronchi reduces in diameter and is rigid due to the presence of resilient bronchial cartilage in the walls. It is this cartilage that prevents the collapse of the airways when a person is breathing in a hyperbaric environment.[281] The muscular and terminal bronchioles are sufficiently flexible to allow them to be constricted or even closed, especially if a vacuum is also created as a result of the rapid flow of gas. The closure of the bronchioles leads to a partial stopping of the gas exchange and, as a consequence, to local hypoxia.

An increase in respiratory resistance can easily cause a disturbance in gas exchange. These effects are not considered when estimating the decompression procedure. Divers should be trained in conscious breathing to avoid disturbing the decompression process. Inhaling should take place slowly in a controlled manner, while during exhalation, resistance can be increased by clenching the mouth and performing a forced exhalation, as recommended in cases of asthma.

Breathing equipment for both diving and inhalation should be checked as often as possible, not only for the tightness of its component parts, but above all, also because of the breathing resistances.

Any form of compression, which usually results in reduced blood flow, can provoke skin symptoms of decompression sickness.[282] With some types of diving gear, local compressions cannot be avoided, such as the neck seals and sleeves of dry-type diving suits, or the neck seals of diving helmets, etc. Research carried out to develop the decompression process does not take this into account.

A diver most often performs underwater work in a suit, which, by limiting his/her movements, increases the workload. It may also cause a tightening so significant that it results in shallow breathing and effects similar to those of asphyxiation.[283] It also often causes localized compressions, which can lead to reduced blood circulation. This, in turn, can lead to local, painful symptoms of decompression sickness. Some local compressions are difficult to eliminate, such as the already-mentioned sleeve seals or the neck seal of the suit; hence, it is necessary to optimize them. Most often, the diver is loaded with a diving apparatus which, similar to the suit, restrains his/her movements and causes local compressions.

Local body compressions can and should be counteracted during decompression in the decompression chamber. Divers often forget about it, so the task of reminding the diver about this lies with the staff.

When carrying out underwater work, the diver, in moving different parts of structures, must often use different tools. The tools are usually designed in accordance with ergonomic principles, but the displaced structural elements are not. Underwater tools are usually heavier than those used on surface.

Performing hard work underwater requires stabilizing the diver's profile in the water, and in this case, no simple methods exist which would take into account ergonomic principles. The stabilization of the diver's position is most often achieved by applying variously distributed additional weights, such as breast weights, weight belts, special footwear, or other local weights. All these elements impose restrictions on movement and the necessity to expend additional effort as well as cause local compression.

The necessity to use some elements of equipment and tools is distracting for the diver, which is why he/she cannot completely focus only on useful work but must also pay attention to elements of equipment and diving parameters. For example, the lack of correct settings of respiratory resistance on the breathing gas supplied to the diver may result in a slight increase in respiratory resistance and, as a consequence, the need for additional breathing effort. Similarly to the lack of sufficient control over buoyancy, such resistance results in unnecessary additional exertion to counter it. This may result in anaerobic metabolism, with all the consequences associated with it.

Exercising control of diving and decompression parameters is an activity which requires attentiveness. Thorough and appropriate attentiveness is necessary, as errors can produce significant consequences. The process maintaining attention is disturbed by the toxic influence of components of the breathing mix. When planning a dive, it is important to pay attention to this difficulty by giving the diver sufficient time to consider and, if possible, control his or her actions.[284]

Negligence in proper fitting and arrangement of equipment may cause local compressions and consequently lead to local, most often dermal, symptoms of *DCS*.

The approval for underwater work should be preceded by preparation involving a properly selected training course which covers possible underwater work scenarios. The commencement of work, on the other hand, should be preceded by preparation consisting in verification of all elements of equipment and current diving parameters.

2.10.16 Mental State

To work as a diver requires performing various jobs in an environment which humans are not adapted to. This is why it is necessary to control the diving process using measuring instruments, focusing on hazards, whilst at the same time providing limited protection based, to a large extent, solely on the possibilities of self-rescue. Diving can take place in situations causing agoraphobia, operating in isolation with limited possibilities of communication, difficult spatial orientation, sometimes in enclosed spaces and often darkness which increases claustrophobia, etc. This causes anxiety, compounded by a feeling of weightlessness, distortion of image and dimensions of objects, exposure to cold, sometimes a feeling of hotness, etc. These factors lead to physiological disorders, usually similar to seasickness. Such factors impede concentration and disturb the diver's emotional state.

Military diving is additionally accompanied by psychological problems associated with explosive ordnance handling, operating in the immediate vicinity of an opponent, awareness of the existence of anti-diver barriers and anti-frogman weapons, etc.

Two extreme types of psychological adaptation are most often made use of. One consists in displacing thoughts related to threat with thoughts about various forms of individual compensation the diver might expect; the other consists in a pragmatic approach to threats, promoting a phlegmatic response to risk. The former gives a high degree of self-confidence, and considerable accumulation of stress sometimes allows one to survive, but when the immunity barrier is exceeded, an anxiety attack is explosive and, in most cases, can no longer be contained. The second style of adaptation also creates the possibilities of working under stress, but exceeding the limit of endurance causes discouragement and resignation. However, it is possible to wake up from this torpor once the right impulse materializes. Depending on the development of any given stressful situation, it is often a case that parallel personalities can actively exist in humans, as in Stevenson's novel *The Strange Case of Dr Jekyll and Mr Hyde*. It is possible to try to push the boundaries between them through adaptive training; however, it is a labor-intensive process.

It seems that psychological training also has a large impact on the safety of decompression. Theoretically, this mechanism can be explained, for example, by its influence on the process of secretion of stress hormones. Their action is based on direct stimulation of the receptors causing the contraction of blood vessels and, hence, an increase in blood pressure. This is accompanied by an acceleration of the heart's action, an increase in its stroke volume, an increase in diastolic pressure in the aorta, and an increase in cerebral and coronary blood flow and pupil and bronchial dilatation. It also mobilizes fat burning, inhibits intestinal peristalsis and the secretion of digestive juices and saliva, and reduces smooth muscle tension. Of course, not all these changes must cause a direct or indirect increase in the risk of symptoms of decompression sickness, but any imbalance in the body is undesirable from the point of view of decompression safety. Controlling the fine line between entering a stressful situation and the outbreak of panic creates a handicap.[285] Such situations hinder in the body's ability to facilitate the decompression process and make it more difficult for the symptoms of *DCS* to be identified.

2.10.17 Training

The scope of diver training, apart from maintaining the diver's general psychophysical and physical health,[286] should also include specialist diving, adaptive and psychological training, as well as care and hygiene procedures.

The conditioning adaptive training consists of a controlled swimming effort together with breath control. It develops a predisposition for underwater swimming. Such training may consist of underwater swimming with one breath, and also with additional buoyancy,[287] and swimming in an apparatus over a distance requiring great effort, etc. Complementary adaptive training comprises pressure expositions conducted in a hyperbaric chamber, allowing for respiratory adaptation to the hyperbaric environment and for undergoing decompression. Such training should be conducted continuously, and the level of training used as a parameter to assess the degree of adaptation to work in a hyperbaric environment. The assessment of the current level of adaptation to work in a hyperbaric environment can be carried out properly only according to the order scale. To assess the fitness of a diving team, it is not enough to simply obtain information about the currently completed training sessions; the underwater operator should plan and personally supervise the training.

The preceding forms of training should be supplemented by adaptation to cold and thermal shock exposure. This training creates conditions for safe decompression by reducing homeostasis disturbances as a result of cooling. The training can be organized in a pool with unheated water, and later in open water. The rule is not to use warm water before entering the pool or after exercise. At a later stage, it is possible to enter the pool by jumping.[288] Gradually, diving training may be added using different options of swimming with breath held. Swimming with basic and diving equipment may also be included. The level of workout load should be controlled by grading it properly, initially without being too strict on the trainee, for instance, allowing the use of head warmers or warmers for any other body parts.

Training in a decompression chamber should be supplemented by elements of psychological training of resistance to claustrophobia. Of course, psychological training can be conducted additionally in an isolated form, for example, in the form of meditation or antistress training. However, it is impossible to separate the elements of psychological training from general development training or specialized diving training. The psychological and diving training should also include elements leading to the following physiological reactions: hypoxia-induced sleep, aerobic blackout, choking, deep transition to anaerobic metabolism, respiratory stress, etc. Such training can be organized in swimming or diving pools, also during regular swimming or general development training.

The spatial orientation training of claustrophobic immunization can be conducted on specialized obstacle courses. In our hydrological conditions, it is difficult to organize a training course adapting to agoraphobia. It can be carried out only in waters with high visibility and with faults occurring at great depths, as an element of diving or swimming in a bathyscaphe. If this is not possible, it can be organized close to a technical infrastructure, such as an oil rig, using a multipoint underwater lighting set up to create an appropriate spatial exposure. Dangerous but exploitable areas would be sites such as flooded quarries.

Diving in a different environment should be preceded by adaptation. Such adaptation does not only mean adaptation when changing elevation, as before diving in alpine areas; there is also a need for adaptation when changes occur in meteorological[289] or hydrological conditions.[290]

2.10.18 Diet

Divers need to observe the general calorific recommendations and supplements for people performing hard work. However, the diet should also take into account specific requirements.

The centers of nucleation of the free gas phase growth can be inclusions from dietary supplements, dietary elements, or medicines taken. Therefore, information on this subject cannot be ignored or trivialized. For example, there is a belief that fruit juices are a source of valuable vitamins and should be recommended for divers. In general, such a conviction is correct, also from the point of view of the desired hydration of the body. However, juices contain a significant amount of acids, which quickly reach the blood, potentially lowering the *pH* of the blood. The organism defends itself by using buffering effects that are potentially unfavorable from the point of view of decompression. The protective mechanism in the form of buffering the *pH* of blood is based on the processes of carbon dioxide emission, causing faster breathlessness or symptoms of wheezing. Similarly, a diet consisting of sour food may cause analogous effects.

A diet rich in fats causes fats to also be present in the blood serum above physiological limits. It is well-known that a high cholesterol content in the blood serum favors its accumulation in blood vessel walls, which increases the risk of arteriosclerosis. It is not known what effect the lipid content of blood has on the safety of decompression.

Plasma lipids are essential to facilitate the absorption of fat-soluble vitamins, by acting as carriers of these vitamins. In relation to inert gases that are better soluble in fats, the lipids contained in blood may act as a carrier and/or storage medium. The gas storage function should change the character of the tissue from fast to slower.

2.10.19 Human Factors

Personal immunity or sensitivity to the effects of decompression may be caused by the natural presence of a free gas phase in the blood, constituting a factor allowing for its growth during decompression. The free gas phase in blood is formed mainly by cavitation accompanying the beating of the heart.[291] The emission of cavitation free gas phase in the heart is largely dependent on the individual heart structure and especially the work of the valves. Experience in ultrasonography and the possibility of continuous performance of such tests still only gives the possibility of subjective evaluation of the influence of this factor on the safety of decompression, and therefore, it cannot be generalized or normalized as yet.

Personal resistance to strenuous decompression may be caused by increased perfusion of fat tissue through blood.[292] In this respect, differences are observed depending on gender and age.

Stress hormones mainly affect the cardiovascular system, thus improving blood circulation. They also increase the secretion of glucose into the blood. Stress hormones cause the whole reaction of secretion of various other substances into the blood. They are potential sources of growth of the free gas phase. Although hyperactivity causing a rapid increase in perception is, in many cases, a positive behavior, during a dive, it can lead to an increased risk.

There is a strong conviction that individual immunity to the effects of decompression decreases with age. However, long studies of the occurrence of free gas phase in venous vessels show that the situation may be more complicated. A certain percentage of divers advanced in age but still in good general physical condition and undergoing systematic hyperbaric training do not show signs of *DCS* despite the presence of a significant amount of free gas phase in their blood.[293] Although this phenomenon is recorded, it is difficult to explain on the basis of the current physiological knowledge. The occurrence of these phenomena does not affect the general trend in age-related burdens, which make it necessary to moderate decompression regimes for older divers.

2.10.20 Gender

The discussion on the impact of gender on the safety of estimated decompression was triggered by publications on the risk of *DCS* in aviation. No differences were observed in the pilots self-studies (Kłos, 2007b). But after further analysis of the literature, it appears that noticeable differences exist in *hypobaria*, whereas in *hyperbaria*, no such differences were observed, although studies were conducted on a substantial number of individuals (Edmonds, Bennett, Lippmann and Mitchell, 2020; Cole, 1993; Brubakk and Neuman, 2003).

However, there are a number of burdens associated with gender differences that have a significant impact on decompression safety.

To date, few scientific studies have been carried out on the impact of gender on the risk of the *DCS*. This is because almost all scientific research is being conducted with a view to military and commercial applications. Currently, not many women are active in these fields. Currently, the *Polish Navy* has started recruiting women for diving roles.

If a woman can endure vigorous exercise during her menstruation, she should also be able to dive without major problems. If a woman alleviates her menstrual symptoms by taking painkillers, she should refrain from diving. Premenstrual stress[294] may cause some women to have a tendency to reduced environmental tolerance or mental breakdowns. In general, the question whether a woman can dive during this time depends on how she feels.

Many women dive in the first weeks of pregnancy, when they are not yet aware of conception. During this time, the fetus is particularly vulnerable to teratogenic factors.[295] The fetus should have an adequate level of oxygenation – neither too high nor too low. There is speculation that a free gas phase that does not cause problems in the mother[296] may “attack” the fetus or placenta by restricting blood flow and thus reducing the fetus’s oxygenation. Part of the procedures for treating *DCS* consists of recompression and administering oxygen for breathing. Some people believe that high oxygen partial pressure can be carcinogenic to the fetus.[297]

2.10.21 Conclusions

The material presented cannot provide an in-depth analysis of the academic literature. Although literature on the safety of decompression is extensive, the reports tend to focus on individual cases without, as yet, making indisputable recommendations. Only very few authors present analyses of the factors that interfere with the decompression process in the literature that is widely available (Lippmann, 1990).

Most of the problems presented stem only from the so-called good diving practice which has been adopted and followed by the Polish Navy specialists. These professional standards were taught on specialist courses, but probably because they have not been developed in a consolidated form, they have gradually disappeared. They are briefly described here and should be taken into account in obligatory analyses of the risk inherent in decompression, although their interpretation has not been undisputedly confirmed in the process of methodically correct and a complete scientific cognitive process.[298]

Diving is only part of combat operation scenarios; however, diving safety translates directly into combat effectiveness. Devising and checking some of the predicted dive operation scenarios is important for finding an optimal way to plan a combat operation.

2.10.22 Summary

The process of developing diving systems in the *Polish Navy* is characterized by considerable attention paid to safety rules. Therefore, underwater works for the *Polish Navy* are performed in accordance with the described operation sequence derived from the research performed. Apart from other things, the following technologies have been implemented: deep, bounce[299] *Trimix* diving technology with *SCR* use, *Trimix* saturation diving technology, and *Nitrox* combat diving system with *SCR* / *CCR* use. The preceding technologies have been repeatedly successfully used. Some analyses of the preceding technologies employed are described further.

2.11 BUBBLING

During experimental diving, the sonography technique can be applied for assessing the decompression risk. The Bayesian approach applied to validation of the Doppler method for intravascular detection of the free gas phase instead of the typical statistical inference has been demonstrated there.

It is almost a commonly held opinion that the Doppler ultrasonic detection of the intravascular free gas phase is not a procedure that can be particularly useful in decompression research. The main objection is that detection of the free gas phase in venous vessels is a weak function to predict the presence of the free gas phase in tissues and arterial blood, so this method is not suitable for assessing the risk of decompression. Only a few countries disagree with this commonly held view and use this method to assess the risk of decompression in decompression studies. France has introduced detection of the free gas phase in venous vessels for diving research and then, together with Canada, improved this method and developed it to a standard form.

Based on the published results of the Canadian research, the technique was evaluated at the Polish Naval Academy using statistical methods. The Academy accepted and adopted the results of this research and started to use this method in its own research on decompression over 25 years ago and continues to use it to great effect.

2.11.1 Context

In ocean engineering and underwater technology, diving technologies still play an important role, especially in offshore industries and search and rescue at sea. Combat and police divers are powerful forces who play a key role as response tactical teams, as reconnaissance units, and in mine countermeasures, submarine rescue, combat search and rescue missions, and so on. Recreational diving is also an important segment of the sports market.

Diving apparatuses are still important appliances on board marine vessels that can be used in emergency situations, such as shipboard damage control, evacuation of people cut off by water in ship compartments and still alive in air pillows, etc. Decompression sickness DCS is dangerous from the depth of $12\,mH_2O$, and prevention of DCS symptoms involves the implementation of effective decompression procedures.

As it follows from the preceding text, despite the huge market for unmanned technologies, it is still important to improve diving technology through scientific progress.

2.11.2 Modelling

Navies constantly want to increase operational security. One possible way is to look for factors that have a decisive impact on the safety of diving. In an analytical approach, the diving schedule model can be divided into several submodels.

Our perception of reality is gained only by building *models*, because the reality is too complicated for us. The model is a maximally simplified *system* reflecting a real *system* that is able to ensure the reproducibility of the *process* we are interested in with the required precision and accuracy.

The *system* is a set of distinct elements with links between them. Elements and connections between them form the structure. In the structure, relations and arrangement can be distinguished. The arrangement can be distinguished by series and order. The subsequent activities that are aimed at achieving a specific target are a *process*. The *process* can occur in a *system* whose elements ensure its sustainability.

The theoretical *model* is a deductive *model* resulting from the knowledge of the *system* and *processes* occurring in it. Theoretical *models* are most often based on structural isomorphism to the known *systems*, homology with the known *processes*, or structural and *process* analogy. Cybernetic logical *models* reflect the basic structure of the *system*, ensuring analogous information exchange inside and outside the modelled *system* (Figure 2.33).

The economic *model* is used to assess the profitability of developing a *model* of the *system* that is of interest to us, for example, in the form of a feasibility study.

The most frequent *models* are statistical or deterministic *models*. Validation *models* are most often the former. Before commencing the validation *process*, its feasibility should be reconsidered, taking into account the economic *model*. A positive result of the validation *process* allows the *model* to be implemented. Validation of the received decompression schedules is based on the stochastic *model*.

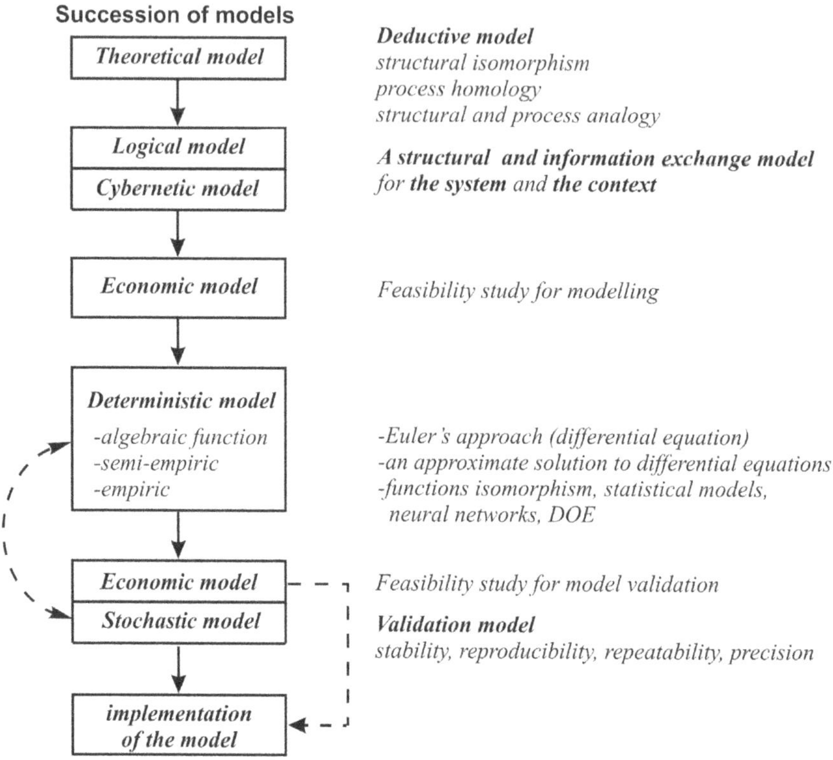

FIGURE 2.33 Scheme of the research project conducted to determine, validate, and approve the model.

2.11.3 Studies

The *processes* accompanying decompression constitute a complex problem for scientific research, because of the lack of precise measurement methods to monitor the processes taking place in body tissues. The mathematical *models* applied to describe the decompression process reflect only a small part of the total phenomena taking place, and they do not reflect the physiological *processes* taking place during decompression. Such mathematical *models* should be treated only as computational methods for the approximation of safe decompression procedures, because they are developed by matching experimental data to relatively simple mathematical functions.

Collecting data used to develop decompression procedures through experimental diving investigations is difficult, consuming much time and money. The data obtained are sometimes controversial and susceptible to various interpretations. However, it is the only way to obtain and validate such procedures.

As mentioned earlier, in the laboratory research phase, new mathematical *models* of decompression are developed, or already-existing *models* are modified. The proposed procedure thus obtained is tested. The experimental *process* is established in accordance with the principles specified in the Declaration of Helsinki. Chamber tests, where most of the *process* parameters can be accurately controlled, are essential for the laboratory research phase, but pool tests and open water tests may also be performed. Such experiments should be continued until the results, based on medical and scientific considerations, justify a transition to the implementation phase. The laboratory phase may be stopped at any time. The number of laboratory tests required to start the implementation phase may vary according to the *model* being tested. If an entirely new *model* is being tested, then the number of trials must be large enough to achieve an adequate confidence level for acceptance of the results. On the other hand, if the model had been previously tested and only minor changes were introduced, then the number of tests needed may be smaller. After completing the test program, the decision to terminate the laboratory research phase and begin the implementation phase is taken.

In the implementation phase, the validation tests are performed under the proposed operational diving conditions together with a high concern for diver safety. Management of the dives is given to the most experienced operational diving teams, with the assistance of well-equipped medical and scientific teams to monitor the medical and physiological status of the dive subjects and the safety of the dive procedures being tested. This is to ensure that a sufficiently large quantity of high-quality data is obtained while maintaining a high level of safety for the experimental divers. This will also form the basis for the decision to subsequently approve the decompression tables for operational use.

However, this is not the end of the research process. During operational use, the decompression procedures and the results of the dives carried out will be monitored. If any problems appear, they should be investigated in the research phase again. If a decision about the withdrawal of the decompression procedure or about its modification is made, the process will have to start anew.

2.11.4 Existing Solutions

Traditionally, the decompression risk assessment has been conducted using statistical inference methods. The classic approach is based on statistical inference based on experimental dives. The collected results are used to assess the risk of decompression sickness *DCS*, expressed as the likelihood of occurrence of *DCS* symptoms in the future by statistical inference based on a binomial distribution or using classical sequential analysis (Wald, 1947).

A conclusion based only on the occurrence or absence of *DCS* symptoms seems to be quite primitive, which is why methods were sought that could be used to determine, according to a more accurate scale, the likelihood of occurrence of *DCS* symptoms. The method of determining the concentration of the free gas phase in venous vessels by Doppler ultrasonography detection was chosen. This popular method was adopted and developed by the French and the Canadians, and the achievements of the latter have been adopted and developed in Poland (Nishi, 1990; Nishi, Brubakk, and Eftedal, 2003).

The simple Doppler Bubble Monitoring *DBM* application[300] is useful in experimental studies of decompression profiles. *DBM* is a standard tool used by several laboratories conducting research on decompression (Eftedal, 2007; Seddon, Thacker,

Gilbert, and Anthony, 2010). Its main advantage is that it can be used outside the laboratory in an actual time under pressure. The Naval Academy has used Doppler ultrasonic detection of the intravascular free gas phase as a method for monitoring decompression safety in the course of diving technology–focused research and development for over 25 years, to great effect (Kłos, 2020). A similar approach was presented earlier (Blatteau, Hugon, Gempp, Castagna, Pény, and Vallée, 2012).

The adequacy of the decompression schedule is assessed under the control of the Doppler ultrasonic detection of the intravascular free gas phase (Kłos, 2002). The symptoms of decompression sickness *DCS* are not necessarily accompanied by the formation of the free gas phase in venous blood as measured by the instrument. Measuring the concentration of the free gas phase in tissues is much more suitable for assessing the development of *DCS* risk symptoms. As for the results of the Canadian research, good statistical agreement between *DCS* symptoms and the concentration of the free gas phase in veins is recorded.

The Polish research on decompression safety is based on measurements performed with the Doppler ultrasonic detection of the intravascular free gas phase carried out to predict the risk of *DCS* in the intravascular free gas phase. These long-term studies seem to have confirmed the usefulness of this method. A description of the statistical inference procedures used for decompression sickness risk and the measurement scores of intravascularly free gas phase was presented earlier (Kłos, Nishi, and Olszański, 2002). In practice, three places are used for determining the free gas phase. These are the precordial and both subclavian veins.

The precordial area should be monitored from the vena cava side, because any vein gas from the venous system should pass through this area before it is removed from the lungs. The free gas phase is rarely observed on the arterial side. Gas bubbles reach the arterial side only when the bubbles of the free gas phase on the vein side are so numerous that they exceed the capacity to filter their entire population through the lungs. By monitoring the precordial area, it is possible to estimate the total number of bubbles entering from the whole of the vein blood at a given time.

In practice, not all gas bubbles can be detected, because the amplitude of the signal coming from smaller bubbles may be smaller than the amplitude of the basic background signal derived from the heart rate and blood flow.

Proper placement of an ultrasound transducer over the heart is very important from the point of view of the techniques for detecting gas bubbles in vein blood. The operation of the tricuspid valve may help locate the optimal placement of the probe, but its movement tends to mask the signals coming from the free gas phase bubbles, so it should not be allowed to dominate in the listening signal.

In some divers, especially those with a massive physique, it may be difficult to get a good signal around the heart area. In this case, the diver should lean forward, causing the heart to move closer to the chest wall. This will increase the signal strength.

Sometimes it is difficult to get a good signal in the precordial area. In this case, the gas bubble may not be detected in this place. Therefore, it is recommended that other places be monitored as well. Subclavian veins are easy to monitor, and gas bubbles are easily detected there. For this reason, it has been routine to study the presence of bubbles in both the subclavian veins and the precordial region. It is often possible to detect the free gas phase in the subclavian veins when it is not possible to detect them in the precordial region. However, monitoring only the subclavian veins is not sufficient, because gas bubbles flowing from other parts of the body could then be undetected. The subclavian vein is located close to the subclavian artery, so it is easy to find by searching for a pulsating sound coming from the artery. The probe should be positioned so as to obtain the maximum venous signal and the minimal signal from the artery. Monitoring of the subclavian veins is accompanied by a characteristic Doppler signal, similar to a blowing wind. This sound becomes louder when the blood flow through the vein increases as a result of clenching of the fist by the subject on the monitored side.

2.11.5 Validation

The Bayesian approach was chosen for the validation process (Kłos, 2014a). In the inference process, the physiological symptoms of DCS and the results of measurements of the presence of free gas phase in the blood veins at 1,726 exposures in DRDC Toronto were compared (Sawatzky, 1991; Eftedal, 2007; Table 2.5).

In the investigations described, the acceptable safe gradient was taken as $< II+$ (Nishi, 1990; Nishi, Brubakk, and Eftedal, 2003). Estimation of the probability of events involving the absence of the signal designated as dangerous $B:\{G\}$ with the subsequent onset of DCS for air and nitrox decompression $A:\{DCS\}$ is $P(\neg G|DCS) \triangleq \frac{n(\neg G \cap DSC)}{n(DSC)} = \frac{5}{41} \cong 0.12$ (Table 2.5). For the case of lack of diagnosis by nondetection of intravascular free gas phase $B:\{G\}$ for air and other nitrogen–oxygen decompression procedures with the DCS occurrence $A:\{DCS\}$, the appropriate probability is $P(DCS|\neg G) \triangleq \frac{5}{1289} \cong 0.004$, and this is small enough for us to recommend this method for use in support of decompression research.

TABLE 2.5
The cumulative Frequency Distribution of *DCS* and Measurements of the Signal from *DBM* for Air and Other Nitrogen–Oxygen Decompression

	G	$\neg G$	TOTAL
DCS	36	5	41
$\neg DCS$	401	1,284	1,685
TOTAL	437	1,289	1,726

where $\neg DCS -$ is the number of the absence of DCS symptoms; $DCS -$ is the number of the onset of DCS symptoms; $\neg G -$ is the number of signals from DBM, suggesting the absence of DCS symptoms; $G -$ is the signal from DBM, indicating conditions for onset of DCS symptoms for the free gas phase in the precordial zone or in the subclavian vein obtained by selecting the larger of the signals.

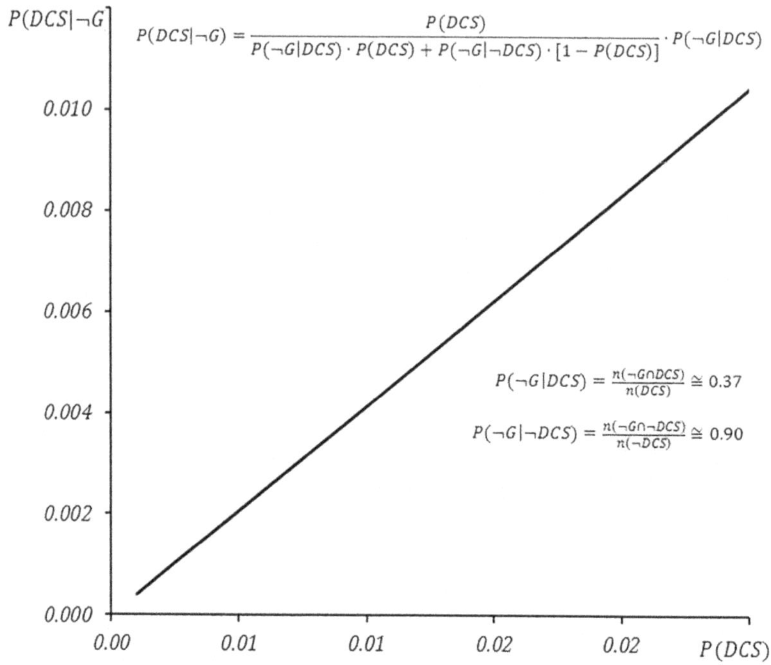

FIGURE 2.34 Probability $P(DCS|G)$ that, despite the fact that a free gas phase was not detected during motion or at rest $B:\{G\}$, a diver will have DCS symptoms $A:\{DCS\}$.

In accordance with the global trends, the Polish Navy has adopted measures to encourage the development of decompression systems for which the *DCS* risk is less than 1% (Huggins, 1992). Hence, for the evaluation of experimental dives, the adoption of this assumption level seems to be $P(DCS)=0.01$. Using this assumption, data from Table 2.5, and the Bayesian approach, it is possible to calculate the probability of an event in which, despite exceeding the boundary for an undetected signal originated from the intravascular free gas phase during motion or at rest, the diver will have the symptoms $P(DCS|\neg G)\triangleq\frac{0.01}{0.370.01+0.900.99}0.37\cong 0.004_1 << 1\%$ (Figure 2.34; Kłos, 2007b).

These results suggest that the diagnostic techniques when using *DBM* are highly reliable and sufficient to assess the *DCS* risks during experimental dives. These results are entirely consistent with the previous observations recorded by the Polish Navy. Other researchers have drawn similar conclusions by using other inference methods (Eftedal, Lydersen, and Brubakk, 2007; Hugon, Metelkina, Barbaud, Nishi, Bouak, Blatteau, and Gempp, 2018).

2.11.6 Measuring

The Naval Academy has used the Doppler ultrasonic detection of the intravascular free gas phase as a method of monitoring decompression safety in the course

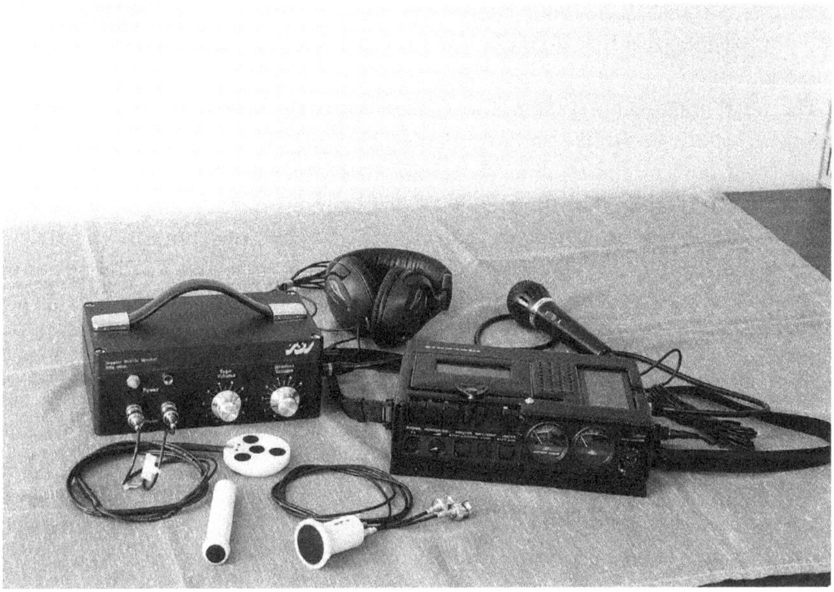

FIGURE 2.35 Doppler Bubble Monitor DBM 9008.

TABLE 2.6
Description of Doppler Bubble Monitor Shown in Figure 2.35

Probe	Two piezoelectric ceramic elements (transmit and receive) are both rigidly and permanently situated inside the head of the sensor assembly. Two styles of probes are a pencil probe for the subclavian region and a precordial type with variable focal lengths for the precordial region.
Operation mode	Continued wave
Operating frequency	$2.5\ MHz$
Acoustical pressure	$15\ mW\,cm^{-1}$
Power supply	Batteries: $4 \times AA$
Mechanical features	Rugged, splash-proof sealer container
Focal length	Precordial: $7.9\ cm$ and $10.5\ cm$; subclavian: $1\ cm$

of diving technology–focused research and development for over 25 years, to great effect.

In the Naval Academy, the Doppler Bubble Monitoring *DBM* system is used in the procedure for ultrasonic detection of the intravascular free gas phase (Figure 2.35 and Table 2.6).

The simple *DBM* application[301] is useful in experimental studies of decompression profiles. *DBM* is a standard tool used by several laboratories conducting research on decompression (DCIEM, 1986; Eftedal, Lydersen, and Brubakk, 2007; Kłos, 2020). Its main advantage is that it can be used outside the laboratory in real time under pressure.

The *DBM* measuring system is a simple and inexpensive device, but using this technique requires a skilled and experienced operator. The operators of the auscultatory *DBM* device require about three months of intensive training, which, in the case of the interpretation of clinical ultrasound imaging, is much longer and depends on the scope of the authorization granted. All the Academy operators completed a course conducted by $DR-DCToronto$.[302] The experimental results are easily understood by other scientists because of the application of a standard *KM* code procedure which has been used for signals classification (Nishi, Brubakk and Eftedal, 2003; Kłos, 2007b).

In practice, certain places are used for determining the free gas phase. During the research described, the precordial and both of the subclavian veins were used for determining the free gas phase.

The precordial area should be monitored from the vena cava side, because any vein gas from the venous system should pass through this area before it is removed from the lungs. The free gas phase is rarely observed on the arterial side. Gas bubbles reach the arterial side only when the bubbles of the free gas phase on the vein side are so numerous that the capacity to filter their entire population through the lungs is exceeded. By monitoring the precordial area, it is possible to estimate the total number of bubbles entering from the whole of the vein blood at a given time.

In practice, not all gas bubbles can be detected, because the amplitude of the signal coming from smaller bubbles may be smaller than the amplitude of the basic background signal derived from the heart rate and blood flow.

Proper placement of an ultrasound transducer over the heart is very important from the point of view of the techniques for detecting gas bubbles in vein blood. The operation of the tricuspid valve may help locate the optimal placement of the probe, but its movement tends to mask the signals coming from the free gas phase bubbles, so it should not be allowed to dominate in the listening signal.

In some divers, especially those with a massive physique, it may be difficult to get a good signal around the heart area. In this case, the diver should lean forward, causing the heart to move closer to the chest wall. This will increase the signal strength.

Sometimes it is difficult to get a good signal in the precordial area. In this case, the gas bubble may not be detected in this place. Therefore, it is recommended that other places be monitored as well. Subclavian veins are easy to monitor, and gas bubbles are easily detected there. For this reason, it has been routine to study the presence of bubbles in both the subclavian veins and the precordial region. It is often possible to detect the free gas phase in subclavian veins when it is not possible to detect them in the precordial region. However, monitoring only subclavian veins is not sufficient, because gas bubbles flowing from other parts of the body could then be undetected. The subclavian vein is located close to the subclavian artery, so it is easy to find it, by searching for a pulsating sound coming from the artery. The probe should be positioned so as to obtain the maximum venous signal and the minimal signal from the artery. Monitoring of the subclavian veins is accompanied by a characteristic Doppler signal, similar to a blowing wind. This sound becomes louder when blood flow through the vein increases as a result of clenching of the fist by the subject on the monitored side.

It is also possible to monitor other places, such as femoral veins and main veins. Femoral veins may be examined for the presence of the free gas phase in the legs. Main veins are hard to access. The lower main vein is available for examination. In this case, the probe must be pushed deep into the abdominal wall. It is important not to confuse signals coming from the intestines with signals coming from the free gas phase in the inferior vena cava. Checking for bubbles in the brain can be achieved by monitoring the jugular veins. If there are plenty of bubbles in the precordial region, the jugular vein is worth examining.

Identifying the bubbles in the precordial region in the course of ultrasonic detection of the free gas phase is the most difficult task because the signals are masked by the signal coming from the movement of the structural elements of the heart. And unfortunately, this skill can only be mastered over a sufficiently long period of practice. It is recommended that people learning the Doppler technique of detecting the free gas phase be instructed by an experienced person. Monitoring should be carried out in two cases: standing at rest and after a well-defined movement. The goal of the movement is to temporarily increase the number of bubbles squeezed out of the tissues into the bloodstream. This is especially useful when there are small numbers of bubbles. In the case of monitoring located in the precordial region, this movement is defined as a deep knee bend and squat. The subject has to bend his knees deeply and then immediately get up, keeping the transducer probe in the selected location so that the probe will not move. In the case of monitoring located in the subclavian region, this movement tightens the fist on the monitored side, followed by rest. If femoral veins or lower veins are examined, the subject is asked to raise first one, then the other leg, or do a calf raise.

It is advisable to record a reference signal for each diver before each dive. If later on the observers are not sure about the presence of the free gas phase, they can listen to the reference signal. Recording the reference signal helps the observer familiarize himself with the characteristic sounds of each diver's heart.

Divers can be monitored in different stages of staying under pressure, but most often monitoring is done after the dive. After the pressure exposure is over, the divers should be monitored at half-hour intervals, for at least three hours after the end of decompression, or until the free gas phase disappears.

As experience shows, the formation of the free gas phase can occur $t \in [1,4]h$ after the end of decompression.

It is advisable that the monitored signals be recorded and recommended that a complete commentary on the recording be made. Parallel to the commentary, written documentation should also be collected. During sound recording, signal level indicators should be observed to ensure that the amplitude of the signal is adequate. In order to make sure that the signal has been saved correctly, it should be played back. If it is possible, investigators should work in pairs. One places the probe and maintains the signal at the appropriate level, while the other one records the signals. Both should interpret the signals and cooperate with each other.

Three parameters are used to describe the signal coming from the free gas phase. Each parameter is classified on a relative scale from 0 to 4. The frequency and

TABLE 2.7
Frequency Parameter for *KM* Code

Code	Frequency (Numbers)
0	No bubbles, 0
1	Single bubbles, 1–2
2	Single bubbles, 3–8
3	Waving and rumbling signal, 9–14
4	Continuous signal

TABLE 2.8
Time Duration Parameter for *KM* Code for Rest

Code	Time Duration (%)
0	No bubbles, 0%
1	1%–10%
2	10%–50%
3	50%–99%
4	100%

amplitude are classified on an identical scale for rest and movement. The third parameter for rest and movement is defined in a different manner. For rest, it is expressed on the percentage scale, while for movement, on the duration scale. The combination of these three code parameters can be reduced to a single-digit *KM* code.

The first parameter is the frequency representing the number of gas bubbles per one heartbeat. The method for determining this parameter is given in Table 2.7, where, for the relative code 4, the bubbles are so numerous that their numbers cannot be estimated.

Time duration represents the ratio of the time of occurrence of the signal from the free gas phase in relation to the duration of the measurement, expressed as a percentage. The method of estimating this parameter is presented in Table 2.8. For example, if 1–2 gas bubbles[303] occurred during 20 out of 100 heartbeats, the time duration parameter would be 2 from Table 2.8.

If the same signal, 3–8 bubbles, coming from the free gas phase occurs during 20 heartbeats per 100, and if during an additional 40 heartbeats 1–2 gas bubbles occur, the frequency parameter can be chosen as 2 from Table 2.7, and the time duration can be chosen as 2 from Table 2.8, because the signal has been observed for 20% of the heartbeats, or the time duration parameter will be 3 from Table 2.8 for the frequency code 1 from Table 2.7, because for more than 50% of the signal occurrence,[304] its frequency code can be assigned as 1 from Table 2.7. As shown in Table 2.11, almost the same final effect of signal classification can be achieved in several ways.

TABLE 2.9
Time Duration Parameter for KM Code for Movement

Code	Duration of Bubbles (Heartbeats, Percent)
0	0 (0%)
1	1–2 (10%–20%)
2	3–5 (30%–50%)
3	6–10 (60%–100%)
4	Continuous, >10 (100%)

TABLE 2.10
Amplitude Parameter for the KM Code

Code	Amplitude
0	No bubbles
1	Barely perceptible, $Ab << Ac$
2	Moderate amplitude, $Ab < Ac$
3	Comparable amplitude, $Ab \approx Ac$
4	Maximum amplitude, $Ab > Ac$

Movement is a transient event, while rest is treated as a quasi-permanent event. Due to this difference, for movement, the second parameter will not be expressed as the percentage of time duration but as the time of cycle duration.

The duration of measurement refers to ten heartbeats starting from the end of the movement and counting as the number of heartbeats attributable when the number of gas bubbles as defined in Table 2.7 is recorded. The measurement cycle consists of ten heartbeats for measurements located in the precordial area or as the percentage of heartbeats when the signal from gas bubbles is recorded for other places (Table 2.9).

For example, if 9–14 gas bubbles are observed in each of the first four heartbeats per ten heartbeats after movement, then the code of the time duration parameter from Table 2.9 is 2 for the frequency parameter code 3 from Table 2.7, provided that the number of bubbles has the tendency to decrease. If the number of gas bubbles for the next four heartbeats falls to 3–8, the duration code will be 3 for the frequency code 2.

The amplitude Ab of the signal produced by the gas bubbles compares to the amplitude Ac of the normal signal coming from the heartbeat or flowing blood. The method for determining the amplitude parameter is given in Table 2.10.

The three designated codes are presented in the order of the parameters: frequency, time duration, and amplitude. The resulting KM code is stored in the form of fpa for rest and fda for the movement case.

TABLE 2.11
Conversion of the KM Code to the Degree of Gas Bubble Gradation

fpa	g	fpa	g	fpa	g	fpa	g
111	I–	211	I–	311	I	411	II–
112	I	212	I	312	II–	412	II
113	I	213	I+	313	II	413	II+
114	I+	214	II+	314	II	414	III–
121	I+	221	II–	321	II	421	III–
122	II	222	II	322	II+	422	III
123	II	223	II+	323	III–	423	III
124	II	224	II+	324	III	424	III+
131	II	231	II	331	III–	431	III
132	II	232	III–	332	III	432	III+
133	III–	233	III	333	III	433	IV–
134	III–	234	III	334	III+	434	IV
141	II	241	III–	341	III	441	III+
142	III–	242	III	342	III+	442	IV
143	III	243	III	343	III+	443	IV
144	III	244	III+	344	IV–	444	IV

Note: fp(d)a = code KM; *g* = degree of gas bubble gradation; *f* = frequency parameter; *p or d* = time duration parameter for rest or movement; *a* = amplitude parameter.

Table 2.11 shows the method for converting the recorded three-parameter *KM* code into a single-parameter code of bubble gradation. This method is convenient, but the original *KM* code contains more information and should also be stored. In addition to the recommended recording of audio signals, the data on diving and descriptions of the symptoms of decompression sickness should be recorded.

It is often possible to assign more than one classification to one signal. To decide which classification to use, bubble gradation should be determined for individual signals, and the highest value should be selected. For example, for codes 232 and 322, the gradation is III– and II+, respectively, so code 232 should be chosen.

For movement, each measurement should be classified, and the most representative code should be selected out of them. If codes 222, 232, and 332 have been assigned to three measurements, then the central value of 232 should be chosen. A deviation from this rule occurs when two measurements give a value of 0, while in the third one, the free gas phase is observed. In this case, the higher classification should be chosen, because the most important effect of measurements is to distinguish between cases when there is free gas phase and when there is no gas phase.

2.11.7 Conclusions

In Poland, the Ethics Committee does not accept any delay when *DCS* symptoms occur; therefore, all potentially dangerous situations identified by the Doppler Bubble

Monitoring *DBM* measuring system have been corrected by oxygen flush procedures – normobaric or hyperbaric.

The DBM9008 system is used to measure signals from the precordial and subclavian regions only with the use of a precordial probe. It is difficult to get a good signal in the precordial area, so it has been routine to study the presence of bubbles in both the subclavian veins and the precordial region. The latest research validates our approach (Kłos, 2007b).

Monitoring always takes place after the defined movement. The measurements are carried out at $30 min$ intervals, up to $3h$ after the end of decompression or longer, until the free gas phase signal disappears. Three measurements are made in the heart area and each subclavian vein during the one test period. The highest value is selected and later converted into the degree of gas bubble gradation according to Table 2.11. A signal above II+ degree of gas bubble gradation is considered dangerous, and the diver needs an additional decompression.[305] A description of the statistical inference procedures used for decompression sickness risk and the measurements scores of the intravascularly free gas phase was presented earlier (Kłos, Nishi, and Olszański, 2002).

Whenever the potential risk of *DCS* was theoretically increased, the bubble score also increased. This relationship was valid below and inside the so-called "gray area," in which it is difficult to clearly determine the risk of decompression. This method has not been tested in Poland for the region with a high probability[306] of *DCS*.

Over 400 experimental dives under the control of the *DBM*9008 system have been carried out without any symptoms of *DCS* in the last 20 years. Thus, the Naval Academy can recommend ultrasonic detection of the intravascular free gas phase as very useful in research on diving.

NOTES

1 Implementation research.
2 For instance, defined by system limitations.
3 If redundant elements are required owing to the way the system is built, for example, for safety reasons, such an approach is also reflected in the term "reasonably minimal set of elements" used here.
4 Joint action is stronger than the sum of separate actions.
5 That lies outside the system but has an impact on its functioning.
6 In zero approach.
7 Taking into account not only financial viability but also social impacts.
8 Systems engineering is used in engineering science; the system analysis used in management sciences is based on the same principles.
9 Adapted.
10 Military diving, record-breaking diving, wreck diving, cave diving, etc.
11 Record 701 mH_2O.
12 Simulated diving.
13 Mainly no-decompression dives.
14 In some countries, decompression procedures are introduced and followed by diving federations and local law.
15 Decompression schedules.
16 Not recommended.
17 In order to rescue the diver if unintentionally the time of dive exceeds the limit.

18 Self-contained air *UBA*.
19 Mine countermeasure.
20 For example, Polish *SCR*-type $APW-6M$.
21 For example, as the failure result.
22 For example, *FGG III SCR* manufactured in Germany (Figure 1.3a).
23 For example, failure of basic gas supply, of external gas supply, or of regeneration systems.
24 That is, attached and connected at the place that enables easy access.
25 Like the diving bell, rescue submersible habitat, or the emergency *UBA* at the decompression station where the diver is decompressed, to the surface, etc.
26 For example, in case of necessity, cut off external gas supply and switch on independent supply or abandon the basic diving equipment.
27 It is possible only if the umbilical is not damaged.
28 Approximate volume $1\,dm^3$.
29 For example, while the diving bell or underwater vehicle is lost or immobilized and there is no possibility to rescue the divers.
30 Basic gas supply system.
31 Post-accident procedures.
32 Particularly at designing of the special *UBAs*.
33 For example, *Drgerwerk AG Lbeck* has published the basic consequences of the assumptions of the *FGG III*-type *premix* – *SCR* ventilation mathematical model in its description.
34 For example, Det Norske Veritas.
35 Or fixed load that enables to keep for a long time the load level constant, or short-distance swimming.
36 Push-on ergometer.
37 Courtesy of the Military Diving School and the Military Sport Centre FLOTA.
38 Courtesy of *DR* – *DC*, former *DCIEM* Canada, advertising materials.
39 Standard.
40 An effort program.
41 Brief initial design.
42 Space of the limited gas exchange.
43 Emergency situations as well.
44 For example, at the author's supervision phase.
45 For example, organic injuries that occur after years.
46 It can occur during diving and decompression as an effect of breathing, with the gas mixture having low or too high oxygen content.
47 For example, once a year.
48 Especially those of the cumulative mechanism character.
49 Dives per one diver.
50 No diving company will employ a diver when rumor is going around that he/she has personal predisposition to develop diving sickness or has poor health.
51 As a rule, it is usually performed just before or during decompression.
52 Heart rate, blood pressure, breathing rate, etc.
53 For providing surface decompression for this type of diving technology, safety divers in the event of an occurrence of decompression sickness or other diseases and so on.
54 Caisson works, hyperbaric (diving bells, flying diving bells, and lockout-type underwater vehicles, like underwater vehicles with a hyperbaric compartment for divers) and normobaric (observation chambers, welding chambers, manned vehicles, one-atmospheric diving suits, and so on) underwater intervention, and so on.
55 For submarine rescue, for protecting victims who have stuck in "air pillows" in sunken surface ships, and so on.

56 Currently, it is believed that hyperbaric treatment of some diseases is cheaper than other therapies, like antibiotic therapy of the diabetic foot, or it is the only relatively effective therapy, for example, in the case of traumatic rupture of the auditory nerve.
57 For example, emergency power generator, *system* of emergency disconnection/isolation/cutoff of the complex from external installations, fire protection *system*, and so on.
58 For example, an automatic *system* for increasing/decreasing pressure in the hyperbaric complex, an automatic *system* for maintaining a constant partial pressure of oxygen in the complex atmosphere, and so on.
59 If the current time is not downloaded from the computer that is currently the main time server.
60 Depending on the scope of the planned diving process and the structure of the hyperbaric system.
61 Powered from main supply, auxiliary mains grid, power generator, self-contained battery, external emergency supply systems.
62 Powered from main supply, auxiliary mains grid, power generator, self-contained battery, external emergency supply systems, flashlights powered by mechanical hand-generated electric current, strobe markers, chemical lights, and so on – some elements are only checked by compatibility with a computer list or a paper checklist.
63 Powered from main supply, auxiliary mains grid, power generator, self-contained battery, external emergency supply systems, chemical heaters, and so on – some elements are only checked by compatibility with a computer list or a paper checklist.
64 Powered from main supply, auxiliary mains grid, power generator, self-contained battery, external emergency supply systems, voice-powered system, and so on.
65 Possibility of recording on the internal memory of the recorder, transfer to the internal server, transfer to the outside; possibility of consolidation with measurement data.
66 Powered from main supply, self-contained battery, hand-operated mechanical drive system, personal filtering masks working in the filtration system of the exhaled breathing mix.
67 Powered by gas and water, personal handheld fire extinguishers, fire blankets, etc. – some elements are only checked by compatibility with a computer list or a paper checklist.
68 Powered by gas and/or electrical supply, personal handheld valve systems, cutoff of umbilical system – some elements are only checked by compatibility with a computer list or a paper checklist.
69 Consolidated with video and voice recording.
70 Usually, 24 *h* pressure tests are carried out for checking the tightness of the seal of a hyperbaric object, which can be combined with a toxicological test of the taken atmosphere sample.
71 The efficiency of the CO_2 removal *system* can be checked by feeding it inside a hyperbaric facility with CO_2, with the simultaneous removal by various elements of the structure of the hyperbaric complex atmosphere regeneration *system*.
72 The efficiency of the O_2 content maintenance *system* can be checked by introducing, for example, nitrogen into the interior of a hyperbaric facility, with the simultaneous automatic dosing of O_2 into the atmosphere of the hyperbaric complex.
73 The test is related to the introduction of toxic gases or vapors that can potentially contaminate the hyperbaric *system* in a permanent manner to the input of the catalytic converter; therefore, such a catalytic converter test should be carried out outside the hyperbaric *system*.
74 Courtesy of *ORP Lech*.
75 Operation under maritime exercise conditions, aboard a vessel.
76 For example, Divers Alert Network.
77 For example, the Association of Diving Contractors.
78 Human Research Ethics Committee.
79 Providing a solid base for statistical deduction.
80 Bernoulli's trials.
81 Interval of probability values of *DCS* onset.

82 Selection criteria for the groups are decided by the researcher.
83 Reject–support situation.
84 Equivalently, the probability α_1 of a type II error must be acceptably small.
85 Reject–support testing.
86 α_k should be as low as possible; in statistical books, it is a common opinion that α_k should be $\alpha_k \leq 0{,}05$.
87 In a lot of computer programs, it is p.
88 Or p value calculated by statistical program.
89 For example, increasing the size of probe N.
90 Comparative distribution.
91 $1-\alpha_1 > 1-\alpha_{1,k}$.
92 Theoretical lines are consistent with previous calculations made for the first step for $N = 40 \mid n = 1$ of the *NMRI* procedure, and the lines from the simulation show the deviation due to the averaged assessment for the two steps of the complete *NMRI* procedure.
93 *Technology* is understood here as system optimization for a given process.
94 Resulting from tactics.
95 By way of various analyses, including feasibility studies, competitiveness or risk analysis.
96 System where more than the necessary elements have been applied due to functional reasons for the purpose of increasing its reliability.
97 Although it does not have to happen.
98 Undersystem.
99 Mine countermeasure.
100 That is, with little risk of an occurrence of even mild symptoms of *decompression sickness DCS*.
101 Slower for the decompression process.
102 Polish literature refers to such a procedure as extended decompression.
103 *Conservatism* will amount to 100% for the gradient equal to 30% of the maximum saturation value and 0% for 70% of the value of this gradient.
104 In accordance with the estimation as for the epidemiological threat for the entire population admitted to diving by checking compliance with medical standards
105 Decompression Illness *DCI*, as a term broader than *DCS*, for instance, by including central nervous syndrome *CNSyn*.
106 Improving the mobility of disabled persons.
107 Enabling people with disabilities to participate in life, social, and sporting activities, at a level similar to that of others with similar disabilities, for example, diving for people lacking limbs or diabetics' diving clubs.
108 For example, failure of a hoist, winch, or underwater vehicle; explosion; displacement of a wreck or its component; etc.
109 This situation occurs when a diver is rescued by their safety diver during any type of diving.
110 Explosive ordnance disposal.
111 Special response team.
112 This applies to operations in peaceful conditions in the fight against terrorism, but when asymmetric actions are undertaken by the opponent, the appetite for risk may reach much higher values.
113 Some countries use procedures for assessing and selecting threats according to their tactical situation assessments.
114 Because the most common way to intensify decompression is to use possibly high oxygen partial pressures.
115 For example, rest time before diving or undisturbed sleep in the night before diving.
116 Irrespective of the time of stay at it.
117 For instance, when entering a diving bell.

118 In the sense of physical or mental phenomena subject to experimental cognition.
119 The divers were conventionally classified as "delicate" and "resistant" to the conditions of strenuous decompression.
120 Current attitude to diving.
121 Resistance to panic, melancholy, depression, or resignation is not comparable to the ability to soberly assess the situation, make clear use of knowledge resources, and take effective action under stressful conditions.
122 The eyes are "the mirror of the soul."
123 Muscle tension or tremor can be checked in passing while checking the equipment by touching the diver's muscles or placing a hand on his/her shoulder.
124 Not only heart rate, but also pulse volume.
125 Verifying that the diver remembers and has easy access to valves, switches, watch, panel, knife, etc.
126 Decompression does not end with reaching the surface, but a state of equilibrium with the respiratory atmosphere is achieved through slow isobaric decompression; after saturated dives, the process is the longest.
127 Up to 1 hour after completing previous decompression.
128 More than 1 hour from the completion of previous decompression.
129 Clearance time.
130 It does not have to be an air atmosphere, because the diver can breathe on the surface, for example, from an oxygen inhaler.
131 It ensures, for instance, being able to reach for a mouthpiece that has escaped from the mouth and having access to valves, possibly floating lines, vegetation that may entangle the diver, etc.
132 In water bodies, the temperature in the water layers can stratify; in the thermocline, the temperature changes quickly with a change in depth.
133 For example, during transport activities or when resting on the beach.
134 Standard operating procedures.
135 Vasoconstriction.
136 $6.0\,N$ oxygen is on the market, which means 99.9999% oxygen in the gas.
137 Although, in this case, too, the content of impurities increases with the increase in "dead spaces."
138 Washout period.
139 Washing of the breathing space of the diving apparatus, while at the bottom is undesirable due to the direct proximity of a mine or diver detection and combating systems.
140 For example, it is easier for a diver to double the predicted time of oxygen decompression and to complete it with the use of an operational mixture in the event of a failure of oxygen supply than to calculate this time according to a different formula.
141 More conservative.
142 The diver omits a part of the decompression and, after the ascent and a limited time on the surface, is compressed again and subject to further decompression in the decompression chamber.
143 Thus making it more *conservative*.
144 Artificial extension of stay time at the bottom.
145 This technology variant takes into account stressful situations, such as when a diver has, for example, to operate in sea currents.
146 Treatment tables.
147 Such as passing the oxygen tolerance test (Kłos, 2023).
148 In the case of recreational dives, the risk is usually specified at the level $p = 1\%$, and for combat dives, at $p = 5\%$, with significance $\alpha = 0.01$.
149 Outside the saturation zone, expending more effort is dangerous.

150 Rest will be necessary for the performance of mine work or caused by tactical situation connected with the need to reduce noise emission, but in such a case, the problem of an increased risk of central oxygen toxicity should be considered.
151 For example, with an open circuit of the breathing mix.
152 For example, in the form of additional devices taken by a diver or suspended in the water depths.
153 Many additional elements of the system can be added, such as a communication subsystem, increasing mental comfort, having a significant impact on decompression processes.
154 Speeding up.
155 For instance, in the case of oxygen supply failure.
156 Additional rebreathers suspended in water or carried by a diver.
157 According to Canadians, but in Poland, an interval of up to 5 *min* was allowed.
158 In such a case, there is a much greater certainty that the diver will reach the decompression chamber on time.
159 Such designs are common for cable-supplied breathing systems, such as *Polycom* 101, manufactured by Drägerwerk AG Lübeck (Kłos, 2000).
160 For example, exposure to nitrogen narcosis.
161 For example, during descent and staying at the bottom, *Trimix* will be used, and in the decompression process, it will be replaced with *Nitrox*, and later with oxygen.
162 The partial pressures of the gas components of the breathing mix $(p_1 \cdot\cdot p_\bullet, p_i)$ do not always have to change smoothly with a change in the depth H, because a rebreather can implement an algorithm to maintain a constant partial pressure of a given gas through an adapted change in its content during a change in depth H, such as devices achieving a constant partial pressure of oxygen $H\ p_i$: *Mk*15/16.
163 This does not apply to dives from the saturation plateau.
164 The supply of oxygen can also be achieved by delivering it to the circuit together with other components of the breathing mix.
165 For example, flinching during a panic attack.
166 An actuating element of an automation system that performs a predefined process of reaction to forced action.
167 Often with electronic checking.
168 Decrease of respiratory action in trained athletes in relation to the average.
169 Heart rate decrease in trained athletes in relation to average; this effect is amplified in the case of immersion in water.
170 What is meant here is not diving training but adaptive training for other activities.
171 In real-life conditions, such verification has thus far been difficult to implement.
172 The technology should have a declared appetite for risk that directly results from the tactical context.
173 *External/auxiliary breathing system*, emergency breathing system: "pony," etc.
174 Due to thermal protection preventing cooling and overheating, and thus influencing the risk of central nervous syndrome caused by oxygen toxicity *CNSyn*.
175 Analyzers of oxygen partial pressure and/or composition, carbon dioxide content, etc.
176 The theory and practice of using equipment and instruments to achieve the intended purpose.
177 Autonomous underwater vehicle.
178 Unmanned undersea vehicles.
179 Remotely operated underwater vehicle.
180 That is, a reconnaissance vehicle operating from the vessel.
181 Network-centric warfare.
182 *MCMV* usually needs shielding ensured by other classes of ships and aircraft.

183 *SRT* divers can, for example, be used to release entangled *MCMV* equipment, to prepare to extract exercise torpedoes fired by submarines, to harvest or destroy undetonated self-propelled or dropped explosive devices, etc.
184 Also secretly, for example, by submarines.
185 For instance, of the infrastructure securing against reconnaissance or landing operations.
186 Unexploded ordnance.
187 Diving technology for such activities should differ from typical military and training technology.
188 For example, by rapid pulling back by a fast boat, shooting a high-velocity stream of air from a special cannon, etc.
189 For example, underwater air curtains or water sleeves, lining with rubber, shock wave–damping plastic, used on land.
190 Activities involving the destruction of electronics or the jamming of internal and external signals.
191 Wooden, plastic, lead bullets, water stream, etc.
192 Assuming, as an alternative variant of the operation, an attack commenced by units deployed in the first line who are capable of using mine clearing line charges under combat conditions, and burning access routes or blocking passageways with missile strikes in order to burn crossing. However, the launch of LCLNs turns the attack into the operation for which the opponent may get prepared. Opponent forces are able to detect the advancing force and launch counteract attacks or aerial attacks. In addition, it is not possible to maintain secrecy until the landing has taken place. In addition, if significant losses are incurred, then despite a successful breakthrough, occupying the defended area or conducting a pursuit may not have a chance of success. Consideration of the use of a *special response team SRT* or *MCM* force to secretly locate and mine obstacles, despite about 100% risk of incapability to recover the team of divers, if there is a probability of completing the task with 70% confidence at 95% confidence level, it seems justified from the point of view of economics of conducting combat operations, provided that the possible detection of underwater activity can be immediately compensated by the implementation of the breakthrough variant.
193 Very shallow water.
194 For instance, in the form of a floating ring.
195 For example, having the capability to move or bury.
196 The use of statistical models was reserved only for the validation process.
197 It was difficult to choose such a tactic, as it is widely held that it is impossible to develop sufficiently accurate and reliable deterministic models for such a complex human–machine interaction system, as is the case with diving.
198 A typical analytical approach was applied, which requires a synthesis of the developed models and validation with the use of a statistical model.
199 Most often related to the volume of ventilated space.
200 *Breathing module* ε represents the breathing module and is the ratio of oxygen consumption to lung ventilation $\varepsilon(H)=\dot{u}/\dot{V}_E$ related to diving depth H.
201 At least several may be recommended: *neo-Haldanian*, *RGBM*, *VPM*, etc.
202 The ratio of consumed oxygen stream $\dot{v}$ to ventilation stream $\dot{V}_E : \varepsilon = \frac{\dot{v}}{\dot{V}_E}$.
203 Repeatability and reproducibility with assumed accuracy at the assumed significance level.
204 Similar, for instance, to the role of ventilated volume in the ventilation model for a submarine (Kłos, 2021).
205 The representativeness of the group is limited to divers meeting the criteria for military divers and those who can perform the role of experimental divers.

206 In relation to oxygen content in the *premix* supplied into the breathing space.
207 For stabilized respiratory parameters of a diver.
208 Biodiversity.
209 Human factors.
210 For anaerobic organisms, it is a lethal toxic agent.
211 Symptoms similar to lung burning with chemicals.
212 In relation to the central nervous system.
213 Reduced red blood cell count, decreased vital lung capacity, oxygen-induced blindness, etc.
214 This theory forms the basis for epidemiological research, research into mortality and biodiversity, machine reliability, risk analysis, etc.
215 A return to the state preceding the oxygen exposure, consisting in deactivation of potentially dangerous reactive oxygen compounds by biochemical mechanisms as a result of exposure to high partial pressures of oxygen.
216 Commonly used in computer programs for planning civil technical dives
217 Considered dangerous to a wide range of divers.
218 Possibility of performing decompression at the last station at the depth of 6 mH_2O, possibility of omission of oxygen decompression, possibility of additional effort load, taking into account previous dives, taking into account the type of transport, taking into account low-temperature dives, etc.
219 Such deviations occur when a diver has a significantly developed muscle tissue as compared to the typical one; for this reason, one diver was rejected.
220 As military dives are carried out in pairs.
221 Reproducibility and repeatability.
222 For example, in the diver's physical preparation, reduction of the effectiveness of flushing the respiratory space of the rebreather, inappropriate preparation of breathing mixes or deterioration of their required quality, inappropriate packing of the scrubber or reduction of its quality parameters, etc.
223 The occurrence of rare medical conditions or diseases.
224 Oxygen toxicity includes *central nervous syndrome CNSyn*, pulmonary toxicity, and somatic oxygen toxicity (Kłos, 2023).
225 Lorrain Smith effect.
226 Short exposures.
227 Optimal conditions for "washing out" the body from nitrogen.
228 Effective reductions in the size of gas bubbles at the decompression disease.
229 Inseparably linked.
230 Residual risk which can no longer be minimized under given conditions.
231 The assessment of the condition presented on the basis of studies and analyses.
232 Activities to test suitability for the assumed level of accuracy and precision.
233 They are most often collected in the form of a case study, and the conclusions of the case study are introduced as pilot recommendations for use or usable patents that make up good diving practice.
234 Knowledge not yet validated by methods and methodologies recognized as scientific.
235 Or a military transport plane.
236 The pressure in a corridor cut off by water can be much higher than the pressure produced only as a result of it being below sea level.
237 The decompression meter typically cannot take into account the residual risk.
238 The most common comment concerns the use of the decompression system being at one's own risk.
239 To avoid pejorative meaning, these skills can be compared to the skills of diamond polishers, master chefs, etc.
240 This refers to professional spacemen, not people who are merely space tourists.

241 At present, it is becoming increasingly common that having servicing instructions requires a franchise agreement with the manufacturer.
242 Discrepancy!
243 Some believe that Jaroslav Hašek achieved such a goal by describing all the possible situations that could happen in the army in a relatively small work titled *The Good Soldier: Schweik*.
244 $J-$factor frequently is an arbitrarily defined function or value, usually abbreviated by a J in equations, and defines an analogy between parameters and thus corrects errors in the calculation process.
245 They are the results of tests carried out under controlled laboratory conditions.
246 Materialization of residual risk.
247 Without changes in total pressure, the diver could not leave the hyperbaric environment.
248 Most of even the simplest dives conducted in real conditions have a more complex structure for the decompression profile.
249 *Syndrome* means a group of symptoms which consistently occur together, or a condition characterized by a set of associated symptoms.
250 For example, decompression acceleration, oxygen therapy, treatment of decompression sickness, etc.
251 This makes it possible to bypass the relatively cumbersome procedure for the production of correctly composed *heliox Hx*, allowing a relatively wide range in terms of composition.
252 Similar disorders occur in distance swimmers and are associated with deep, rapid inhalations, causing the collapse of the bronchioli.
253 Containment of carbon dioxide in the body.
254 Oscillatory circuit device.
255 Owing to the induction of skin symptoms of decompression sickness, there may also be direct gas exchange between the atmosphere and tissues through the skin.
256 For example, via hemoglobin or blood lipids.
257 Physically dissolved gas in blood plasma.
258 *Depth* is the unit of measurement of length, although in this case it is a specific unit of pressure.
259 The reference to fresh water is unambiguous; the reference to the depth expressed in units of seawater is ambiguous because it depends on the salinity of the body of water. In the past, a reference value to water density was given on the manometer; now, in order to standardize the units of measurement, it is assumed that $10\ mH_2O\ f.w. \triangleq 100\ kPa$.
260 The suspension in water of certain minerals, such as barite, is known as the process of producing so-called "heavy water." For example, such suspensions are used in the drilling industry to exert counterpressure on the walls of a borehole. This name may be misleading, as *heavy water* is also called *deuterated water*. In this case, we are dealing with the development of permanent suspensions that increase the density of water. Diving at the same depth expressed metrically in water with suspended solids is an exposure to higher pressure compared to clear fresh water.
261 A pneumatometer is used to measure pressure at depth (it differs in design from the device of the same name used in medicine to measure the inhalation negative pressure and expiration overpressure) – it is a tube which is fed with gas so that it flows slowly and freely into the water at its end, at the same time measuring the pressure in the tube.
262 For example, depending on the diver's position in the water, as regards decompression in the diver's vertical position, there is a significant difference in depth between the feet and the head; the most common assumption is to measure the depth around the diver's lung centroid.
263 The accuracy of the method and the place of measurement may cause the error for the measured pressure to be comparable to the distance between the decompression stations.

264 In airliners, cabin pressure can be reduced for robustness reasons – this way, the aircraft can be constructed for smaller pressure variations; hence, their construction can be lighter.
265 In accordance with generally accepted guidelines, the pressure inside the cabin of a passenger aircraft must be maintained at a level not lower than that corresponding to $2438\ m$ above sea level ($0.75\ atm$).
266 Just like in sport, where not only individual predispositions count but also fitness at the time of competition. However, meeting these conditions does not guarantee success, because a lack of mental resilience during the competition itself can reduce the importance of predisposition and training.
267 Lactic acid present in muscles can provoke aching effects.
268 It is connected with a gradual change in temperature, especially when, during transport to the surface, an additional process of lowering the pressure in the bell takes place.
269 During a saturation dive, the bell usually does not function as a thermal habitat; therefore, even while in the bell, the divers use active thermal protection measures.
270 For example, when chopping wood for firewood in freezing conditions.
271 In some regions, special operations forces divers still prefer wet suits, which are equipped with Velcro systems for attaching the straps of an unbuttoned suit; this allows for partial unbuttoning of the wet suit underwater and clarifying it in such a way that it does not increase swimming resistance.
272 It is widely believed that after $12\ h$, the effect of saturation is unnoticeable even after deepwater diving.
273 At present, such measurements usually do not present major difficulties.
274 It is not possible to precisely define the significance of the premise related to the increased risk of decompression sickness for serial dives, because sometimes the safe interval between dives will be $12\ h$, and in another case two days; one will apply to the second dive and in another case to the subsequent dives.
275 The maximum amount of oxygen that the body is able to uptake during intensive effort.
276 From *Latin*.
277 For example, health issues within the family unit, financial situation, recent setbacks, etc.
278 The collapse of their chest with depth does not cause mechanical damage to the body.
279 Of course, lightweight breathing medium components are used, but usually for diving at considerable depths; hence, there is an increase in its density compared to atmospheric air. Even under normal conditions, the Dräger company has also advocated the use of lightweight breathing medium components for filling firefighting breathing apparatuses.
280 Adoption of an existing decompression system is possible but should be validated.
281 Ca. $20\ kPa$.
282 More serious symptoms in the form of joint pain may also be possible, usually in decompression after saturation due to a bad body position during sleep; therefore, decompression during nighttime is not recommended, and divers are not allowed to sleep during daytime.
283 Similar to the suffocation effects caused by some snakes strangling their victims.
284 It is good practice to ask the diver questions, as it not only gives him/her control over their actions but also gives the diver time to adjust their gear and equipment and to reflect on the current situation.
285 Facilitating one process and, at the same time, making the other difficult.
286 Understood here as the current physiological state of the organism undergoing changes under the influence of external environmental factors.
287 For example, immersion to a few meters with or without a swimming board and with or without support of the diver's fins, or distance swimming at a shallow depth whilst wearing fins, with or without breath-holding and with a swimming board.
288 Of course, adequate measures should be taken to deal with incidents such as cardiac arrest.

289 Diving in a different climate or meteorological condition.
290 Differences in the occurrence of thermocline, tides, or diving in flowing water.
291 Also between muscles and tendons, in tendon sheaths, or in periarticular areas.
292 As in marine mammals, humans may also have fat tissue that is well perfused through blood.
293 Despite an extremely high buildup of free gas phase in their blood, they do not show signs of DCS, which would have occurred with a high probability when they were younger.
294 Premenstrual syndrome, PMS.
295 Factors from the external environment that affect the organism during its development inside the womb.
296 The main blood vessels are larger in the mother than in the embryo.
297 The organ of vision can be particularly easily damaged, as shown by observations made in infant incubators (Bartosz, 2008).
298 However, the interpretation of the examples quoted here is based on current knowledge.
299 Beyond the saturation zone.
300 Ultrasound detection of free gas phase signals transformed using the electronics into an audible beep.
301 Ultrasound detection of free gas phase signals transformed using the electronics into audible beeps.
302 Former Defence and Civil Institute of Environmental Medicine, Toronto.
303 Frequency parameter is 1 from Table 2.7.
304 Ratio is 60 to 100.
305 Normo- or hyperbaric, with the use of oxygen or not.
306 Over 10%.

REFERENCES

Alonso J.V., Chowdhury M., Borakati R., Gankande U. 2017. Swimming-induced pulmonary oedema an uncommon conditio diagnosed with POCUS ultrasound. *American Journal of Emergency Medicine*. 35, 1986.e3–1986.e4. DOI: 10.1016/j.ajem.2017.09.029.

Arieli R., Yalov A., Goldenshluger A. 2002. Modeling pulmonary and CNS O2 toxicity and estimation of parameters for humans. *Journal of Applied Physiology*. 92, 248–256. DOI: 10.1152/japplphysiol.00434.2001.

Bartosz G. 2008. *Druga twarz tlenu—wolne rodniki w przyrodzie*. Warszawa: Wydawnictwo Naukowe PWN. ISBN: 978-83-01-13847-9.

Bennett P.B., Elliott D.H. 1993. *The physiology and medicine of diving*. London: W.B. Saunders Company Ltd. ISBN: 0-7020-0538-X.

Betts E.A. 1994. *Introduction to enriched air diving*. Freeport: American Nitrox Divers Inc.

Blatteau J.E., Hugon J., Gempp E., Castagna O., Pény C., Vallée N. 2012. Oxygen breathing or recompression during decompression from nitrox dives with a rebreather: effects on intravascular bubble burden and ramifications for decompression profiles. *European Journal of Applied Physiology*. 112, 2257–2265.

Bøe J., Hartung K.-H. 1983. Employment of the polycom 101 gas mixing unit for divers in major project in Norway. *Dräger Review*. 51, 26–28.

Boussuges A., Ayme K., Chaumet G., Albier E., Borgnetta M., Gavarry O. 2017. Observational study of potential risk factors of immersion pulmonary edema in healthy divers: exercise intensity is the main contributor. *Sports Medicine-Open*. 35, 1–8. DOI: 10.1186/s40798-017-0104-1.

Brubakk A.O., Neuman T.S. 2003. *Bennett and Elliott's physiology and medicine of diving*. Philadelphia: Saunders Ltd. ISBN: 0-7020-2571-2.

Bühlmann A.A. 1984. *Decompression-decompression sickness*. Berlin: Springer-Verlag. ISBN: 3-540-13308-9; ISBN: 0-387-13308-9.

Bühlmann A.A. 1995. *Tauchmedizin*. Berlin: Springer-Verlag. ISBN: 3-540-58970-8.
Butler F.K., Thalmann E.D. 1986. *CNS oxygen toxicity in closed-circuit SCUBA divers III*. Panama City: USN Experimental Diving Unit. Report No. 5–86.
Castagna O., de Maistre S., Schmid B., Caudal D., Regnard J. 2018. Immersion pulmonary oedema in a healthy diver not exposed to cold or strenuous exercise. *Diving and Hyperbaric Medicine*. 48, 40–44. DOI: 10.28920/dhm48.1.40-44.
Cole B. 1993. *Decompression and computer assisted diving*. Dive Information Company. ISBN: 0-9520934-0-5.
Comex Marseille. 1986. *Medical book*. Marseille: Comex.
DCIEM. 1986. *Procedures for Doppler ultrasonic monitoring of divers for intravascular bubbles*. Toronto: Defence and Civil Institute of Environmental Medicine. DCIEM No. 86-C-25.
DCIEM. 1995. *Diving manual*. North York: Defence and Civil Institute of Environmental Medicine. DCIEM No. 86-R-35A.
DMW. 1982. *Tabele dekompresji i rekompresji nurków. Praca zbiorowa*. Gdynia: Dowództwo Marynarki Wojennej. Sygn.Mar.Woj. 860/81.
Donald K.W. 1947. Oxygen poisoning in man part I. *British Medical Journal*. 17, 667–672.
Donald K.W. 1992. *Oxygen and the diver*. Harley Swan: The SPA Ltd. ISBN: 1-85421-176-5.
Edmonds C., Bennett M., Lippmann J., Mitchell S. 2020. *Diving and subaquatic medicine*. Boca Raton, FL: CRC Press. ISBN: 9780367575557.
Eftedal O.S. 2007. *Ultrasonic detection of decompression induced vascular microbubbles*. Trondheim: Norwegian University of Science and Technology Faculty of Medicine Department of Circulation and Medical Imaging. ISBN: 978-82-471-3697-3.
Eftedal O.S., Lydersen S., Brubakk A.O. 2007. The relationship between venous gas bubbles and adverse effects of decompression after air dives. *Undersea and Hyperbaric Medicine*. 32, 99–105.
European Committee for Standardization. 2014. *Respiratory equipment—open-circuit self-contained compressed air diving apparatus—requirements, testing, marking*. Brussels: CEN-CENELEC Management Centre. European Standard. EN 250:2014 E; ICS 13.340.30.
Fire Research, Test, Development and Education Centre. 1993. *Test report: safety at 8% oxygen—15 minutes*. Copenhagen. Fire Research, Test, Development and Education Centre.
Frånberg O. 2015. *Oxygen content in semi-closed rebreathing apparatuses for underwater use: measurements and modeling*. Stockholm: School of Technology and Health. ISSN: 1653-3836.
Gardette B., Massimelli J.Y., Gortan C., Delauze H.G. 1993. *Hydra 10—a 701 msw onshore record dive using "heliox"*. Proceedings of the XIXth Annual Meeting of EUBS, Trondheim, Norway.
Gernhardt M.L., Abercromby A.F., Conkin J. 2007. *Potential fifty percent reduction in saturation diving decompression time using a combination of intermittent recompression and exercise*. Dunkirk: Undersea and Hyperbaric Medical Society, Inc. Abstract of the Undersea and Hyperbaric Medical Society, Inc. Annual ScientScientific Meeting, 14–16 June.
Gerth W.A. 2002. Overview of survival functions and methodoology. In *Survival analysis and maximum likelihood techniques as applied to physiological modeling*, ed. Gerth W.A., Weathersby P.K. Kesington: Undersea and Hyperbaric Medical Society Inc.
Gerth W.A., Ruterbusch V.L., Long E.T. 2007. *The influence of thermal exposure on diver susceptibility to decompression sickness*. Panama City: Navy Experimental Diving Unit. NEDU TR 06-07.
Gussmann J. 1984. *Człowiek zdobywa głębiny*. Gdańsk: Wydawnictwo Morskie. ISBN: 83-215-4047-3.
Hamilton R.W. 1989. Tolerating exposure to high oxygen levels: repex and other methods. *Marine Technology Society Journal*. 23, 19–25.

Harabin A.L. 1993. *Human central nervous system oxygen toxicity data from 1945 to 1986*. Bethesda: Naval Medical Research Institute. NMRI 93-03; AD-A268-225.

Haux G. 1982. *Subsea manned engineering*. London: Bailliére Tindall. ISBN: 0-7020-0749-8.

Huggins K.E. 1992. *The dynamics of decompression workbook*. Ann Arbor: The University of Michigan.

Hugon J., Metelkina A., Barbaud A., Nishi R., Bouak F., Blatteau J.E., Gempp E. 2018. Reliability of venous gas embolism detection in the subclavian area for decompression stress assessment following scuba diving. *Diving and Hyperbaric Medicine*. 48, 132–140.

Imbert J.-P., Bontoux M. 1987. *Proposition d'un manuel de procedures de decompression a l'air*. Marseille: Comex Services. Final Report to Centre d'Etudes Petrolleres et Marines. Fiche No. 4723.

Johnstone R.E., Thring M.W. 1957. *Pilot plants, models and scale-up methods in chemical engineering*. New York: McGraw-Hill Book Co.

Kenyon D.J., Butler G. 2014. *Crude neon with nitrogen and oxygenasa hyperbaric intervention breathing mixture*. Pub. No.: US 2014/0290651 A1 USA, 11.2, 2014. 128/203.12: 96/134.

Kłos I., Kłos R. 2004. *Polish soda lime in military applications*. Oświęcim: Chemical Company DWORY S.A. ISBN: 83-920272-0-5.

Kłos R. 2000. *Aparaty Nurkowe z regeneracją czynnika oddechowego*. Poznań: COOPgraf. ISBN: 83-909187-2-2.

Kłos R. 2002. Metabolic simulator supports diving apparatus researches. *Sea Technology*. 12, 53–56.

Kłos R. 2007a. *Mathematical modelling of the normobaric and hyperbaric facilities ventilation*. Gdynia: Polskie Towarzystwa Medycyny i Techniki Hiperbarycznej. ISBN: 978-83-924989-0-2.

Kłos R. 2007b. *Zastosowanie metod statystycznych w technice nurkowej—Skrypt*. Gdynia: Polskie Towarzystwo Medycyny i Techniki Hiperbarycznej. ISBN: 978-83-924989-26.

Kłos R. 2008. Principles of work of different types of underwater breathing apparatus. *Polish Maritime Research*. 58, 72–84. DOI: 10.2478/v10012-007-0100-3.

Kłos R. 2009. *Wapno sodowane w zastosowaniach wojskowych*. Gdynia: Polskie Towarzystwo Medycyny i Techniki Hiperbarycznej. ISBN: 978-83-924989-5-7.

Kłos R. 2011. *Możliwości doboru dekompresji dla aparatu nurkowego typu CRABE—założenia do nurkowań standardowych i eksperymentalnych*. Gdynia: Polskie Towarzystwo Medycyny i Techniki Hiperbarycznej. ISBN: 978-83-924989-4-0.

Kłos R. 2012. *Możliwości doboru ekspozycji tlenowo-nitroksowych dla aparatu nurkowego typu AMPHORA—założenia do nurkowań standardowych i eksperymentalnych*. Gdynia: Polskie Towarzystwo Medycyny i Techniki Hiperbarycznej. ISBN: 978-83-924989-8-8.

Kłos R. 2013. Oxygen tolerance test. *Polish Hyperbaric Research*. 45, 69–78.

Kłos R. 2014a. *Helioksowe nurkowania saturowane—podstawy teoretyczne do prowadzenia nurkowań i szkolenia*. Gdynia: Polskie Towarzystwo Medycyny i Techniki Hiperbarycznej. ISBN: 978-83-938322-1-7.

Kłos R. 2014b. Inherent unsaturation: the risk of central nervous system oxygen toxicity part 1. *Polish Hyperbaric Research*. 46, 37–64. DOI: 10.13006/PHR.

Kłos R. 2014c. The pathophysiology related to the toxic effect of oxygen: the hazard of central oxygen toxicity part 2. *Polish Hyperbaric Research*. 47, 15–34. DOI: 10.13006/PHR.47.2.

Kłos R. 2014d. Survival analysis: the risk of central oxygen toxicity part 3. *Polish Hyperbaric Research*. 48, 33–48. DOI: 10.13006/PHR.48.3.

Kłos R. 2014e. The hazard of central oxygen toxicity occurrence: the risk of central oxygen toxicity part 4. *Polish Hyperbaric Research*. 49, 19–31. DOI: 10.13006/PHR.49.2.

Kłos R. 2016. *System trymiksowej dekompresji dla aparatu nurkowego typu CRABE*. Gdynia: Polskie Towarzystwo Medycyny i Techniki Hiperbarycznej. ISBN: 978-83-938322-5-5.

Kłos R. 2017. Pollutions of the hyperbaric breathing atmosphere. *Scientific Journal of Polish Naval Academy*. 208, 31–44. DOI: 10.5604/0860889X.1237621.

Kłos R. 2019. Context analysis in the development of diving technologies. *Polish Hyperbaric Research*. 62, 7–58.

Kłos R. 2020. Ultrasonic detection of the intravascular free gas phase in research on diving. *Polish Maritime Research*. 27, 176–186. DOI: 10.2478/pomr-2020-0039.

Kłos R. 2021. *Ventilation of normobaric and hyperbaric objects*. Boca Raton, FL: CRC Press (Taylor & Francis Group, LLC). ISBN: 978-0-367-67523-3 (hbk); ISBN: 978-0-367-67524-0 (pbk); ISBN: 978-1-003-13164-9 (ebk).

Kłos R. 2023. *Oxygen diving*. Boca Raton, FL: CRC Press (Taylor & Francis Group, LLC). ISBN: ISBN: 978-1-032-31389-4 (hbk); ISBN: 978-1-032-31390-0 (pbk); ISBN: 978-1-003-30950-5 (ebk). DOI: 10.1201/9781003309505.

Kłos R., Konarski M., Olszański R. 2004. The implementation of factor analysis for the evaluation of selected blood parameter changes induced by hyperbaric exposure. *International Maritime Health*. 55, 87–101.

Kłos R., Nishi R., Olszański R. 2002. Validation of diving decompression tables. *International Maritime Health*. 53, 1–4.

Lambertsen C.J. 1986. *Studies in isobaric counterdiffusion*. Filadelfia: Institute for Environmental Medicine.

Lepawsky M., Wong R. 2001. *Empirical diving techniques of commercial sea harvesters*. Richmond: Undersea and Hyperbaric Medical Society Inc. Fiftieth Workshop of the Undersea and Hyperbaric Medical Society. ISBN: 0-93406-21-4.

Lewbel G.S. 1984. *The decompression workbook*. New York: Pisces Book Co. ISBN: 0-86636-023-9.

Lippmann J. 1990. *Deeper into diving*. Carnegie: J.L. Publications. ISBN: 0-9590306-3-8.

Lippmann J., Mitchell S. 2009. *Deeper into diving*. Aschburton: J.L. Publications. ISBN: 0-9752290-1-X.

Lundgren C.E.G., Pendergast D.R. 2010. Improve your exercise stamina by training your breathing muscles. *Faceplate*. 14, 10–12.

Natural Environment Research Council. 1974. *Code of practice for scientific diving*. London: Natural Environment Research Council.

NEDU. 1994. *U.S. Navy unmanned test methods and performance goals for underwater breathing apparatus*. Panama City: US Navy Experimental Diving Unit. Report No. 01-94.

Nishi R.Y. 1990. Doppler evaluation of decompression tables. In *Man in the sea*, ed. Shida K.K., Lin Y.C. San Pedro: Best Publishing Co. ISBN: 0-941332-12-8.

Nishi R.Y., Brubakk A.O., Eftedal O.S. 2003. Bubble detection. In *Bennett and Elliott's physiology and medicine of diving*, ed. Neumann T.S., Brubakk A.O. Edinburgh: Saunders. ISBN: 0-7020-2571-2.

NOAA. 1976. *Vertical excursions breathing air from nitrogen-oxygen or air saturation exposures*. Rockville: National Oceanic and Atmospheric Administration. U.S. Government Printing Office 1976-210-801/366.

NOAA. 2001. *NOAA diving manual—diving for science and technology*. Flagstaff: Best Publishing Co. ISBN: 0-941332-70-5.

NOAA. 2017. *NOAA diving manual—diving for science and technology*. Flagstaff: Best Publishing Co. ISBN: 9781930536883.

PN-EN-14143. 2005. *Sprzęt do nurkowania. Autonomiczne aparaty do nurkowania z obiegiem zamkniętym*. Warszawa: Polski Komitet Normalizacyjny. ICS 13.340.30; 97.220.40.

Predictive Studies IV. 1978. *Work capability and physiological effects in He-O_2 excursions to pressures of 400-800-1200-1600 fsw*. Philadelphia: Institute for Environmental Medicine University of Pennsylvania. Report No. 78-1.

Risberg J., Møllerløkken A., Eftedal O.S. 2019. *Norwegian diving and treatment tables*. Bergen: Personal Publisher. www.dykketabeller.no/Western Norway University of Applied Sciences. ISBN: 978-82-690699-6-9 (printed version); ISBN: 978-82-690699-7-6 (digital/PDF version).

Rutkowski D. 1990. *Nitrox manual*. Key Largo: Hyperbarics International, Inc.

Sawatzky K.D. 1991. *The relationship between intravascular Doppler-detected gas bubbles and decompression sickness after bounce diving in humans*. Toronto: York University.

Schipke J.D., Tetzlaff K. 2016. Why predominantly neurological decompression sickness in breath-hold divers? *Journal of Applied Physiology*. 120, 1474–1477.

Seddon F.M., Thacker J.C., Gilbert M.J., Anthony T.G. 2010. *Doppler monitoring of compressed air workers; Belfast sewer project*. Norwich: Health and Safety Executive Offshore Technology. RR788 Research Report.

Shykoff B. 2007. *Performance of various models in predicting vital capacity changes caused by breathing high oxygen partial pressures*. Panama City: Navy Experimental Diving Unit. NEDU Report TR 07-13.

Siedow L.I. 1968. *Analiza wymiarowa i teoria podobieństwa w mechanice*. Warszawa: Wydawnictwo Naukowo-Techniczne.

Stone W.C. 1995. Designing a redundant life-support system. *AquaCorps*. 12, 29–34.

US Naval Sea Systems Command. 2016. *US Navy diving manual*. Washington: The Direction of Commander, Naval Sea Systems Command. SS521-AG-PRO-010 0910-LP-115-1921.

Vann R.D. 1982. *The effect of exercise during decompression*. Norfolk: Undersea and Hyperbaric Medical Society, Inc. Abstract of the Undersea and Hyperbaric Medical Society, Inc. Annual Scientific Meeting held 1–5 June in Norfolk, VA.

Wald A. 1947. *Sequential analysis*. New York: John Willey & Sons, Inc.

Wienke B.R. 1993. *Diving above sea level*. Flagstaff: Best Publishing Co. ISBN: 0-941332-30-6.

Wienke B.R. 2003. *Basic decompression theory and application*. Flagstaff: Best Publishing Co. ISBN: 1-930536-14-3.

Wienke B.R. 2008. *Diving physics with bubble mechanics and decompression theory in depth*. Flagstaff: Best Publishing Co. ISBN-13: 978-1-930536-33-3; ISBN-10: 1-930536-33-X.

Wienke B.R. 2016. *Biophysics and diving decompression phenomenology*. Sharjah: Bentham Science Publishers. ISBN: 978-1-68108-320-9.

Zlokarnik M. 1991. *Dimensional analysis and scale-up in chemical engineering*. Heidelberg: Springer-Verlag. ISBN-13: 978-3-540-54102-8. DOI: 10.1007/978-3-642-76673-2.

3 Principles of *UBA* Operation

The kind of breathing gas used exerts a large effect on the construction of the *UBA*. The kind of breathing gas that should be used depends on the diving depth range and the construction of the *UBA*. These three principles constitute the basic criterion for diving *UBAs* into types and should be taken into account simultaneously. The design of *UBA* is determined primarily by the type of breathing circuit; hence, discussed here are the principles of operation of general types of engineering solutions,[1] according to the following division:

- Open circuit *UBA*
- Semi–closed circuit rebreather *SCR*
- Closed circuit rebreather *CCR*

3.1 DIVING APPARATUSES

Traditionally, diving equipment can be divided into heavy[2] and lightweight equipment (Clarke, 1992; Kłos, 2008b). The classification of diving equipment according to this criterion is based on the diving technique used.

The diving gear having similarity to *August Siebe's* construction, patented in 1836, can be regarded as heavy diving equipment. The diving equipment was further developed and improved (Figure 3.1).

It evolved toward the helmet systems equipped with diving regulators. An improved diving equipment with the free flow of breathing gas was also developed (Figure 3.2).

The features of the heavy diving equipment are presented in Table 3.1.

Assuming the classification criteria presented in Table 3.1 distinguishing the heavy diving equipment is sufficient, it should be recognized that the *Henry Fleuss's* $oxy-CCR$ from 1879 was an independent version of the heavy diving equipment, as the dive involved placing an especially loaded diver on the bottom (Figure 3.3).

Changes in diving methods have led to further development of the $oxy-CCR$ and their transformation into light diving equipment.

Commander Jacques-Yves Cousteau is generally considered to have invented the air open circuit SCUBA in 1940. It was called *Aqua – Lung* (Figure 3.3). It is well-known that a prototype of such apparatus had been designed earlier. In 1860, Benoit Rouquayrol and Auguste Denayrouze constructed a similar *SCUBA* (Figure 3.4).

However, Cousteau's gear was a real landmark in the history of diving and led to the development of free diving (Rawlis, 1989; US Navy diving manual, 2016).

Paul Bert's physiological research and Elihu Thomson's theoretical works contributed to the first experimental mixed-gas diving performed in 1924. In 1940,

DOI: 10.1201/9781032649207-3

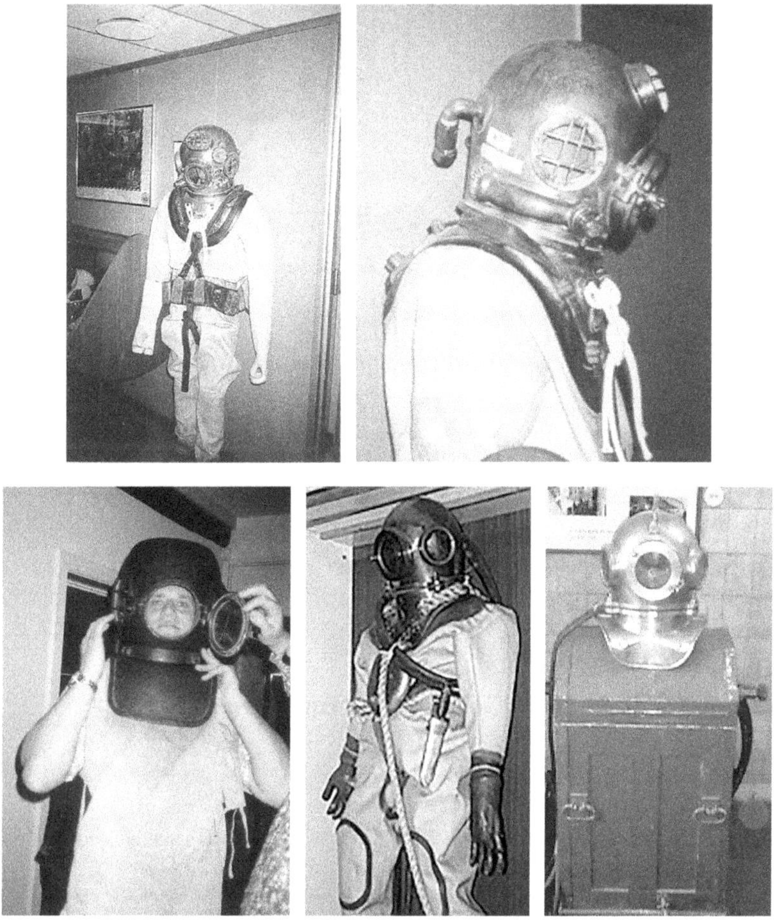

FIGURE 3.1 *Top*, the diving helmet Siebe Gorman; *bottom left*, the author with probably one of the oldest, still ready-to-use diving helmet in the world; *bottom right*, the diving helmet *UWS* – 50*M*.[3]

Source: Kłos (2000).

experiments resulted in the development of the independent *Nx-SCUBA* proposed by Lambertsen (Figure 3.5).

However, the first diving apparatus type *DM* 40 was developed and manufactured by *Drger* in 1915 (Figure 3.6).

Yet many historians accept that it was Lambertsen's gear that began the real development of *UBAs* (US Navy diving manual, 2016). Apart from the breathing gas used, the features of *UBAs* are as shown in Table 3.2.

At present, the differences between the heavy- and lightweight diving equipment are dependent upon the additional gear chosen; the basic set can function as the classic or lightweight diving equipment.

FIGURE 3.2 *Left*, the author ready to dive with diving equipment type DM 200/2; *right*, diving equipment type $AH-3$.

Source: Kłos (2000) and author's own photos, respectively.

TABLE 3.1
Characteristics of the Heavyweight Diving Equipment

The heavyweight diving equipment	
During normal work, an equipped diver has high negative buoyancy. The buoyancy can be changed, however, only to change the diver's work plane or while descending or ascending.	
Disadvantages	• The diver can perform work only in a fixed area, such as a trap, a diving platform, the bottom and the decks of sunken vessels, etc.
	• The diver's mobility is limited. Considerable effort must be done by the diver to move around as required by work situations.
Advantages	• The diver's position is stable. Heavy work can be performed without large effort to keep the diver's position.

Here, the *diving equipment* is generally defined as a technical equipment that enables a man to safely descend, stay at depth for a period of time, and then ascend. There are some tendencies to extend the preceding definition by normobaric dives. In the accepted definition, it is essential to indicate that the diving equipment enables the man to perform direct work at depth. As it follows from the earlier text, the water environment exerts pressure on the diver's equipment that is directly transferred on the man.

This feature distinguishes the diving equipment from the other kinds of submersibles which enable normobaric dives, for example, atmospheric diving suits[4] or manned underwater vehicles (Figure 3.7).

Descent and ascent in the water environment should be safe; therefore, the diving equipment requires that the diver, except possessing operating skills, meet standards for divers and have undergone specialized training.

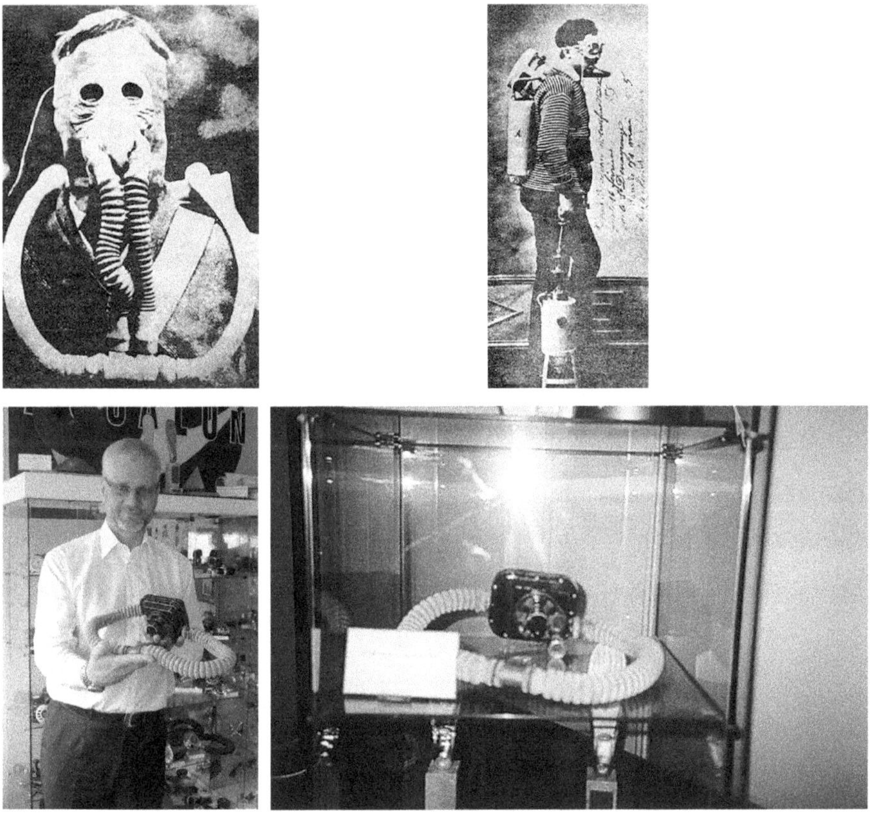

FIGURE 3.3 *Top, Henry Fleuss oxy – CCR*; *bottom*, an author with Cousteau's diving regulator.

Source: Gussmann (1984) and author's own photos, courtesy of AQUA-LUNG France, respectively.

FIGURE 3.4 Rouquayrol's *UBA*.

Source: Bachrach, Desiderati, and Matzen (1988).

FIGURE 3.5 Lambertsen's *UBA*.

Source: Kłos (2000).

FIGURE 3.6 *DM* 40 *SCUBA* manufactured by Dräger, 1915.

Source: Bachrach, Desiderati, and Matzen (1988).

3.1.1 The *UBAs* Classification

Except for the traditional classification as heavy- and lightweight diving equipment, there are many other classification criteria. Most of those presented in the literature are based on the *UBA* constructional criteria (e.g., Haux, 1968; Rawlis, 1989; Clarke, 1992). This book will present a classification that seems to be more accurate and useful than those offered so far. Furthermore, it will deal with the *UBA* as part of the diving equipment (Kłos, 2008a).

TABLE 3.2
The Features of the Lightweight Diving Equipment

The lightweight diving equipment	
During normal work, the equipped diver has neutral buoyancy that can be adjusted within a certain range.	
Disadvantages	• The diver underwater is almost weightless. Even if the diver is supported underwater by immovable elements, weightlessness makes the diver's work (drilling holes, surface cleaning, shooting pegs, welding, etc.) difficult.
Advantages	• Compared to heavyweight diving equipment, the lightweight equipment enables relative long-distance diving without significant effort and free movement in all directions underwater or remaining at a given place underwater. This feature permits the inspection of the technical conditions of large underwater objects, underwater sweeping, etc.

FIGURE 3.7 Underwater scientific and rescue vessels: (a) one-atmospheric diving suit Newtsuit; (b) general view of the underwater vehicle *SeaCliff*; (c) the historical underwater vessel of Comex; (d) the author in the *DSRV Mystic* cockpit; (e) general view of *DSRV Avalon* at the technical service station, with the hull covers removed; (f) a historical underwater submersible rescue vehicle; (g) the author and a general view of *URF*, a submarine rescue vehicle; (h) the author and a manned underwater robot.

Source: Kłos (2000, 2008a).

FIGURE 3.7 (Continued)

UBAs are defined as parts of the diving equipment responsible for supplying the diver with breathing gas. The breathing gas pressure is adequate for the diving depth. The breathing medium can be stored outside the *UBA* and supplied through a suitable hose, the containers holding the breathing medium can be an integral part of the *UBA*, or the medium can be obtained[5] in a chemical reaction.

To classify the *UBAs* for the tasks assumed, the following criteria are the most convenient:

- Kind of breathing gas
- Operational depth range
- Principle of operation

The *UBAs* with chemical preparation or complementation of the breathing gas are not discussed here. The classification of *UBAs* according to the aforementioned criteria is presented in Table 3.3.

This classification of the *UBAs* used for underwater work enables collecting, systematizing, and analyzing data on trends in construction developments.

Although it is not perfect, it is the most complete classification that has been met in professional literature up to date. Its usefulness has been confirmed for systemizing and analyzing the *UBAs*' data.

3.1.2 Summary

The *UBA* classification based on the three criteria, that is, the type of breathing gas, the operational depth range, and the principle of operation, makes the classification complex enough for use in analyzing the direction trends in the development of *UBAs* and forecasting the demands to be met by *SCRs*. The breathing gas used is the most important criterion. The other two basic classification criteria are derived from the first one; therefore, it should be treated as one complex criterion.

3.2 OPEN CIRCUIT *UBA*

Open circuit *UBA* is basically known as air self-containing *UBA* – *SCUBA* (Rawlis, 1989; Kłos, 2000, 2008b). The engineering variations of this type of *UBA* are

TABLE 3.3
Classification of *UBAs*

Specification		Classification of the *UBAs*									
1. Breathing medium		Air		Oxygen	Mixed gases (most often nitrox, heliox, trimix, hydrox, hydreliox, neox)						
2. Depth range		$0-50 mH_2O$		Up to $6 mH_2O$	$0-200 mH_2O$				Most often deeper than $200 mH_2O$		
3. Principle of work	a. Group	Of the open gas circuit		Closed gas circuit	Semi-closed gas circuit				Closed gas circuit		
	b. Subgroup				Premix		Preparing breathing gas during diving process				
	c. Type	Self-contained	Hose-supplied	Self-contained	Self-contained	Hose-supplied	Self-contained	Hose-supplied	Small circulation	Increased (enlarged) circulation	High circulation
4. Examples	(Haux, 1968)	PA-38/3600	PL-70	LAR V	—	FGG III	SM III	SM III-S	Elektrolung	—	GAK-600
	Polish constructions	PR-27	UAN-82	—	APW-6M	GAN-87	APW-3	—	—	—	—
	Other examples	AGA MkII	KMB-10	COBRA	DC-55	Sealab III (Fink Rig)	VIPER+	—	Mk-15/16	DOLPHIN 7	CCBS
5. Remarks		Not recommended for underwater works	Recommended for underwater works at shallow and mean depths	Military diving	Mostly used for diving work performed beyond the saturation zone (or the bailout *UBAs* during saturation) and for operational diving at large depths – standard approximately $150 mH_2O$				Military diving at large depths	*UBAs* used for underwater works during saturation diving	

WARNING! The *UBAs* with chemical preparation and complementation of the breathing gas are not taken into account here.

Source: Haux (1968); Kłos (2008a).

many. Apart from *hose – UBA* and *SCUBA* types, they can also be divided according to the number of stages of pressure reduction to be obtained at the demand valve:

- Single-stage pressure reduction regulator
- Two-stage pressure reduction regulator

Systems of a higher number of pressure reduction stages are not found.[6] The two-stage reduction systems may be divided into thus:

- Designs with separated stages of reduction
- Designs with combined stages of reduction

Recommended for diving work is a hose apparatus with two-stage pressure reduction (Haux, 1968, 1982). Regardless of the type of the automatic pressure reduction allowing for the ambient depth of diving, the general principle of the open circuit *UBA* is similar. This general principle is displayed in Figure 3.8.[7]

A more detailed discussion focused on constructions of this type can be found in literature (e.g., Hiszpański, 1966).

SCUBA[8] consists exclusively of the B equipment, as marked in Figure 3.8. In its version with hose, it additionally contains external supply unit A and supply line C. The major interest of the construction of *UBA* is the demand valve, which reduces the pressure in the supply hose[9] to the ambient pressure of the diving depth. The demand system is divided into thus:

- Water space 4 with the escape valve 5
- Gaseous area 7 containing the pressure reduction[10] unit 8
- First stage of pressure reduction[11] unit 10

The first-stage reduction for *SCUBA* unit 10 is controlled by hydrostatic pressure via the diaphragm it cooperates with. Thanks to that, the breathing medium remains under a constant overpressure in relation to the hydrostatic pressure of a given diving depth. The reduction unit of the second stage 8 is controlled by hydrostatic pressure via the diaphragm 6. As a result, it can deliver the breathing medium into the diver's lungs at a pressure equal to the hydrostatic pressure at the depth of diving. In order to diminish inhalation resistance, this unit is often equipped with a device aiding suction – not shown in Figure 3.8. In order to ensure the emergency supply of the breathing gas, the external supply unit should also include reserve holders. For *SCUBA* cylinders, set 15 should also have a pressure valve of the breathing medium reserve marked in the Figure 3.8 as 13. During regular work, the diver receives the breathing medium from external supply unit A – distribution valve 3 is set as indicated in Figure 3.8, and valve 11 from autonomous cylinder set 15 is open.

It is recommended that the breathing medium be administered at a pressure reduced earlier[12] but allowing the second stage of the demand valve to work properly. Owing to that, hoses resistant to lower pressure may be used. *SCUBA* cylinders set 15, together with its instrumentation, when cut off with distribution valve 3, serve as

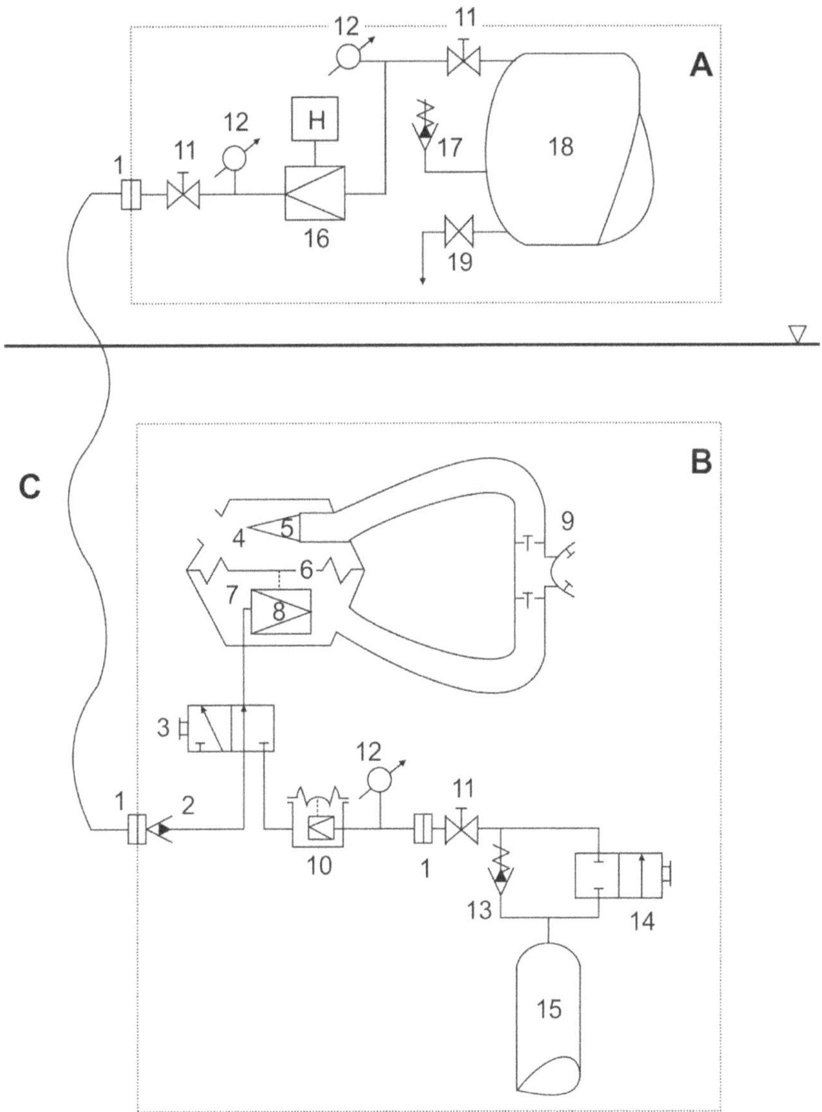

FIGURE 3.8 Open circuit *UBA*: (a) External supply of hose-version *UBA* with open breathing circuit. (b) Open circuit *SCUBA*. (c) Supply line: (1) junction; (2) nonreturn valve; (3) selector valve; (4) water space of demand valve; (5) nonreturn scape valve; (6) diaphragm; (7) gaseous space of demand valve; (8) unit of breathing medium pressure reduction cooperating with the diaphragm – second-stage pressure reduction (second-stage regulator); (9) mouthpiece device with valves and hoses; (10) reducer holding overpressure of breathing medium constant relative to hydrostatic pressure (first-stage pressure reduction – first-stage regulator); (11) cutoff valve; (12) manometer; (13) throttle–stop valve of the reserve unit; (14) valve actuating reserve of the breathing medium; (15) unit self-contained holders; (16) manually controlled reducer; (17) safety valve; (18) external holders with the breathing medium stored; (19) drainage valve.

TABLE 3.4
Basic Features of Open Circuit UBA

Open circuit *UBA*	
This type of *UBA* is known primarily as an air *SCUBA*. The air *SCUBA* version is recommended for recreational and sports purposes. For commercial diving, recommended is exclusively the hose version with two separate-stage pressure reduction regulator.	
Disadvantages	• Limited maximum diving depth for air as the breathing medium, not suitable for deep exploration and operation. • The system with an artificial breathing medium is unacceptable for economic reasons, which relates to helium mixtures in particular. • Relatively short time of protective action of the autonomous version, air used as the breathing gas, long decompression time, and long time needed for recovery afterward. • Complicated construction of the regulators when compared to other technical solutions, affecting reliability of operation and requiring personnel having special skills.
Advantages	• The breathing medium used does not get mixed with the fresh one. • An inexpensive and readily available breathing medium in the air is used. • Diving depth, with air as the breathing medium, sufficient to carry out harbor underwater work, repair damage to ships, and provide maintenance of hydroengineering objects.

an emergency reserve of the breathing medium. The moment the breathing medium from the external supply unit has been cut off, the user switches valve 3, triggering an autonomous supply. The reserve of the breathing medium should be sufficient to let the diver leave the working site safely and surface or reach a safe place, for example, a diving bell.

The circuit of breathing gas from an external supply is a simplified version of what happens in it in a *SCUBA* ; hence, taking the latter case as a model, it is discussed. During the regular work of a *SCUBA* version, reserve valve 14 is shut in position as in Figure 3.8. From cylinder set 15, the breathing medium is pushed through the throttle–stop valve of reserve device 13, through first reduction stage 10 to reduction unit of the demand valve 8,[13] where, through a hose, the inhale valve and mouthpiece device 9 go right into the diver's lungs. Then, from the lungs through the exhale valve of the mouthpiece, hose, and nonreturn escape valve 5, through the water section of demand valve 4, into the water. The moment the pressure in cylinders set 15 has fallen to the value at which valve 13 gets cut off, the diver feels the inhalation resistance rising, and then a shortage of the breathing medium. This is a signal that the reserve of the breathing medium[14] has been cut off. To trigger this supply, the user must switch on reserve valve 14.

A description of the types of open circuit *UBA* is presented in Table 3.4.

Examples of the open circuit *UBAs* listed in Table 3.3 are shown in Figure 3.9.

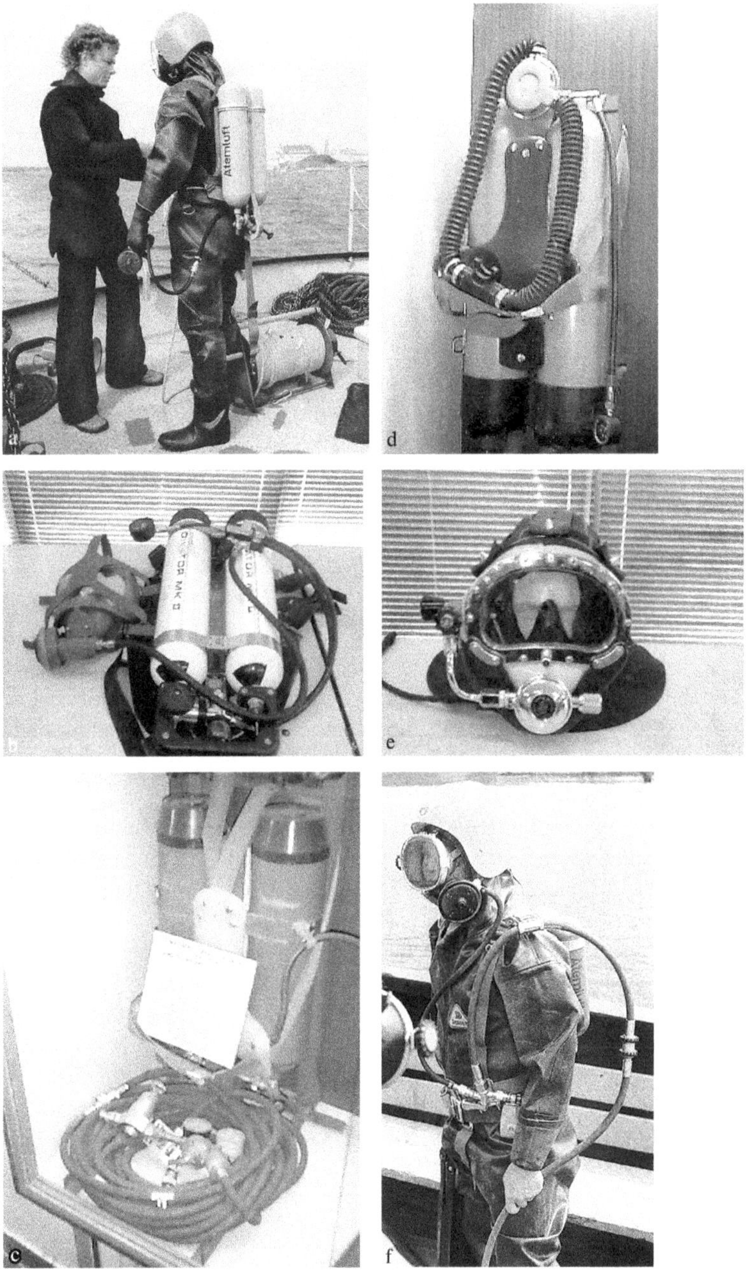

FIGURE 3.9 Examples of open circuit *UBAs*: (a) $PA-38/3600\ SCUBA$; (b) $AGA\ Mk\ II\ SCUBA$; (c) $UAN-82\ UBA$; (d) $PR-27\ type\ SCUBA$; (e) full-face mask $HeliOx-18B/KMB-10$; (f) $PL-70\ UBA$.

Source: Kłos (2000, 2008a).

3.3 SEMI–CLOSED CIRCUIT UBA

In a semi–closed[15] circuit *UBA* (referred to as *SCR*), exclusively, artificial breathing media are used (Table 3.3):

- *SCR* using ready breathing medium: *premix – SCR*
- *SCR* producing breathing medium during a dive: *automix – SCR*

The *automix – SCR* offers the following possibilities:

- Producing breathing medium during a dive at a constant ratio of oxygen content and inert gases
- Producing breathing medium during a dive with a changeable ratio of content between oxygen and inert gases according to depth

Taking the mode of supply as a division criterion, we can distinguish between thus:

- Self-contained *SCR*: *SC – SCR*
- External supply *SCR*: *hose – SCR*

In *SCR*, the role of the system supplying the breathing medium under the ambient pressure of diving depth has been taken over by elastic bags or breathing bellows. Such an *SCR* can be equipped with one bag or two bags. A bigger number of bags are seldom to be met. Again, considering the arrangement of the bags in relation to each other, the two-bag system can be classified as designed with two separate bags or one bag contained in the other.

A characteristic of diving systems which use premix is that they administer breathing gas at a constant volume or actuated according to the breathing rhythm of the user. The technical design of the metering device allows grouping them as follows:

- *SCR* with nozzle reduction metering system
- *SCR* taking advantage of evacuation of the metering cylinder
- *SCR* with system of breathing gas metered with a set of bags arranged one in the other

The classification of various technical designs of *SCR* in regard to the breathing bag construction is shown in Table 3.5.

3.3.1 PREMIX *SCR*

The general principle of *two separate – bag premix – SCR* with the nozzle reductive metering system is shown in Figure 3.10.

The *SC – SCR* part of the *one – bag SCR* is illustrated in Figure 3.11.

Their principles of work being so much alike, it is the *SCR* presented in Figure 3.11 that will be discussed in detail. When discussing the principles of operation of the *SCRs*, the term "metering" will be used, that is, mass flow per time unit passing through the bore of the nozzle of the *SCR*.

TABLE 3.5
Classification of SCRs Related to Type of Construction of Breathing Bags

Classification of SCRs		
One – bag SCR	*Two – bag SCR*	
	With separate bags	With bags arranged one in the other
Disadvantages	• In the *single – bag SCR*, the breathing medium designed for rebreathing is partly lost. • All the *SCRs* are less-hygienic than the open circuit breathing gas system. • All types of *SCR* construction are less-economical in use than *CCRs*. • The complex construction of breathing bags in the *SCR* type with bags one in the other may result in smaller reliability of the device.	
Advantages	• In the *two – bag SCR* construction, the losses only involve the breathing medium which has not been purified from carbon dioxide, thus economizing on filling of the absorbent and prolonging the duration of its protective action (vital for the hose version). • In the *SCR* with separate bags, it is easier to arrange the bags in the casing, making the whole unit more compact. • With the *SCR* having bags contained one in the other, we are able to precisely determine the ratio between the amount of the breathing medium removed and the total amount of this medium in the circuit. Metering the breathing gas is carried out on a periodical basis and adjusted to the breathing rhythm of the diver. To set the relief valve pressure for the opening in this type of *SCR* turns less important than in other systems with semi-closed circuit of the breathing medium.	

The part that is crossed and marked as B represents *SCUBA-SCR*. Again, connecting parts A and B with supply hose C creates hose supply *SCR* version *house – SCR*, where breathing gas supply stored in unit 15 is the emergency supply.[16] In the *hose – SCR* version, gas from cylinder 17 is passed via reducer 12 and selector valve 4, metering nozzle 11, inner supply unit into supply hose C through the supply unit, return valve 2, valve selecting mode of supply 3, into inhalation bag 10. In the autonomous version, the breathing medium from unit of cylinders 15 passes through cutoff valve 14 and manometric pressure control 13 to reducer 12. Then, through selector valve 3[17] and nozzle 11, which was selected with nozzles selector valve 4, into breathing bag 10.

The breathing bags are made of an elastic, gastight fabric; therefore, the pressure in the bags approximates the summarized value of the ambient hydrostatic pressure at a given depth and the pressure relief valve opening.[18] The breathing medium from inhalation bag 10 is lost through the hose and inhalation valve of mouthpiece device 9 into the diver's lungs. Exhalation runs through the exhalation valve and the exhalation hose into exhalation bag 8, wherefrom the breathed-out gas passes into carbon dioxide absorbent 7. Relief valve 6 that is mounted on the exhalation bag is meant to lose the excess of exhaled gas into the water. The expired medium running through the absorbent is purified by carbon dioxide removal and joins the clean breathing medium circuit entering inhalation bag 10, mixing here with the fresh breathing medium. Worthy of note is that the value of the reduced pressure of the breathing medium should remain at a level permitting the selected nozzle to

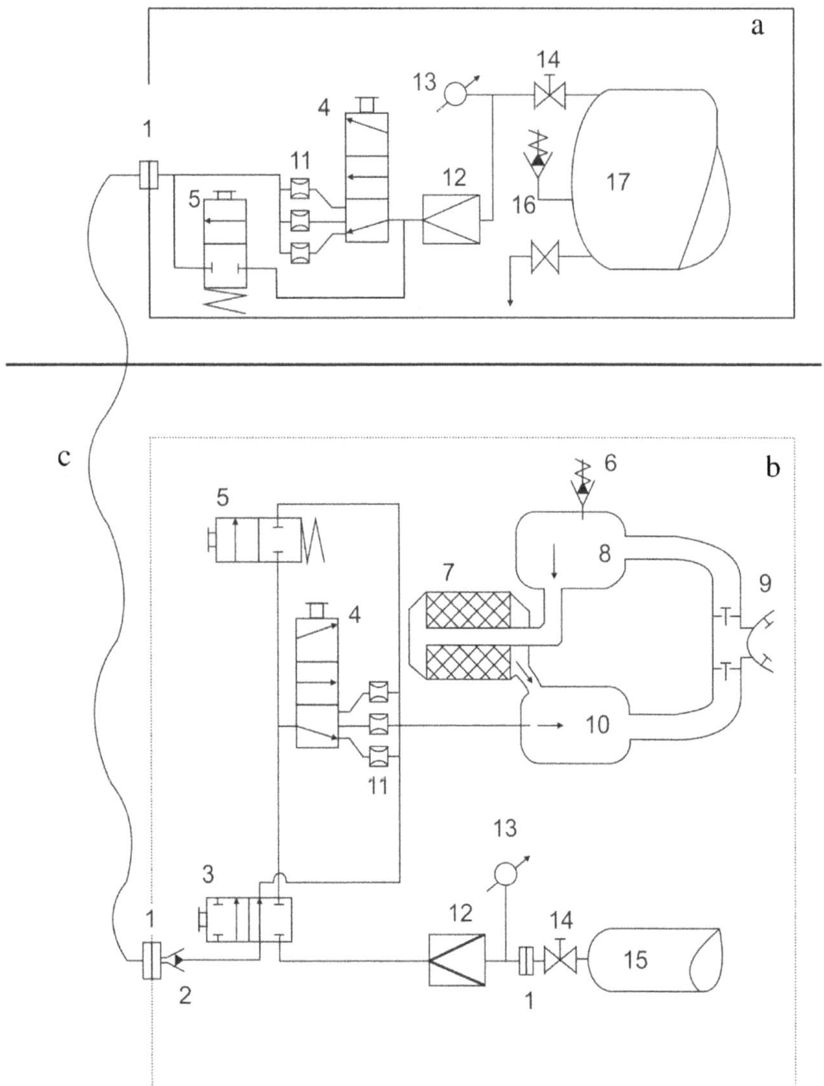

FIGURE 3.10 Two-bag version of the *premix – SCR*: (a) External supply unit of hose version of *premix – SCR*. (b) *premix – SCR SCUBA*. (c) Supply hose: (1) coupling; (2) return valve; (3) supply mode selector valve; (4) metering nozzles selector valve; (5) bypass valve; (6) relief valve; (7) carbon dioxide absorbent; (8) exhalation bag; (9) mouthpiece device with valves and hoses; (10) inhalation bag; (11) metering nozzles; (12) reducer; (13) manometer; (14) cutoff valve; (15) autonomous set of cylinders; (16) safety valve; (17) external cylinders with breathing medium stored; (18) drain valve.

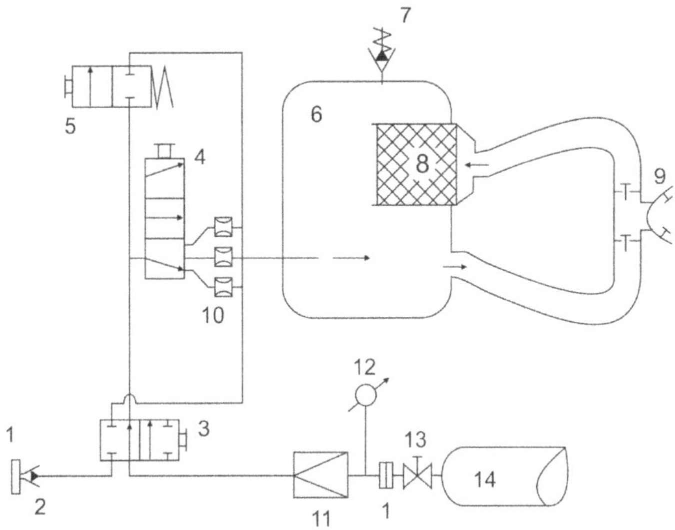

FIGURE 3.11 One-bag *premix* – *SCR SCUBA*: (1) coupling; (2) return valve; (3) selector valve; (4) selector valve of metering nozzles; (5) bypass valve; (6) breathing bag; (7) relief valve; (8) carbon dioxide absorbent; (9) mouthpiece unit with valves and hoses; (10) metering nozzles; (11) reducer; (12) manometer; (13) cutoff valve; (14) autonomous set of cylinders.

work within the range of its supercritical flow, maintaining it at the whole range of the diving depths allowed. This means that its mass expenditure is kept at a constant level with an assumed tolerance. This type of hose supply *premix* – *SCR* is equipped with special bypass valves 5, designed for filling and washout[19] of the breathing area. In the *hose* – *SCR* version, the bypass valve[20] is steered by external supply A team – communicating with the diver by means of technical aids. A basic feature of *premix* – *SCR* is shown in Table 3.6.

Apart from the nozzle metering system, other solutions are also to be met. They operate on the principle that ventilation of the diver's lungs is a function of the oxygen stream consumed (Kłos, 2021). The solution applied in Swedish *ACSC* type *SC* – *SCR*[21] lies in the evacuation of the special metering cylinder. In French, *FENZY* – 68 and *DC* – 55 metering of fresh breathing medium is regulated by a set of bags contained one in the other (Figure 3.12).

The metering system making use of evacuation of the metering cylinder is presented in Figure 3.13.

Every slight movement of the breathing bellow 1 induced by breathing is counted by the pawl mechanism of the system regulating the metering valves 2. Filling and evacuation of the metering cylinder 4 is carried out by switching valve 3 by steering system 2.

TABLE 3.6
Characteristics of *premix – SCR*

Apparatus of semi-closed circuit of ready breathing medium *premix – SCR*	
Premix – SCR is known as a system designed for using gas mixtures as a breathing medium, though oxygen is possible. Among the breathing media used most frequently, one may find the mixtures helium–oxygen, *Heliox*; helium–nitrogen–oxygen, *Trimix*; or nitrogen–oxygen, *Nitrox*.	
Disadvantages	• The choice of breathing medium and its composition and metering is confined to a relatively narrow range of depths. • Relatively complicated preparation of *SCR* for work, e.g. preparing breathing mixture, checking its composition, checking correctness of the absorbent's filling, checking metering and positioning of the relief valve, checking leak tightness etc. • Technical difficulties resulting from prolonged storage of the breathing medium, e.g. likelihood of stratification of the components, with *Heliox*, in particular. • Despite a partial re–use of the exhaled medium it gets wasted through the relief valve, which adversely affects the efficacy of diving, especially with the high costs of the breathing medium components. • Technical problems connected with manufacturing of metering nozzles and changes in their characteristics with the elapse of time—loss of proper nozzle calibration. • High sensitivity to of the breathing medium to dust contamination likely to cause clogging of the metering nozzles. • Due to a partial closure of the breathing medium circuit all *SCRs* are less hygienic than open circuit systems.
Advantages	• Simplicity of the design providing its reliability. • Partial reuse of the exhaled breathing gas allows for expansion of the protection period of the *SCR*, thus economizing the process of diving relative to diving with open circuit *UBA*. • In majority of cases, it offers a choice of parameters of work, allowing for the effectiveness of diving operations and shortening the diver's decompression time by selecting a range of oxygen partial pressures and the reduction of breathing medium density – by application of light gases, etc.

In Figure 3.13, valves 3 are shown in the position of evacuation of metering cylinder 4. To maintain the constant mass rate of the fresh medium metered, the supply overpressure in cylinder 4 is kept at a constant level by pressure reducer 5. This system can sense the outer pressure via an elastic diaphragm. When the sum of the bellow's motions achieves the certain definite number, valves 3 get switched and the content of cylinder 4 is forced through into the breathing loop of the *SCR*. This solution gives the possibility of economical management of the fresh breathing gas stock, as its metering is carried out on demand according to the breathing rate. It allows the precise development of decompression procedures. For the designs with constant metering of fresh breathing medium, this relationship is much more distinct. *ACSC* is equipped with a number of very complicated mechanical safety devices, such as emergency supply of breathing medium, a device preventing subsidence of breathing bellow during a rapid submersion, etc. For this reason, it turns a very complicated

FIGURE 3.12 *premix – SCRs*: (a) *SC – SCR* type *APW – 6M*; (b) *SCR – FGG III U*; (c) *SCR* applied in *Sealab III* project; (d) *SCR* type *DC – 55 SCUBA*; (e) *SCR / CCR – ACSC SCUBA*; (f) *SCR – GAN – 87 UBA*; (g) *SCR – CRABE SCUBA*.

Source: Kłos (2000, 2008a).

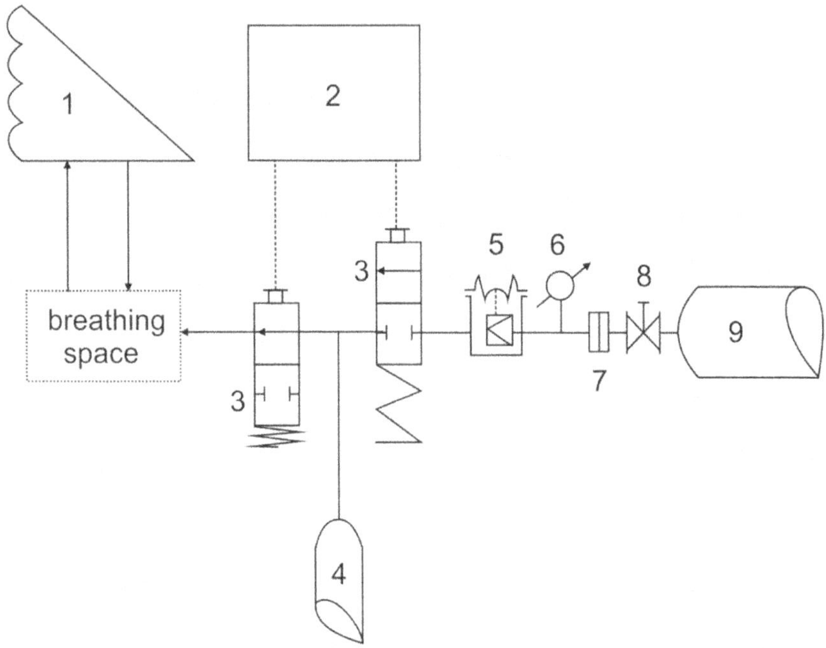

FIGURE 3.13 The metering system in the *premix – SCR* making use of the evacuation of a constant capacity holder and constant overpressure of loading: (1) breathing bellow; (2) system-regulating metering valves; (3) metering valves; (4) metering cylinder; (5) reducer maintaining constant overpressure of breathing medium supply relative to ambient pressure; (6) manometer; (7) coupling; (8) valve; (9) supply cylinder.

and a delicate design; first costs and operating costs are high. This *SCR* is known only in nitrox version, although several attempts have been undertaken to use it in deep diving as well (Frånberg, 2015). This type of *SCR* is employed exclusively for military purposes, that is, in mine countermeasure operations. To this end, it is made of nonmagnetic material and is equipped with a device diffusing breathing medium released through the relief valve into the water.

French *SCRs* : *FENZY* – 68, *DC* – 55, and *CRABE* metering of the fresh breathing medium is solved by means of a special unit of bellows arranged one in the other (Figure 3.14).

CRABE and *DC* – 55 are used, as much as *ACSC* are used only in the navy, in $Nitrox^{22}$ version to about $50\,mH_2O$.

When discussing the operating principles of *CRABE*, a clear distinction must be made between the exhale and the inhale phases. One-way circuit of the breathing medium is maintained with nonreturn valves 1. During the exhale phase, the escape valve of the mouthpiece device gets open (Figure 3.14a).

The exhaled gas is let through mouthpiece 12, exhale hose, and escape valve into CO_2 absorbent 11, and then to large bag 2, therefrom via nonreturn valve into small bag 3. On inhalation, the escape valve is closed (Figure 3.14b). In turn, the

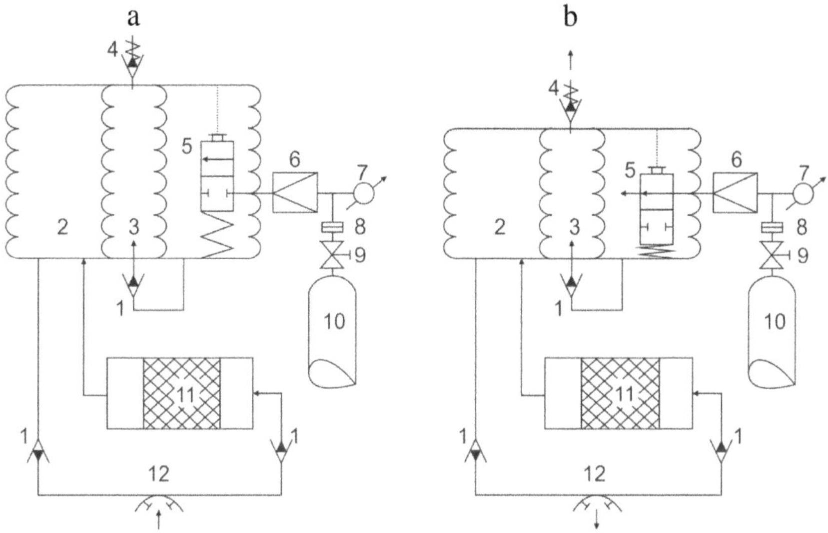

FIGURE 3.14 Engineering principles of *SCR* with breathing bags arranged one in the other: (a) Exhalation phase. (b) Inhalation phase: (1) nonreturn valve; (2) large bag (outer); (3) small bag (inner); (4) relief valve; (5) metering valve; (6) reducer; (7) manometer; (8) coupling; (9) valve; (10) cylinder with breathing medium; (11) CO_2 absorbent canister; (12) mouthpiece device.

inhale valve opens. From large bag 2, the breathing medium goes into the lungs, through the inhale valve, inhale hose, and mouthpiece device 12. On inspiration, the capacity of bag 2 diminishes. When collapsing, large bag 2 is pressing the metering valve 5 lever, triggering it. It causes the fresh breathing gas from cylinder 10 to flow, passing cutoff valve 9, coupling 8, reducer 6, and metering valve 5, into large bag 2, to get mixed with the recycled breathing medium. And now, via the nonreturn valve, the breathing hose, and mouthpiece device 12, right into the lungs. This solution is economical because the breathing medium is administered on demand, actuated by the breathing action and the capacity relation between the large and the small bags. Such a system, however, does not provide sufficient metering at greater diving depths and longer expositions. For this reason, at greater depths[23] and at longer expositions,[24] this construction must be supported by a nozzle system with constant dosing of the breathing mix in order to intensify the ventilation process.

3.3.2 Automix *SCR*

The operation principle of one-bag breathing *automix – SCR* is illustrated in Figure 3.15.

The system produces breathing medium characterized by constant oxygen-to-inert-gases ratio. Designs of this type in the *hose – SCR* version are equipped with installations used for mixing the breathing medium outside the system.[25] The entire

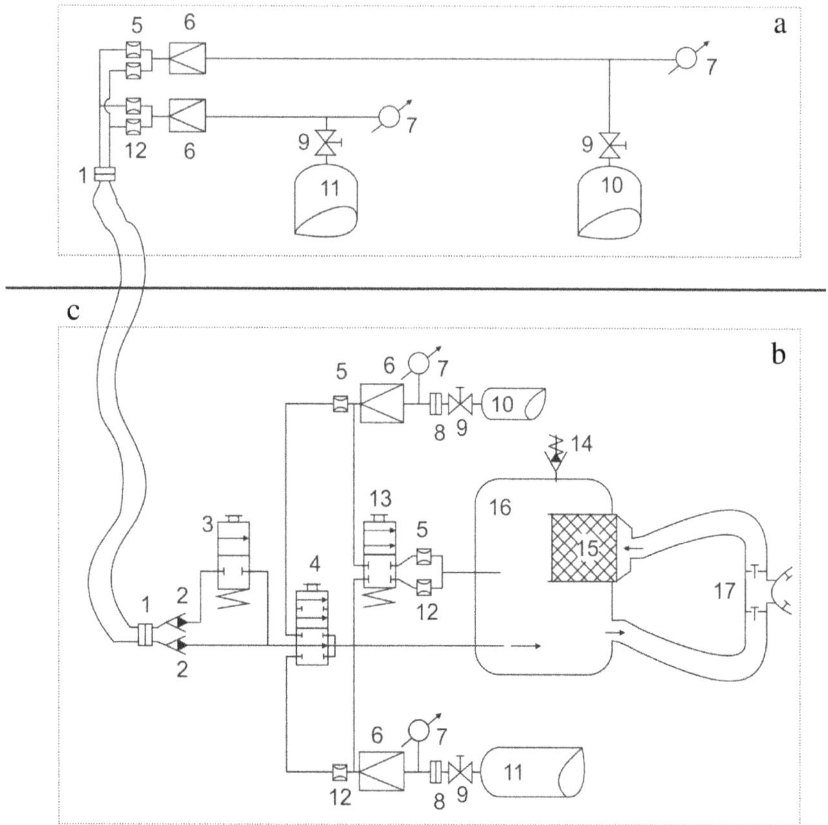

FIGURE 3.15 One-bag version of *automix – SCR*: (a) External supply unit of *hose – automix – SCR*. (b) *SC – automix – SCR*. (c) Group of supply wires and pipes – umbilical: (1) coupling; (2) nonreturn valve 3, bypass valve cooperating with the external supply unit A; (4) selector valve; (5) oxygen metering valve; (6) reducer; (7) manometer; (8) cylinder coupling; (9) cutoff valve; (10) cylinder of oxygen; (11) cylinder of inert gas; (12) metering nozzle of inert gas, for example, helium; (13) bypass valve cooperating with independent supply unit; (14) relief valve; (15) carbon dioxide absorbent canister; (16) counterlung; (17) mouthpiece device with valves and hoses.

system producing the breathing mixture is localized in a diving station or diving bell. It may additionally be fitted with sensors of oxygen content.[26] In its *hose – SCR* version, the apparatus may consist of a combination of several units engaged in *automix – SCR* and other *– SCR*.[27]

Marked as B, the *SCR* shown in Figure 3.15 is a *automix – SCR*. The *SCR* part B is equipped with one breathing bag 16, carbon dioxide absorbent canister 15, gas cylinders 10 and 11, and a unit generating the breathing medium. The breathing medium is produced by mixing a stream of oxygen and a stream of inert gas. The metering of gases is carried out by means of nozzles metering within their

TABLE 3.7
Basic Features of *automix – SCR*

Semi–closed circuit breathing medium mixed during a dive *automix – SCR*	
Automix – SCRs have never been widely used due to the problems with their manufacturing and running. It is known only in experimental or limited runs.	
Disadvantages	• Homogeneity and stability of the breathing medium contained in the bag much more difficult to maintain than in the designs using a premixed breathing medium. • When the *SCR* produces breathing medium having constant content ratio of oxygen and inert gas, each time, the composition of the breathing medium as well as the mode of metering must be carefully selected, sufficing only to a relatively narrow range of diving depths. • Prior to each dive, the *SCR* must be subjected to a quite-troublesome preparation and check procedure. • Despite partial reuse of the breathing medium, it gets wasted at relief valves, which affect the cost-effectiveness ratio of diving with breathing mixture made of expensive components. • High sensitivity to dust pollution likely to chock metering nozzles. • Due to partial closure of the breathing medium circuit, all these designs are less-hygienic than the open circuit systems.
Advantages	The same as those characteristics of the premix *SCR*. Potential for operating on clean gases instead of their mixtures. This allows avoiding problems connected with producing and storing precisely composed breathing mixtures. It enhances the cost-effectiveness ratio of running this apparatus in comparison with the *premix – SCR*. • A change of breathing medium for a selected range of depths requires merely a change of the metering nozzles. • In the case of the *SCR* producing a breathing medium having variable oxygen-to-inert-gas ratio of the potential dynamic change in oxygen-to-inert-gas ratio, making the breathing medium according to the depth of diving.

supercritical ranges. Appropriately reduced overpressure of gases is maintained by pressure reducers 6. Fullness of the cylinder is controlled by manometers 7.

Oxygen and inert gas get mixed in the supply hose and enter counterlung 16, wherefrom, via the inhale hose of the mouthpiece, the ready mixture is directed into the diver's lungs. The exhale is released through mouthpiece 17, through its valve and the exhale hose, passing carbon through dioxide absorber 15, into the counterlung. The surplus of the breathing medium escapes from the bag through relief valve 14 into the water.[28] The breathing medium is also administered into the lungs of the diver at a pressure whose value equals the sum of ambient hydrostatic pressure of the given diving depth and of the relief valve opening pressure. To preliminarily fill the breathing loop and to wash it out with fresh breathing medium, bypass unit 3[29] or 13[30] is used. The fundamental features of *automix – SCR* are summarized in Table 3.7.

Examples of *automix – SCR* discussed in Table 3.3 are presented in Figure 3.16 and Figure 1.8.

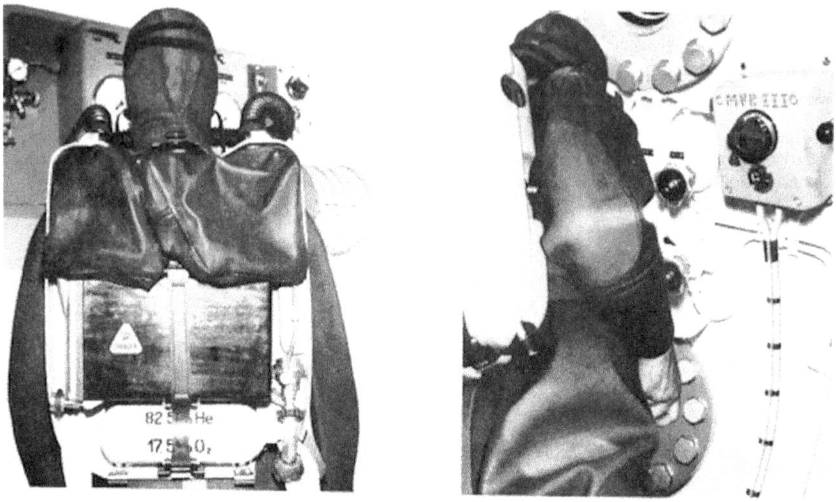

FIGURE 3.16 Examples of *automix – SCR SCR SMS — III.*

Source: Kłos (2000, 2008a).

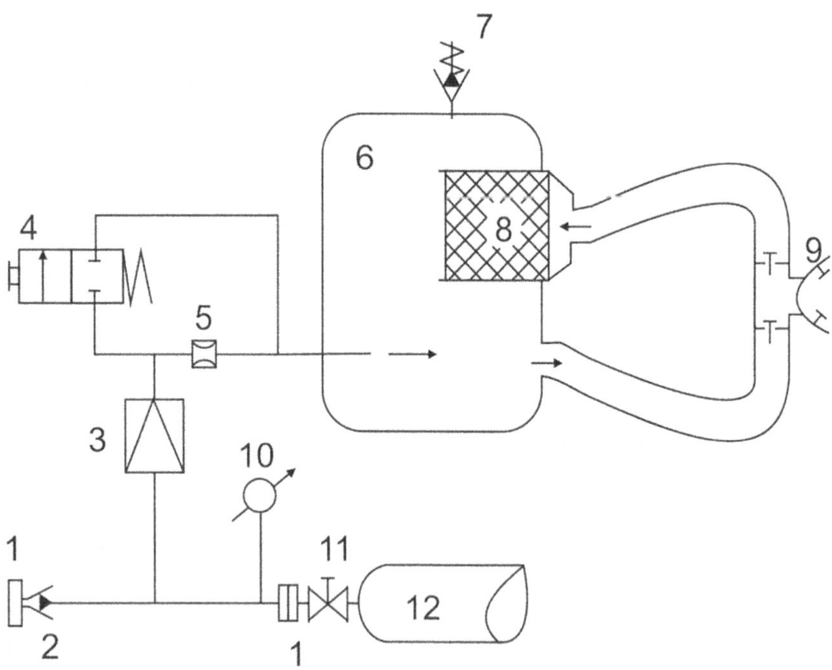

FIGURE 3.17 *oxy — CCR* of constant oxygen metering: (1) joint; (2) nonreturn valve; (3) reducer; (4) bypass valve; (5) oxygen metering nozzle; (6) the breathing bag; (7) relief valve; (8) cylinder coupling; (9) mouthpiece assembly with valves and hoses; (10) manometer; (11) cutoff valve; (12) oxygen cylinder.

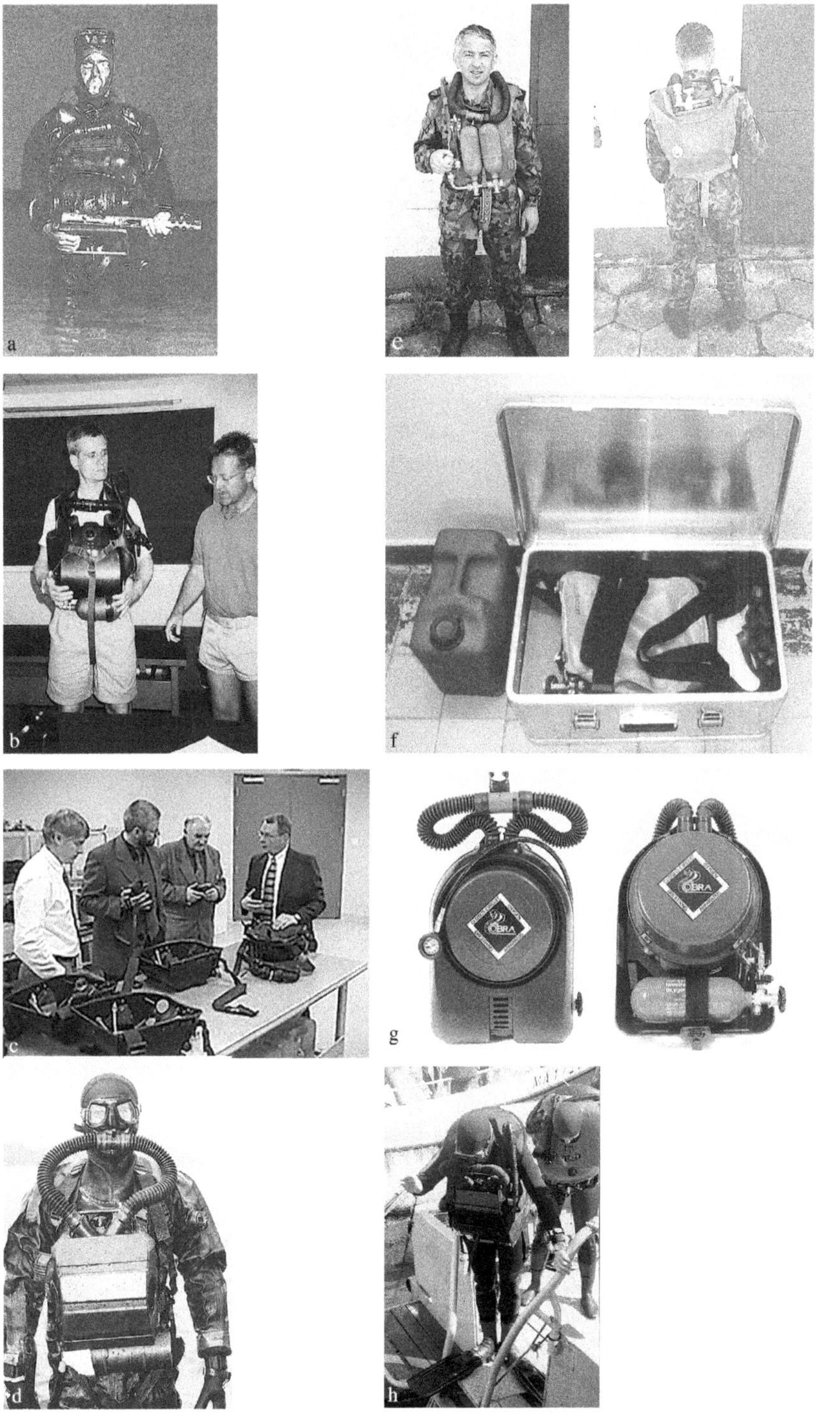

FIGURE 3.18 Examples of $oxy-CCR$: (a) $AGA-Oxydive$; (b) the author with $LAR\ V$; (c) the author during $S-10$ presentation; (d) $Oxy-NG$; (e) $LT\ II$; (f) $LAR\ VII$; (g) $Cobra$; (h) the author before dive with the use of $Oxy-NG$.

Source: Kłos (2000, 2008a).

3.4 CLOSED CIRCUIT BREATHING APPARATUS

The designs of closed circuit breathing apparatus CCR can be divided into thus:

- Oxygen CCR: $oxy - CCR$
- Breathing mixtures CCR: $mix - CCR$

3.4.1 Oxygen CCR

Oxygen-supply UBA is the most often mentioned group of CCR. As for $oxy - CCR$, in its breathing cycle, oxygen is inhaled from the breathing loop of CCR. As it is exhaled, the gas passes through a canister with carbon dioxide absorber to the breathing loop. Oxygen consumption during dives is not high. It is not affected by the diving depth but is dependent only upon the work performed.

When compared to open circuit UBA and $oxy - CCR$ does not require a large supply of breathing gas and offer extended action radius range. Therefore, the construction of $oxy - CCR$ enables an extended operating duration; however, due to oxygen toxicity, it can be used only at shallow depths. As there is no need for periodical release of the breathing medium, there are no bubbles escaping to the water, and $oxy - CCR$ can be used for diving in performed special operations. During normal work, this system requires only periodical washout of the breathing loop to eliminate accumulating contaminants and washout of nitrogen that saturates the human body under normal conditions (Bennett and Elliott, 1993; Brubakk and Neuman, 2003). There is also a need for release of breathing medium excess during ascent of the diver. Due to oxygen toxicity, an oxygen dive is limited by the depth and exposure time. Moreover, an oxygen diver should satisfy the requirement for special physical fitness and mental predisposition. For these reasons, oxygen diving is prohibited in many countries; however, it can be performed in civilian underwater operations.

$Oxy - CCR$ constructions have developed in two main directions. The first is an improvement in the traditional *Fleuss*' construction of the CCR. It consists of the breathing bag connected with a carbon dioxide absorbent canister. The canister may be filled with carbon dioxide absorbent[31] or with various oxygen-generating masses.[32]

$Oxy - CCRs$ may have a manual or constant oxygen metering system, the latter aided with manual or automatic periodical adding of oxygen. An example of $oxy - CC$R is presented in Figure 3.17.

Improvements in this $oxy - CCR$ design consist in the use of patents leading to its better tactical–technical properties. Elimination of buoyancy change effects due to the breathing rate and changes in depth of dive, application of the constructional solutions leading to reduced breathing resistance, application of compact designs, elimination of the magnetic field generated by the $oxy - CCR$, elimination of the noise generated during breathing and $oxy - CCR$ work, application of the $oxy - CCR$ equipment that enables camouflage during transport and covered storage, application of constructions that do not require special tools for daily servicing, corrosion- and trying conditions–resistant materials, water temperature change resistant constructional and exploitation materials, maximization of the apparatus mass to its action radius range, application of streamline shapes of low resistance during swimming, etc.

This group of $oxy-CCRs$ is represented by Swedish $AGA-Oxydive$, German $LT\ II$, or later version $LARV$ and $LARVI$, or Canadian $S-10$ (Figure 3.18).

Another group consists of $oxy-CCRs$, where instead of a breathing bag, a large rubber diaphragm is used, arranged on top of a shallow box, with the lever of the bypass valve on its bottom. If the oxygen volume in the breathing loop falls below a certain established minimum, the diaphragm will press the lever, triggering oxygen bypass valve. The diaphragm is strapped with limiters, adjusting the maximum breathing loop volume to the diver's preferences and mass of his/her equipment. These solutions of $oxy-CCR$ are employed in American *Cobra* or French $OXY-NG$ (Figure 3.18).

3.4.2 Mix *CCR*

Closed circuit gas mixture *UBA*, referred to in this book as $mix-CCR$, can be grouped as follows:

- Small circuit, $mix-CCR$
- Increased[33] circuit, $mix-CCR$
- Large circuit, *CCR*

$Small\,circuit\,mix-CCR$ are called the $SC-mix-CCR$, for example, $Mk15/16$ (Figure 3.19).

$Increased\,circuit\,mix-CCR$ denotes a circuit between the CCR and the diving bell[34] equipped with measuring instruments, systems of regeneration, and enrichment of the breathing gas with oxygen, and so on, for example, $Dolphin7$ (Figure 3.19b). A $large\,circuit\,of\,mix-CCR$ is the circuit between the surface[35] and the diver. The measuring equipment, systems of regeneration, and enrichment of the breathing gas with oxygen are placed on the surface, for example, $CCBS450$ with the surface system $GAK600$, the diver's equipment, and the main table of the system (Figure 3.19c).

The general principle of the $SC-mix-CCR$ operation is presented in Figure 3.20.

Elektrolung, popular name of this design, comes from the apparatus manufactured by Oceanic Equipment Co. (Figure 3.21).

The breathing bag, together with the carbon dioxide absorbent canister, is connected with the mouthpiece assembly by means of breathing hoses (Figure 3.20). The breathing medium circulates within, as the direction arrows indicate in Figure 3.20. Inhale and exhale valves ensure one-way direction. These constructional junctions operate in exactly the same way as *SCRs* discussed earlier. From cylinder 1, via cutoff valve 2, inert gas flows into the breathing bag, running on its way through the manometric pressure control 4 and automatic assembly, allowing for ambient hydrostatic pressure at the depth of diving 5. The oxygen from oxygen cylinder 16 runs through cutoff valve 2, manometric control 4, oxygen reducer 15, and flow controlling valve 13 into the breathing bag. Oxygen sensor 9 placed in the breathing bag constantly measures the oxygen partial pressure. The signal is simultaneously displayed on the diver's control desk and is transmitted to the current–pressure

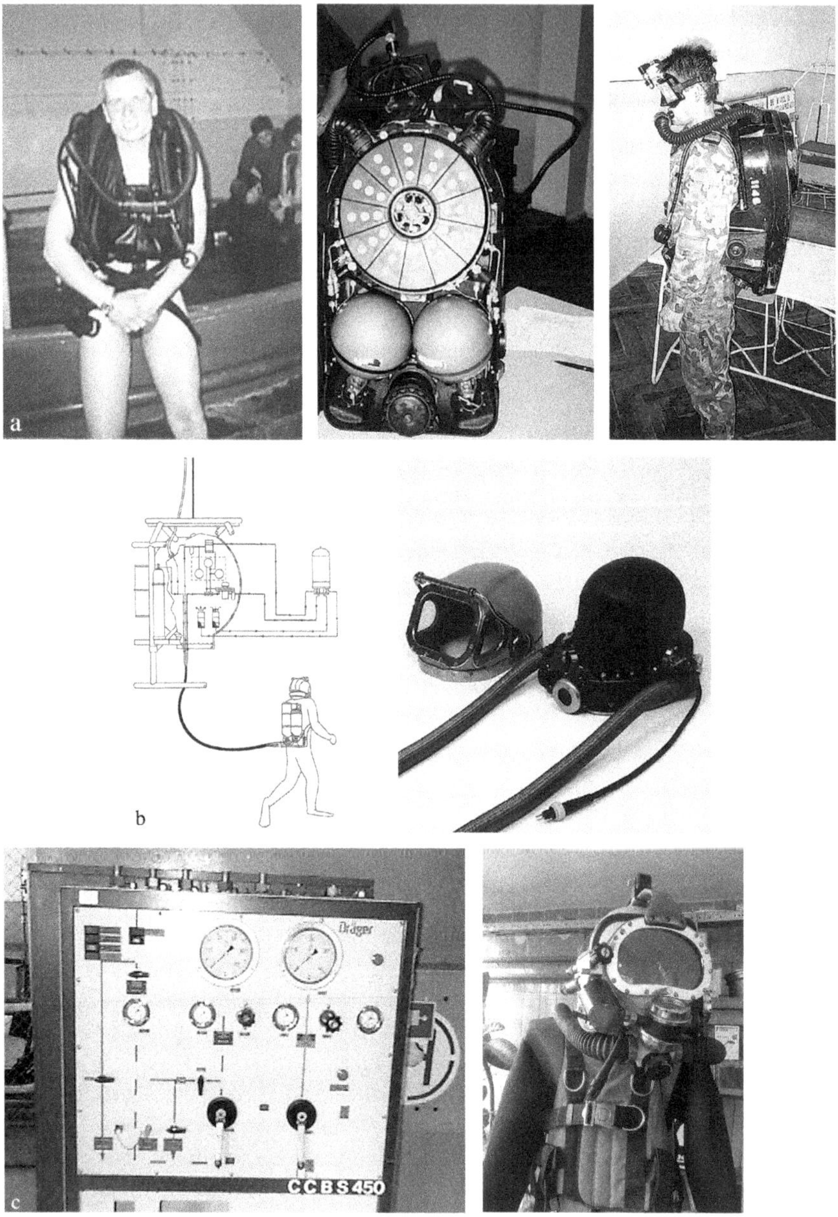

FIGURE 3.19 Groups of $mix-CCR$: (a) the author during mk-15/16 presentation; (b) *Dolphin7*; (c) main board and diving equipment $CCBS\,hose-mix-CCR$.

Source: Own photos, courtesy of the Military Diving School/Explosive Ordnance Disposal Mobile sUnit 11, Oak Harbor, USA; advertising materials, Dolphin 7 Diving System, Reimers Engineering Inc., Alexandria; and the Polish Naval Academy, respectively.

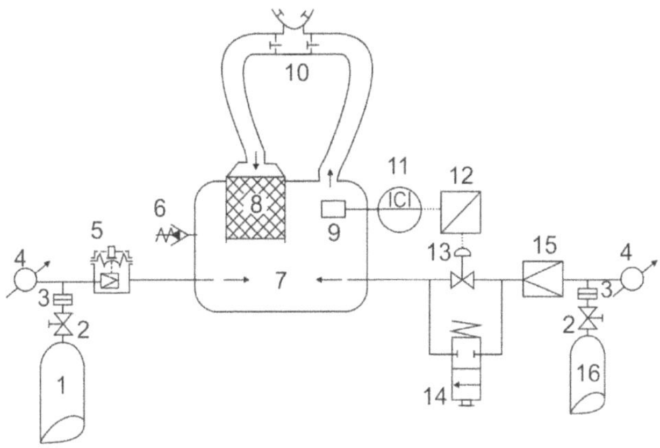

FIGURE 3.20 *mix – CCR SCUBA*: (1) inert gas cylinder;[36] (2) cutoff valve; (3) coupling; (4) manometer; (5) reduction unit maintaining the filling of the breathing bag at the demanded level; (6) relief valve; (7) breathing bag; (8) carbon dioxide absorbent; (9) sensor of oxygen partial pressure; (10) mouthpiece device with valves and hoses; (11) measuring-controlling unit of oxygen partial pressure; (12) converter of electric signal into pressure response; (13) valve with pneumatic servo; (14) oxygen bypass valve; (15) oxygen reducer; (16) oxygen cylinder.

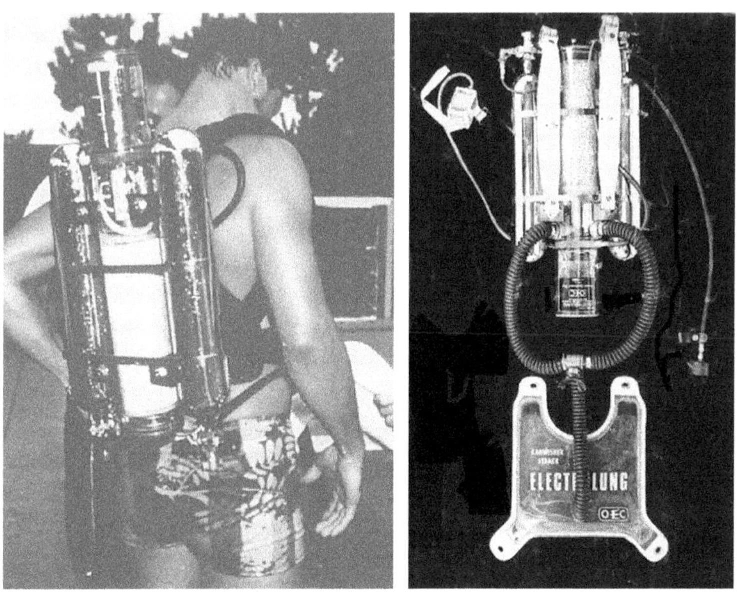

FIGURE 3.21 *mix – CCR SCUBA* manufactured by Oceanic Equipment – CoElektrolung.

Source: Bachrach, Desiderati, and Matzen (1988), and own photo from instruction manual.

transducer 12. The transducer uses the oxygen line to supply the executive element regulating the opening of valve 13. If the oxygen partial pressure is too small, valve 13 will be opened by the executive pneumatic element. Additional bypass valve 14 is used for preliminary filling of the breathing bag and its washout during oxygen compression. Instead of one, three sensors of the oxygen partial pressure were applied in this type of *CCR*.[37] The indications of each one were displayed on the diver's control board. Additional sensors, not shown in Figure 3.20, served also as sensors of low or high partial pressure signal in the breathing bag. They were specifically calibrated so that their sensitivity should be maximal at the established maximum oxygen partial pressure *IIAH* and the established minimum oxygen partial pressure *IIAL*. They signaled any exceeding of these values. Apart from measuring and transmitting the signal on the diver's board, main sensor 9 controlled the opening of the oxygen valve.

Enlarged and *large circuit mix – CCR* are complete systems supporting dives. An example of such a system is presented in Figure 3.22.

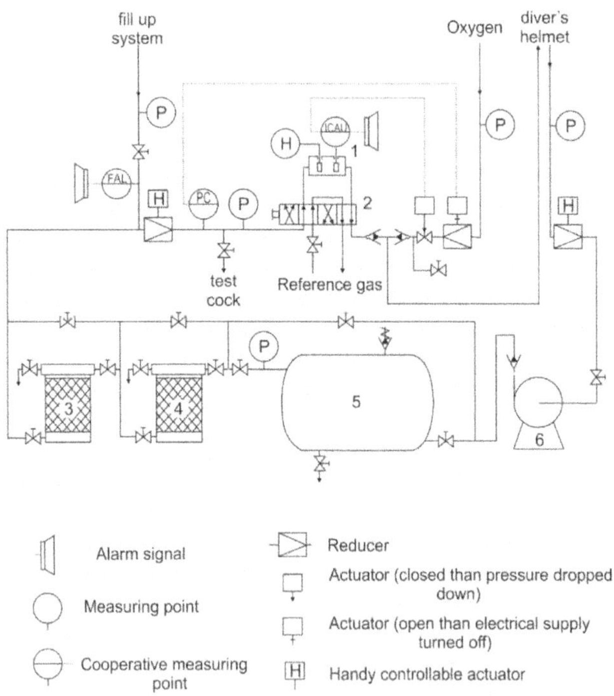

FIGURE 3.22 Supply unit of *large closed circuit of breathing medium mix – CCR*: (1) unit for measuring oxygen partial pressure and breathing medium humidity; (2) group of selecting valves for control of measuring unit in operation; (3) drier; (4) carbon dioxide absorbent; (5) storage reservoir; (6) compressor unit. *Note:* *P* = point of measuring pressure; *PC* = point of measuring pressure and automatic control; *H* = measuring point of humidity or manual control; *ICAU* = measuring of sensor current and automatic control with a signaling of low and high level; *FAL* = flow indicator.

The diver's helmet is equipped with a switch of operation mode that can be set on operation either in closed or open system. It can also be set in the position of the *mix —CCR*,[38] provided the system has one. In case of system failure, *bail – out UBA* cannot be used if the diver switches on the diver's helmet emergency valve to the open circuit system, being supplied then with emergency breathing medium.

Breathing medium is stored in the surface store room or in gas cylinders placed on the diving bell. In the example of the breathing circuit system presented, the circulation breathing medium is forced by pump 6 (Figure 3.22). The pump sucks the exhaled gas through a manually operated reducer that is being set on sucking. Then the gas is pumped to the system, from where, after purification, it is repeatedly pumped to the diver through the pump reducer.[39] The settings on the reducers on the sucking and pumping line are connected with each other and depend upon the range of the diving depth and ventilation rate required. Together with the valves of the mouthpiece assembly, they ensure proper circulation of the breathing medium. They prevent the pump from sucking out the whole breathing gas form the circuit and the breathing gas from being all suddenly pumped into the diver's helmet at too high pressure. The breathing medium, when sucked by the pump, passes through storage reservoir 5, carbon dioxide absorber 4, dryer 3, and the earlier-mentioned reducer to the measurement chamber 1. In the measurement chamber arranged on the gas line, the breathing medium humidity H is measured and the oxygen contents $ICAU$. The oxygen bypass valve is controlled by the measurement. In the event of failure, oxygen can be added manually, and the supply unit is fit to work in an emergency mode of supplying the breathing medium through the valve, which enables filling in the system. In this case, the gas exhaled by the diver is directed into the water.[40] The rate of

TABLE 3.8
Characteristics of Mix CCR

Closed circuit breathing gas mix underwater breathing apparatus *mix – CCR*	
Mix – CCR can be classified in two groups that differ with the breathing medium applied. There are two types: *oxy – CCR* and *mix – CCR*. *Oxy – CCR* is primarily used for military purposes. *Mix – CCR* is mostly used in large or enlarged circuit version in saturation diving operations.	
Disadvantages	• Large circuit *mix – CCR* requires an extensive and technically complicated system of gas supply operated by highly skilled personnel. • Small circuit *mix – CCR* is a very complicated construction and requires expert personnel. Due to the lack of good oxygen partial pressure sensors, *mix – CCR* can be unreliable. • *Oxy – CCR* should be used by especially trained divers of special predisposition due to the possibility of carbon dioxide poisoning. • Due to closure of the breathing gas circuit in all *CCR*-type constructions, accumulation of contaminants can occur.
Advantages	• Due to economy of the *CCR*, they are the most economical types of *UBA*. • *CCR* offers the possibility of rapid and optimum adjustment of its working parameters to each of the diving depth.

breathing gas dehumidification is operated manually by the dryer bypass valve. The appropriate alignment of this valve with the dryer inlet valve separates the stream of breathing gas that enables control of flowing breathing gas dehumidification. The system must offer a possibility of calibrating its measuring instruments with standard gases[41] and periodical breathing gas sampling[42] for gas analysis. The characteristics of *CCRs* are given in Table 3.8.

3.4.3 Conclusion

Beyond any doubt, *air – SCUBA* is the most often used construction of those that have been described. The intensive work on these constructions has led to the development of a *UBA* which is highly reliable despite its relative complexity.

The other types of *UBA* described exemplify the elementary production of specialized companies. Some types of *UBA* have never been implemented or have been manufactured occasionally and are not in common use. Undoubtedly, except the *air – SCUBA*, constant metering *premix – SCR*,[43] *mix – CCR*,[44] and *oxy – CCR*[45] are the most often used *UBAs*.

Despite their unquestionable disadvantages, *premix – SCRs* are relatively simple designs. This makes them technically reliable. Recently, great interest has been focused on special construction diving apparatuses. *Premix – SCR* is employed for independent deep diving operations. *Technical diving* is a special hazardous kind of diving that attracts wide groups of divers looking for strong sensations from all over the world.

NOTES

1 Not of any model in particular.
2 Classic.
3 Courtesy of the *Military Diving School – Gdynia*.
4 Normobaric diving suits, single-atmosphere diving suits.
5 Or complemented.
6 Although installation of the so-called pilot valves is claimed to serve as an additional reduction stage by some.
7 Construction with demand valve of two separate pressure reduction stages.
8 Used as escaping *UBA*.
9 Or in integral set of pressure cylinders 15.
10 Second stage of pressure reduction.
11 For *SCUBA* version.
12 Reducer 16.
13 Selector valve 3 is now positioned different from that shown in Figure 3.8.
14 Indicated by the manufacturer of the apparatus.
15 Semi-open.
16 Escape supply.
17 Set differently than shown in Figure 3.10.
18 Certain fluctuations of the pressure in the breathing area should be expected according to the positioning of the *SCR* on the diver's body and his very position in the water.
19 Ventilation.
20 For the design described earlier.

21 Alternatively closed/semi-closed demand breathing apparatus (Figure 3.12).
22 It is its intention to use *CRABE* with *Trimix* to deeper dives.
23 For example, for $DC-55$ greater than $25mH_2O$.
24 For example, for *CRABE* longer than $30min$.
25 External supply unit marked as A in Figure 3.15.
26 Partial pressure.
27 $SC-SCR$ emergency supply.
28 Just like in the case of the *SCR* previously discussed.
29 Is used at work with hose supply.
30 At *SCUBA* operations.
31 Different forms of soda lime and baralime are most often used.
32 Sodium, potassium, or lithium superoxides.
33 Enlarged.
34 Or other submersibles.
35 Via, for example, the diving bell.
36 For example, helium.
37 Similar to $mk15/16mix-CCR$.
38 $bail-out\ SCR$ for *CCBS*.
39 Supply reducer.
40 Open circuit work – after the appropriate switch of the valves in the diver's helmet.
41 Valve 2.
42 Test cock.
43 Used during dives beyond the saturation zone, especially dedicated for *MCM* divers.
44 Used during saturation deep diving.
45 Military application.

REFERENCES

Bachrach A.J., Desiderati B.M., Matzen M.M. 1988. *A pictorial history of diving*. San Pedro: Best Publishing Co. ISBN: 0-941332-09-8.

Bennett P.B., Elliott D.H. 1993. *The physiology and medicine of diving*. London: W.B. Saunders Company Ltd. ISBN: 0-7020-0538-X.

Brubakk A.O., Neuman T.S. 2003. *Bennett and Elliott's physiology and medicine of diving*. London: Saunders. ISBN: 0-7020-2571-2.

Clarke D.W. 1992. The history of breathing apparatus and current state of the art. In *Lung and physiology and divers breathing apparatus*, ed. Flook V., Brubakk A.O. Aberdeen: BPCC-AUP Aberdeen Limited. Proceedings from the International Workshop, Ballater, Scotland, 1–4 November 1991, Department of Biomedical Sciences Marshal College.

Frånberg O. 2015. *Oxygen content in semi-closed rebreathing apparatuses for underwater use: measurements and modeling*. Stockholm: School of Technology and Health. ISSN: 1653-3836.

Gussmann J. 1984. *Człowiek zdobywa głębiny*. Gdańsk: Wydawnictwo Morskie. ISBN: 83-215-4047-3.

Haux G. 1968. *Diving technics*. Lübeck: Drägerwerk AG.

Haux G. 1982. *Subsea manned engineering*. London: Bailliére Tindall. ISBN: 0-7020-0749-8.

Hiszpański J. 1966. *Samodzielne, powietrzne aparaty nurkowe*. Gdańsk: Wydawnictwo Morskie.

Kłos R. 2000. *Aparaty Nurkowe z regeneracją czynnika oddechowego*. Poznań: COOPgraf. ISBN: 83-909187-2-2.

Kłos R. 2008a. Classification of the underwater diving equipment. *Polish Maritime Research.* 55, 80–85.

Kłos R. 2008b. Principles of work of different types of underwater breathing apparatus. *Polish Maritime Research.* 58, 72–84. DOI: 10.2478/v10012-007-0100-3.

Kłos R. 2021. *Ventilation of normobaric and hyperbaric objects*. Boca Raton, FL: CRC Press (Taylor & Francis Group, LLC). ISBN: 978-0-367-67523-3 (hbk); ISBN: 978-0-367-67524-0 (pbk); ISBN: 978-1-003-13164-9 (ebk).

Rawlis J. 1989. The history of commercial, military and sport diving. *Trans.IMarE.* 10, 161–170.

US Naval Sea Systems Command 2016. *US Navy diving manual.* Washington: The Direction of Commander, Naval Sea Systems Command. SS521-AG-PRO-010 0910-LP-115-1921.

4 Breathing Medium

Artificial breathing medium diving means mixed-gas diving.[1] It can be found in the literature dealing with liquid breathing species. They, however, will not be taken into consideration in this book. *Mix-diving* refers to diving operations that are conducted using a breathing medium other than air. The medium may consist of nitrogen and oxygen in proportions from those found in atmospheric air. The breathing medium can be a gas mixture or pure oxygen. Mixed-gas diving does not exclude the use of air, which may be used in some phases of mixed-gas diving or as emergency medium. The use of *SCRs* in diving operations requires using mixed gases.

4.1 HISTORY

Surface air–supplied classic diving equipment was invented in the nineteenth century by August Siebe. In 1860, Benoit Rouquayrol and Auguste Denayrouze developed an independent-air open circuit *UBA*; however, the invention was not disseminated successfully. In 1876, Henry Fleuss began developing $oxy-CCR$ (Bachrach, Desiderati, and Matzen, 1988). Thus, pure oxygen became the first artificial breathing medium. The practical limit 93 mH_2O for air diving beyond the saturation zone was first established in 1915 (USN, 2016). As a result of nitrogen narcosis and a high density of the breathing medium, the diver was barely able to work at this depth for even a very short time – less than 10 *min*. A few years later, it was reported that there were some attempts to apply air in diving operations at greater depths, generally considered as dangerous. Probably, Brent Gilliam was the first man to set up the depth record of 140 mH_2O using air in 1990. After a short exposition period at this depth, he had some problem solutions of the simplest arithmetical formulas. Gilliam prepared himself to find the solutions of these tests during the arduous and time-consuming adaptation training. In one year, he had no break in diving longer than 6 *days*. He had performed 627 dives, of which 103 were at a depth greater than 90 mH_2O (Palmer, 1994).

Elihu Thomson theorized that helium might be used as a component of a breathing medium. In 1924, the Bureau of Mines performed a series of experiments with the use of helium–oxygen mixtures as a breathing medium. The initial work was done at the Bureau of Mines Experimental Station in Pittsburgh, Pennsylvania. In 1927, the US Navy shifted further experiments to Washington. The first experiments were performed on test animals and humans as well. In 1937, a diver who was wearing deep-sea classic diving equipment and helium–oxygen breathing mixture was compressed into a habitat[2] to a simulated[3] depth of approximately 152 mH_2O. The diver was not informed of the actual depth, and when asked to make an estimate of the depth, the diver reported that it felt as if he were at approximately 30 mH_2O.

DOI: 10.1201/9781032649207-4

FIGURE 4.1 Salvage operation of the USS *Squalus*, and the diving and rescue equipment used during the salvage operation: *left*, the author and rescue diving bell used during the salvaging of the USS *Squalus*, Naval Base Coronado, San Diego; *right*, diving equipment used during the salvaging of the USS *Squalus*, Naval Base Coronado, San Diego.

Source: Kłos (2000).

The first practical test of using a helium–oxygen mixture as a breathing medium was done in 1939 when the *USS Squalus* was salvaged from the depth of 74 mH_2O (Figure 4.1).

In 1940, Chris Lambertsen, from the *US Navy*, proposed using oxygen–air mixtures[4] or oxygen–helium mixtures for filling up *UBAs* and performing dives to oxygen toxicity limits (see Figure 3.5 previously and Figure 4.2). In 1957, André Galerne used 50% oxygen in nitrogen in diving operations performed in France at the depth of $H \in [189, 20]$ mH_2O. These experiments are regarded as the first practical underwater works with the use of nitrox Nx (Betts, 1994). However, now it is well-known that in 1915, Dräger manufactured the first nitrox-supplied $UBA\,DM-40$ type (see Figure 1.1). In 1959, nitrox was introduced in the US Navy.

It was not only military divers, however, that used mixed gases in diving. In 1937, Engr. Max Gene Nohl (Figure 4.2) reached the depth of 128 mH_2O in Lake Michigan using Hx.

In 1946, a civilian diver, Jack Brown, carried out simulated diving to the depth of 168 mH_2O using Hx (Figure 4.2). In other countries, where the availability of helium was restricted,[5] divers performed experiments with the use of other gas mixtures, that is, Swedish engineer Arne Zetterström worked on the application of a hydrogen–oxygen mixture[6] as a breathing medium. In 1945, after the successful completion of diving tests at the depth of 110 mH_2O, Zetterström reached the depth of 160 mH_2O. Unfortunately, as a result of a misunderstanding of his surface-support personnel, he was brought to the surface too rapidly. He died as a result of this experiment.

In 1959, Hans Keller reached the depth of 120 mH_2O. In 1960, he reached the depth of 156 mH_2O, with a sensationally short decompression time. In 1961, he carried out a series of simulated dives[7] and reached the depth of 300 mH_2O (Bühlmann, 1984). Keller performed dives in the base of French divers in Toulon. For his next

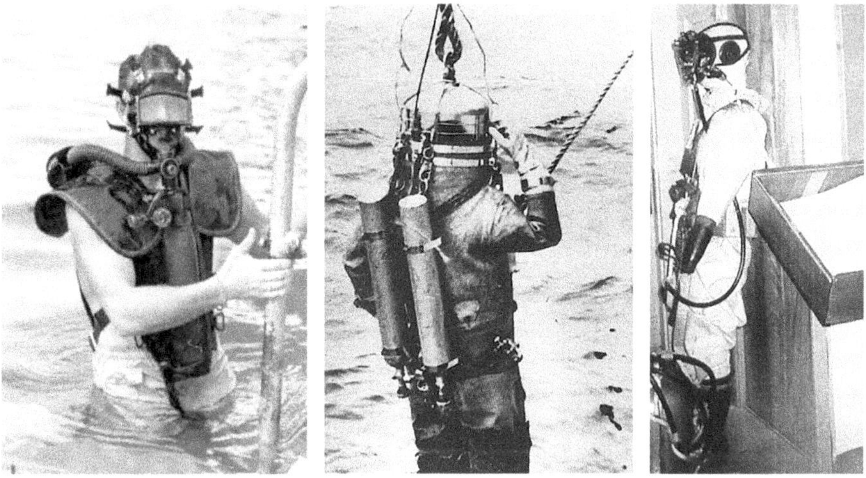

FIGURE 4.2 *Left*, Lambertsen's *UBA*; *center*, Max Gene Nohl during experimental descent; *right*, mask designed by Brown.

Source: USN (2016); Bachrach, Desiderati, and Matzen (1988); and Kłos (2000), respectively.

experiments, Keller employed volunteers. In 1962, after the first successful dives,[8] unfortunately, the diving bell lost leak-tightness, and Keller's companion and the attendant diver died (Gussmann, 1984).

Saturation diving techniques were developed in the late 1950s (Miller and Koblick, 1984) and can be recognized as a new chapter in mixed-gas diving. Different gas mixtures were used in saturation diving experiments. They were followed by diving experiments performed beyond the saturation zone[9] that have been described earlier. Actually, it is the basic way for diving at great depths[10] and long-duration diving operations at medium and large depths. Hx is most often used in saturation diving.

4.1.1 Polish Research

Between the 1960s and the 1970s, the Polish Navy conducted research on the implementation of $automix - SCR$. The research on $automix - SCR$ implementation was ordered by the Rescue and Salvage Operation Department, Polish Navy HQ. $SCRs$ were designed for military purposes. Two versions of $automix - SCR$ were manufactured: $automix - SCR\,APW - 3\,CZAJKA\,SCUBA$ and $automix - SCR\,APW - 6\,KSZYK\,SCUBA$. Underwater breathing apparatuses were produced by the manufacturer of rescue equipment and miner's lamp FASER in cooperation with the Maritime Medicine Department; the Military Medical Academy; Capt. (Navy) Prof. Tadeusz Doboszyński, MD, former head of the Diving Gear and Underwater Technology Department, Naval Academy; and former head of the department Capt. (Navy) Medard Przylipiak, MSc.

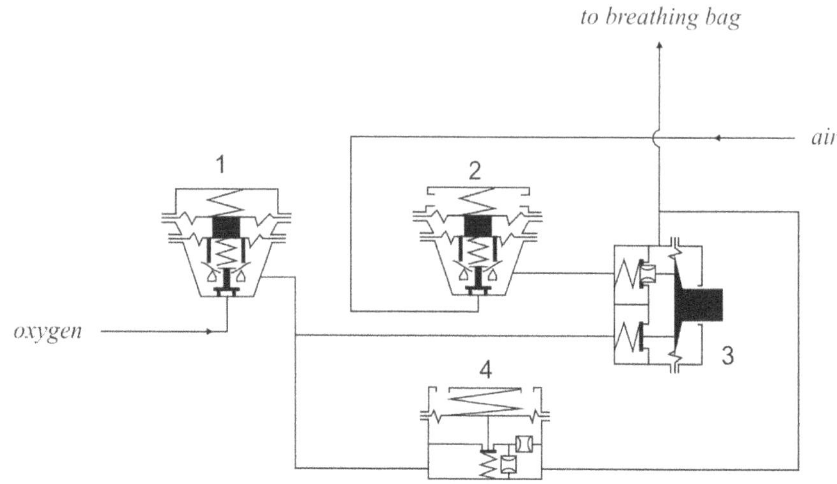

FIGURE 4.3 Principle of APW-3 and APW-6 SCR operation: (1) oxygen pressure reducer; (2) air pressure–reducing cutoff valve (air dosage valve); (3) metering bypass valve; (4) depth valve.

The principle of operation of both $automix-SCRs$ was the same, but the number of breathing bags and the construction of the carbon dioxide absorber differed in both constructions. $SCR\,APW-6\,KSZYK\,SCUBA$ was a two-bag version, and $SCR\,APW-3\,CZAJKA\,SCUBA$ a one-bag version of SCR (see Figure 3.16b–c). The whole construction of $APW-6$ was contained in a knapsack made of plastic. Oxygen and air cylinders were fitted in both SCR versions. Oxygen metering in closed circuits of both $SCRs$ was maintained at a level of 1.6 $dm^3 min^{-1}$ to a depth of $8\,mH_2O$ because oxygen reducer 1 worked as a simple reducer, maintaining constant oxygen overpressure (Figure 4.3 and Figure 4.4).

Depth valve 4 was closed,[11] and oxygen was metered to the system by one metering nozzle. The metering nozzle worked within its supercritical flows because reduced pressure was well adjusted. Therefore, oxygen stream was maintained at a constant level. Valve 2 in Figure 4.3 was closed because the springs of the air pressure–reducing cutoff valve were appropriately adjusted to the depth of $8\,mH_2O$ so that the air was not metered to the system (Figure 4.4). Valve 2 was opened after the depth limit of $8\,mH_2O$ was exceeded, then metering bypass valve nozzle 3 started air metering. It is marked in Figure 4.4a–c. Depth valve 4 was also opened after the depth limit of $8\,mH_2O$ was exceeded, then two metering nozzles of assembly 4 started oxygen metering. It caused a jump in metering to 2.2 $dm^3 min^{-1}$ (Figure 4.4c). As oxygen overpressure was maintained at a constant level by reducer 1 while pressure increase in the breathing system due to increasing depth of diving caused an increase in back pressure of the oxygen metering nozzles (Figure 4.4e), the nozzles therefore worked within their subcritical flows. This caused the oxygen metering to the system to decrease because of the increased depth of diving. The effect is shown in Figure 4.4c–d. The design of the air dosage valve should be discussed separately (Figure 4.5).

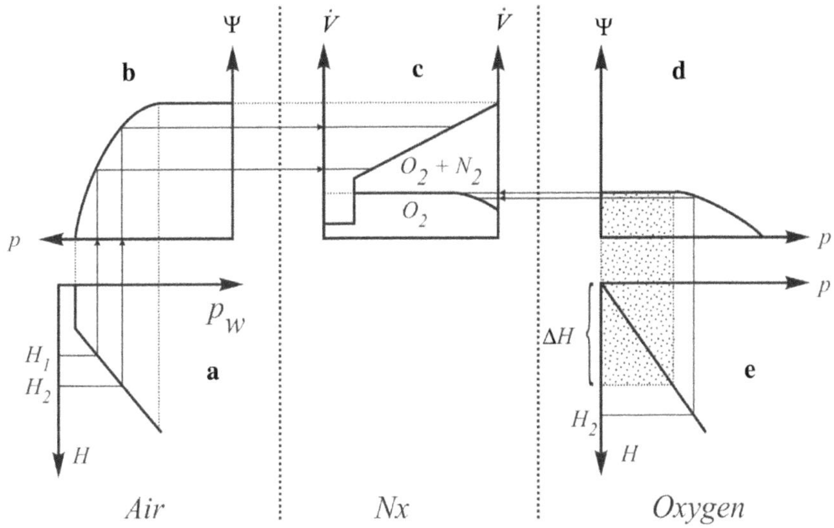

FIGURE 4.4 Principle of APW-3 and APW-6 SCR operation: (a) the relationship between the pressure exerted on the pressure-reducing cutoff valve and the depth of diving; (b) air metering as the function of the pressure exerted on the pressure-reducing cutoff valve; (c) oxygen and air metering as the function of the depth of diving; (d) the relationship between oxygen metering (after increase in metering value) and the depth of diving; (e) the relationship between hydrostatic pressure and the depth of diving.

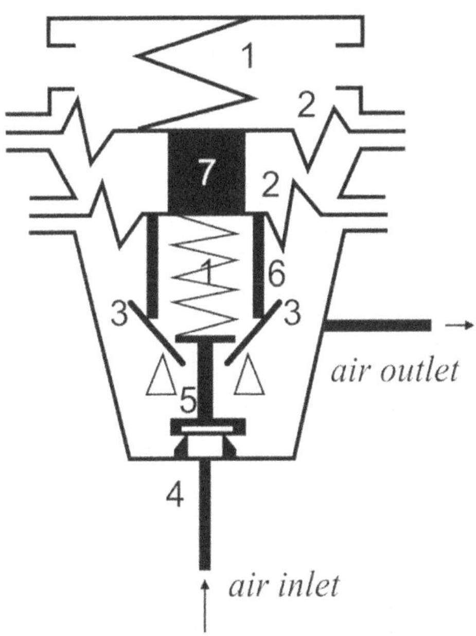

FIGURE 4.5 Principles of operation of air dosage valve: (1) spring; (2) diaphragm; (3) levers lifted valve head; (4) valve seat; (5) head; (6) push rods of the valve head; (7) pusher.

As mentioned earlier, the valve performed two functions:

- The depth valve cut off airflow below the depth of $8\,mH_2O$.
- An air pressure regulator reduced nozzle supply pressure in the function of the current diving depth.[12]

A diagram of the air dosage valve operation is presented in Figure 4.6.

Two diaphragms 2 are a part of the pressure-reducing cutoff valve. The pressure between the diaphragms equals atmospheric pressure p_0. External forces exerting pressure on the diaphragms are reciprocally transmitted by pusher 7. Force exerted from the side of the water on the first, larger diaphragm comes from the force exerted by the upper spring F_s and the water ambient pressure. Force exerted on the lower, smaller diaphragm comes from the force exerted on the downward spring $F_s^{'}$ and air outlet pressure (Figure 4.6). The ambient pressure makes diaphragms move downward and, therefore, displace pusher rods 3 (Figure 4.6). It makes the valve head lift. The formula describing air outlet pressure p_w is given in Figure 4.6.

The theoretical gas mass rate flowing out of the nozzle[13] can be given by the following equation (Kłos, 2015):

$$\dot{m} = A_0 \cdot \Psi \cdot \sqrt{\frac{p_1}{v}} \tag{4.1}$$

where $\dot{m}$ – mass rate of gas; A_0 – surface area of the orifice; p_1 – the nozzle supply pressure; v – molar volume of gas before the nozzle; and Ψ – the outflow number.

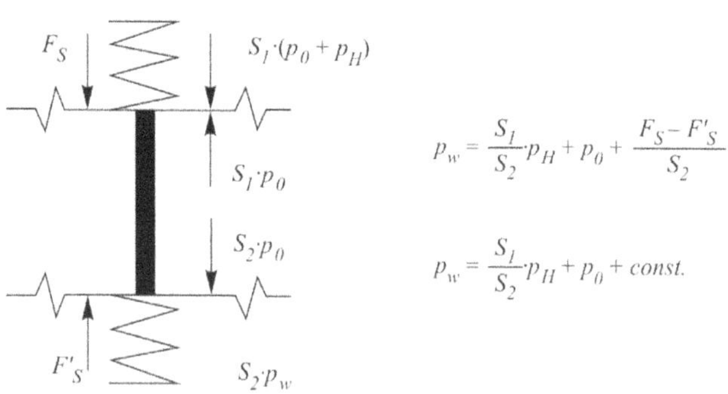

FIGURE 4.6 The force distribution on the diaphragms of the air dosage valve. *Note:* F_s = force exerted by the upper spring; $F_s^{'}$ = force exerted by the low spring; S_1 = surface area of the upper diaphragm; S_2 = surface area of the upper diaphragm; p_w = outlet air pressure; p_0 = atmospheric pressure; p_H = overpressure at a diving depth.

Number Ψ is defined as follows:

$$\Psi = \begin{cases} \sqrt{2 \cdot \frac{\kappa}{\kappa - 1} \cdot \left[\left(\frac{p_2}{p_1} \right)^{2/\kappa} - \left(\frac{p_2}{p_1} \right)^{\kappa+1/\kappa} \right]} = f\left(\frac{p_2}{p_1}; \kappa \right) for \beta \le \beta_c \\ \sqrt{\kappa \cdot \left(\frac{2}{\kappa + 1} \right)^{(\kappa+1)/(\kappa-1)}} \quad for \beta > \beta_c \end{cases} \tag{4.2}$$

where $\kappa -$ adiabatic exponent; $\beta -$ ratio $= p_2 / p_1$ of the pressure behind p_2 and in front of p_1 of the nozzle; $p_1 -$ pressure in front the nozzle (supply pressure); $p_2 -$ pressure behind the nozzle back pressure; $\beta_c -$ critical pressure ratio $\beta_c = \left(\frac{2}{\kappa + 1} \right)^{\kappa/(\kappa-1)}$.

The relationship $\Psi = f(\beta)$ is called the nozzle characteristics. The characteristics of the SCR nozzles are presented in Figure 4.4b and Figure 4.4d. According to the equation in Figure 4.6:

$$\beta = \frac{p_H + p_s}{\frac{S_1}{S_2} \cdot p_H + p_0 + const} \tag{4.3}$$

where $S_1 -$ surface area of the upper diaphragm; $S_2 -$ surface area of the upper diaphragm; $p_0 -$ atmospheric pressure; $p_H -$ overpressure at the diving depth $const = \frac{F_s - F_s^{'}}{S_2}$.

For an oxygen nozzle, it can be written as follows: $\beta = \frac{p_H + p_0}{p_z}$, where p_z is the supply pressure of an oxygen nozzle. The equation for an oxygen nozzle can change as follows: $\forall_{\beta < \beta_c}\, p_H \uparrow \wedge \beta \downarrow \Psi \downarrow \Rightarrow \dot{V} \downarrow$, where $\downarrow$ denotes the value decreasing and $\uparrow$ denotes the value increasing.

The design of the air dosage valve should enable an increase[14] in air rate metering due to the increasing depth of diving. Air and oxygen streams should be established at a level which ensures maintaining the oxygen partial pressure within certain limits. The preceding analyses are presented in Figure 4.4. The areas marked as (d) and as (e) show the working area of the oxygen metering nozzle within its supercritical flows $\beta > \beta_c$, which means constant metering. Figure 4.4c shows the relationship between metering rate $\dot{V}$ for $automix - SCR\,APW - 3CZAJKA\,SCUBA$ and $automix - SCR\,APW - 6\,KSZYK\,SCUBA$ and diving depth H.

Making the previously described SCR pre-production batch was a great success from a technical and medical point of view. However, research on these underwater breathing apparatuses has not been completed. The Polish Navy removed these *UBAs* from its inventory. Between the 1980s and the 1990s, some attempts were made to modernize *automix – SCR APW – 6 KSZYK SCUBA* type. Finally, *premix – SCR APW – 6M SCUBA* was designed. The same group of Polish scientists began research on the problems concerned with automix SCR. Following the unexpected death of Capt. (Navy) Medard Przylipiak, MSc, Cmdr. Marian Pleszewski, MSc, and Capt. (Navy) Stanisław Skrzyński, PhD, successively took over his duties at the Polish Naval Academy. They continued the research according to Polish Navy requirements. The FASER company made a pre-production batch of *SCRs*, which were employed in training divers to use special *UBAs*. Further research on using *SCR APW – 6M SCUBA* as a combat *UBA* was planned to be carried out. This, however, did not happen, and the research will probably never be completed. Some experimental results of the research on these automix SCR–related problems will be discussed here.

4.1.2 Breathing Gases

The most important reasons for using mixed gases as a breathing medium in diving are as follows:

- Oxygen toxicity influence on nervous system
- Oxygen toxicity influence on the lung tissue
- Nitrogen toxicity
- Increase in density due to increased diving depth
- Optimization of diving and decompression
- Increase in therapy effectiveness of decompression sickness *DCS*
- Research on and development of new optimum diving technologies, etc.

Until recently, using a breathing medium other than air had been generally considered to be something extraordinary. This opinion is wrong. To provide the answer to the question concerning the goals of mixed-gas use requires making some medical explanation. Here, only the most important technical aspects will be discussed. Oxygen and/or its gas mixtures used for treatment are needed even during typical air diving. Oxygen is the basic and acknowledged remedy used in case of decompression sickness *DCS* symptoms in divers. Oxygen and its mixtures can be used in hyperbaric *DCS* treatment. The diver placed in a hyperbaric chamber breathes compressed oxygen or oxygen mixtures. The possibility of application of *DCS* treatment in water in case of lack of a chamber is a subject of spirited discussions. Atmospheric pressure oxygen can be used as well; however, its action is less-effective. It can be used not only in therapy but, additionally, also as a supportive gas in decompression. Oxygen washes out the other gases from the human body by physical process. Oxygen is rapidly metabolized.[15] It does not have to be removed by physical desaturation. In other words, the application of oxygen leads to the optimization of diving by shortening decompression time (Bennett and Elliott, 1993; Brubakk and Neuman, 2003). Mixtures enriched with oxygen in relation to the oxygen content in the basic breathing gas are used by the diver in the moment when they can be used for the same purpose as the basic breathing gas. Oxygen-enriched mixtures can be used at a greater depth than at the depth where pure oxygen can be applied. Therefore, decompression time at deeper decompression stops can be shortened without putting the diver at risk of oxygen poisoning. Using mixed gases at the earlier phases of decompression does not exclude using oxygen at shallower depths. The technical possibilities are as follows:

- After exposure to pressure, the diver can be transported to a decompression chamber in a closed bell.[16] Decompression is carried out in a decompression chamber, where the diver can breathe gas mixtures and/or pure oxygen.
- The diver can break in water[17] decompression and then, after rapid ascent, is again compressed in a decompression chamber. Further decompression is carried out in the chamber,[18] where gas mixtures and/or pure oxygen can be used.
- The diver is subjected to in-water decompression, with the possibility of using gas mixtures and/or pure oxygen, for example, inside an open-bottom wet bell, on a diving platform, in an underslung *UBA*, or using an additional *UBA* or cylinder for breathing.

The first method is used in well-equipped professional diving operations. The second one is used as a standard procedure in mine countermeasure operations *MCM*, lately as a standard professional deep diving procedure, and as an emergency procedure today. Although the second method does not require advanced technical equipment, the diver is exposed to additional stress caused by the break in in-water decompression, as the allowable break in decompression time is $t \in [5,7]\,min$. During decompression, the diver has to rapidly ascend to the surface and then partly undressed and again compressed to the depth pressure at which the break occurred. Even exceeding slightly the allowable break time may cause even the death of the diver. At present, such decompression is abandoned; however, the procedures are still in effect as emergency and standard military method. The third method is the most often used in long-duration recreational and technical diving. It is also used in oceanography research (Gilliam and von Maier, 1992; "The twilight zone: the potential, problems and theory behind using mixed gas, surface-based SCUBA for research diving between 200 and 500 feet," *Diving for Science*, September 24–27, 1992). Also, if anybody cannot afford expensive, additional diving equipment[19] but would like to dive to greater depths, they have to use this method.

As it follows from the previous text, the use of mixed gases is common even for recreational diving. Some diving physicians recommend oxidizing before and after a dive by inhaling oxygen from an oxygen inhalator. The aim is to reduce hypoxia during a dive. Hypoxia can be caused by the following:

- Increase in breathing resistance caused by increased mechanical resistance of *UBA*, breathing medium density, pressure exerted on the chest by ambient water and diving suit, etc.

- Decrease in lung ventilation, usually caused by increased effort, pressure of ambient water and diving suit, etc.
- Increase in concentration (partial pressure) of carbon dioxide in the breathing space due to relative decrease in ventilation and the existence of dead space,[20] etc.

As a result of the occurrence of the listed causes of hypoxia, a majority of divers suffer from post-diving headache. It can be stopped after pure oxygen inhalation at the surface.

Oxygen is an indispensable component of a breathing medium. There have been some attempts to use a breathing medium in which oxygen is chemically bound to fluids. Such compounds should be easily decomposed in the human body. This idea has been explored in scientific research and has been used to develop artificial blood.

Only pure oxygen or its mixtures with other gases have been used in diving so far. However, oxygen having high partial pressure is hazardous to the central nervous system CNS. Oxygen is also dangerous to the lung tissue when breathed for a long time, when oxygen partial pressure exceeds $50\,kPa$ (Brubakk and Neuman, 2003; Kłos, 2023). It is well-known that exposure time to oxygen and its partial pressure should be limited due to the effect of oxygen toxicity on the central nervous system CNS. Pure oxygen diving should be limited with respect to its duration and diving depth. According to permissible oxygen exposures listed in $NOAA$, oxygen diving is limited to a depth of $6\,mH_2O$, in case of emergency[21] to $10\,mH_2O$. As it follows from $NOAA$, the use of air as a breathing medium in diving should be limited to a depth of $20\,mH_2O$ and to exposure of $720\,min$ due to the effect of oxygen toxicity on the central nervous system $CNSyn$.[22] The practical meaning of the preceding limitation is unimportant, as in respect of the other factors, exposure time is usually limited to lower values (Kłos, 2023).

Long-duration exposures at increased oxygen partial pressure $pO_2 \geq 50\,kPa$ may lead to pulmonary oxygen toxicity. As it follows from research, even long-duration exposures at atmospheric pressure when the diver breathes pure oxygen may be dangerous for the lung tissue (e.g., Brubakk and Neuman, 2003; Kłos, 2023). Pulmonary oxygen toxicity is particularly dangerous during saturation diving, as it usually lasts approximately four weeks. It is generally acknowledged that during saturation diving, partial pressure of oxygen pO_2 should be less than $pO_2 < 50\,kPa$. Many systems used in the world recommend diving partial pressure of oxygen to be maintained at a level of $pO_2 \in [32, 44]\,kPa$. As it follows from the preceding text, the depth limit for air saturation is approximately $15\,mH_2O$. Saturation diving at such shallow depths is unprofitable from an economic point of view. As regards pulmonary oxygen toxicity, it is necessary to use mixed gases in saturation diving.

At increased pressure, nitrogen present in air is toxic and causes narcosis.[23] It is assumed that the first symptoms of nitrogen toxicity occur during air diving at a depth of approximately $30\,mH_2O$. The symptoms are accentuated at a depth greater than $60\,mH_2O$ (Table 4.1; Edmonds, Lowry, and Pennefather, 1992).

For this reason, many countries have limited commercial air diving to a depth of $50\,mH_2O$ and recreational diving to a depth of $40\,mH_2O$ (Bennett and Elliott, 1993). As it follows from research, the individual susceptibility of divers to nitrogen

TABLE 4.1
Symptoms of Nitrogen Narcosis in Humans Caused by Partial Pressure of Nitrogen During Air Diving

Nitrogen Partial Pressure (*kPa*)	Symptoms of Nitrogen Toxicity (Nitrogen Contained in Air)
300–450	The first symptoms of nitrogen toxicity, small euphoria, delay in reaction
450–600	Drowsiness, hallucination, weakening of judgment, laughing, being talkative, losing control of self
600–800	Joyful orientation, acute weakening of intellectual skills, uncontrolled laughing or panic, fear
800	Shocking, abnormal behavior; euphoria; loss of almost all intellectual skills

narcosis may be improved through adaptation training. However, it is impossible to have an organism adapt to the conditions of nitrogen partial pressure and maintain normal intellectual skills (Narcosis, 1991; Brubakk and Neuman, 2003). This creates a real hazard. The diver is expected to control a series of technical processes, such as measuring depth, time, decompression course, etc. The diver has to control his/her behavior during a dive. Therefore, sometimes, change from air, beginning from a depth of $30\,mH_2O$, to mixed gases is considered (Rutkowski, 1990).

Other gases, regarded as inert gases, are toxic under increased pressure. Some authors show the toxicity of these gases in relation to nitrogen. Helium and neon are considered potentially four times less-toxic[24] than nitrogen, and hydrogen twice less toxic than nitrogen. However, argon is approximately twice more narcotic than nitrogen. The narcotic properties of the components of gas mixtures are given in Table 4.2 (Brubakk and Neuman, 2003).

Research carried out by the US Navy has shown that an increase in breathing air density at a depth greater than $90\,mH_2O$ causes significant breathing resistance. It makes divers unable to perform any work except work of breathing. An increase in breathing medium density leads to an increase in breathing resistance. To support breathing, the chest muscles have to perform increased amount of work. Additionally, alveolar gas exchange is impaired as a result of increased gas density (Flook, 1992). Increased density makes it difficult for the inhaled gas containing oxygen to flow via the bronchial tree to the alveolus, and for oxygen-deficient gas in the lungs to flow outside via the bronchial tree. This stops gas exchange, and the person affected starts to feel dyspnea. Consequently, increased breathing rate is needed. Increased energy consumption related to an increase in breathing rate comes from combustion reaction, where oxygen is needed.[25] In the case of low gas exchange, the diver feels oxygen deficiency. As it follows from the previous text, an increase in density contributes to a significant limitation of the diving depth. Therefore, light gas components are used in deep diving, for example, helium, which is 7 times lighter than nitrogen, or hydrogen, which is 14 times lighter than nitrogen (Table 4.2).

TABLE 4.2
Properties of Some Gases Used as Components of the Breathing Medium

Gas	Chemical Formula	Atomic Number	Molecular Weight $(g \cdot mol^{-1})$	Density (STP Conditions) $(g \cdot dm^{-3})$	Thermal Conductivity Related	Thermal Conductivity Related to Air at 0°C	Relative Partition Ratio between Oil and Water phase	Potential Narcotic Properties Related to Nitrogen	
Helium	He	2	4.003	0.179	5.95	0.015	1.7	4.26	
Neon	Ne	10	20.18	0.90	1.89	0.019	2.07	3.58	
Hydrogen	H_2	1	2.016	0.090	6.88	0.036	2.1	1.83	
Nitrogen	N_2	7	28.01	1.251	7.86	0.067	5.2	1	
Argon	Ar	18	39.95	1.784	0.68	0.14	5.3	0.43	↓
Krypton	Kr	36	83.8	3.74	0.36	0.43	9.6	0.14	Increasing of narcotic reaction
Xenon	Xe	54	131.3	5.89	0.57	1.7	20.0	0.039	
Oxygen	O_2	8	32.00	1.429	0.99	0.11	5.0		
Carbon dioxide	CO_2	–	44.01	1.977	0.59	0.34	1.6		

Solutions of numerous technical and medical problems are found through searching for optimal diving technologies. The choice of a breathing medium has an effect on the following parameters:

- Safety of diving operations
- Decompression rate
- Heat loss rate
- Communication in the system used
- Breathing comfort, etc.

The choice of breathing gases is one of the many factors that influence the safety of diving operations. The toxicity limits for the breathing gas have a significant impact on the diving technology developed. The decompression rate applied is dependent on the breathing gas composition. An increase in physical desaturation rate during decompression can be achieved by using inert gas components having low solubility or being relatively easy to remove from the diver's tissues. Helium, hydrogen, and neon are gases which are removed from tissues faster than nitrogen through the physical process. Thus, by using gases other than nitrogen as breathing gas components, the required decompression time can be reduced. The use of oxygen-enriched gas mixtures as diving and/or decompression mixtures and pure oxygen at shallow decompression stops additionally shortens the required decompression time. Decompression could be accelerated when a special sequence of breathing media is used (Keller and Bühlmann, 1965; Bühlmann, 1975; Wendling, Nussberger, and Schenk, 1999). Oxygen replaces inert gas through washing out the other gases, causing changes in physical equilibrium. Oxygen is removed faster than the other inert gases, not only by physical desaturation, but also through a chemical[26] process. It enables the diver to ascend faster to the surface and avoid remaining longer in an unfriendly water environment. As it follows from an earlier text, the use of pure oxygen or oxygen-enriched mixtures shortens the required decompression time.

Except for some advantages,[27] the use of light gas components of breathing gas mixtures also causes some difficulties. Due to the high thermal conductivity of inert gases,[28] greater heat dissipation from the human body occurs. Additional thermal protection of divers is needed when the aforementioned gas mixture components are used. In many cases, passive thermal protection[29] is not sufficient; therefore, it is necessary to use heated diving suits and to heat the inhaled breathing gas.[30]

The use of light gas components in a breathing medium makes breathing more comfortable due to a decrease in gas density. As gas density decreases, sound propagation in gaseous environment deteriorates, and so does voice communication. The speed of sound propagation in hydrogen is 3.5 times greater than in air and 3 times greater than in helium. This causes voice distortion and its low intelligibility. To improve voice intelligibility, electronic speech unscramblers should be used.

A variety of artificial breathing media are used in commercial diving operations. They can be earlier prepared or prepared according to the special mixing procedures.[31] The breathing media can be specified as follows:

- Medical, used for *DCS* in-water treatment before and post-recompression
- Operational

- Used for emergency breathing
- Bailout
- Used in compression phase before saturation plateau has been reached
- Used in case of emergency decompression, for example, accelerated decompression
- Used in case of fire in habitat
- Used in case of habitat atmosphere contamination, etc.

Common terms applied to some breathing mixtures used in diving operations are specified in Table 4.3.

In *DCS* treatment of divers who have had a diving accident, mixed gases are used. Air–oxygen therapy employed by the US Navy to *DCS* sickness improved medical care provided for military, commercial, recreational, and sport divers (USN, 2016). Air medical treatment applied earlier at present plays a role in an emergency procedure, when a diver is susceptible to oxygen poisoning (DMW, 1982). The introduction of the air–oxygen therapy made it possible to reduce high pressures and the length of time in medical treatment. According to air decompression tables, the required diagnosis-dependent pressure should be maintained at a level between depth $H \in [50,100]\, mH_2O$; however, in the case of oxygen air decompression tables, $H \in [50,100]\, mH_2O$. According to the air decompression tables, diagnosis-dependent required treatment time t should be maintained at a level between $t \in [13,86]\, h$; however, in the case of oxygen–air decompression tables $t \in [3,39]\, h$, the air medical treatment required applying high pressures. To provide proper care for the diver during medical treatment in habitat, the physicians had to be in good condition and well-trained to dive to $H = 100\, mH_2O$. It was also necessary to keep helium stored, because compression higher than $H = 70\, mH_2O$ was performed with pure helium.

French *Comex* was the company that began to use mixed gases in therapy (Comex Marseille, 1986). Some navies try to introduce medical treatment table $CX - 30$, prepared by *Comex*, for practical use (USN, 2016). The methods described earlier bring about many objections. First of all, many cases of the return of the disease have been

TABLE 4.3
Terms Applied to Some Diving Breathing Mixtures

Term	Description of Breathing Mixture
Nitrox, *Nx*	Nitrogen–oxygen
Heliox, *Hx*	Helium–oxygen
Neox, *Nex*	Neon–oxygen
Argonit, *ArOx*	Argon–oxygen
Hydrox, H_2Ox	Hydrogen–oxygen
Trimix, *Tx*	Helium–nitrogen–oxygen
Hydreliox, H_2HeOx	Helium–hydrogen–oxygen

Note: Small (trace) amounts of nitrogen are available in heliox, hydrox, and hydreliox.

recorded after recompression therapy (Hills, 1966, 1997). Now, it seems that the application of the saturation methods will be a solution to the problem (NOAA, 2001, 2017).

Operational mixtures are used during dives starting from the saturation plateau. These mixtures contain more oxygen than does the gaseous atmosphere at a depth of the saturation plateau. This makes it possible for the diver to dive deeper than the saturation plateau without decompression stops. It is worth noticing that it is possible only within a limited range of depth and time related to the depth of the plateau. As a rule, the only allowed changes are toward greater depths. The diver is allowed to work at depths smaller than the saturation plateau only in case of emergency, and as a rule, excursions to nearby shallower depths are allowed. Sometimes, mixtures used in deep-sea diving beyond the saturation zone are referred to as operational mixtures too. To distinguish between operational mixtures and the breathing medium used in decompression, operational mixtures are also called transportation mixtures. Transportation mixtures are used during the descent and bottom stay of the diver. Sometimes it is necessary to use two operational mixtures (Bühlmann, 1975). It occurs when, at the bottom, it is necessary to use mixed gas that is hypoxic at the surface—it is caused by oxygen toxicity. Decompression breathing media replace operational mixtures from a certain depth at decompression. One or two oxygen-enriched Nitrox mixtures and pure oxygen are the most often used decompression gases.

Emergency mixtures can be administered in case of failure of the basic gas supply system. This enables the diver to safely leave the worksite; using a tight mask for breathing during adjustment of gas atmosphere in the habitat, to survive in case of a lost bell incident (switching on the independent gas supply system from the cylinders of the diving bell), etc., he/she has to take into account specificity of the diving phase. For these reasons, it is often necessary to keep several emergency breathing mixtures stored. Different emergency gas mixtures are used to supply the diver in the habitat at saturation plateau, and to supply the diver working in water starting from the saturation plateau, and to support the diver's survival in the lost diving bell.

Bailout mixtures are a special kind of mixed gases. These gas mixtures are stored in a *SCUBA*. The divers and habitats are equipped with a *bailout SCUBA*. For a short time, it allows the diver to breathe during escape. For instance, when the breathing gas supply system fails, the diver escapes from the worksite to a safe place, for example, a diving bell, habitat, or underwater vehicle. *Ponny* are small pressure cylinders usually filled with air. Such cylinders, attached to a rebreather, are used in recreational diving. Also, combat vehicle crew[32] are equipped with bailout *SCUBA* while moving across a water obstacle on the bottom. Usually, there is $oxy-CCR$, where oxygen is obtained through a chemical reaction.

Emergency gas mixtures can be supplied using leakproof masks when, for example, habitat atmosphere is contaminated or there is fire in the habitat. Additionally, in order not to support the spread of fire, when there is fire in the habitat, its gaseous atmosphere should be modified as soon as possible. According to a theory, three elements are necessary to start ignition: oxidizer,[33] increased temperature, and flammable material. The elimination of one of these elements stops the fire. Helium added to the habitat atmosphere reduces oxygen content in the gas mixture and additionally improves its thermal conductivity, which makes cooling of the habitat more efficient.

When contaminated habitat cannot be abandoned in a sufficiently short time – for example, during saturation diving, when the diver cannot move from one habitat to another – the habitat should be ventilated. Emergency gas mixture is very useful in such a case because ventilation of the habitat will not cause significant changes in the gas mixture components at the saturation plateau. In many cases, a gas mixture used to pressurize to the depth of the plateau can perform such a task. To obtain the atmosphere required at the saturation plateau, pure gases or especially composed gas mixtures[34] can be used. The remaining mixed-gas reserve can be used for replenishing the habitat gaseous atmosphere required due to gas leakage, transfer of details into or outside the habitat, separation or connection of the diving bell, etc.

Breathing media designated for emergency and emergency accelerated decompression are primarily used in saturation diving. As a rule, they are not stored as a gas mixture made especially for these purposes. To conduct this kind of decompression, there are special procedures to be applied to modify the atmosphere in a hyperbaric complex. To perform emergency decompression during diving beyond the saturation zone, air is almost always used. In some diving systems, special mixed gases are used as well. Frequently, the design of diving systems enables switching on emergency air breathing in case of failure of the basic gas supply system during decompression. In some systems, it is possible to switch back from breathing air to the basic decompression medium. Sometimes, after emergency air breathing, some systems require completing decompression with air or nitrox and the possibility of using oxygen in the last phase of decompression. In the case of failure during saturation diving, when divers should decompress as soon as possible,[35] accelerated emergency decompression is conducted. This may happen when fire has broken out in the hyperbaric complex and standard decompression cannot be conducted, and also when divers cannot abandon the complex in the rescue transfer chamber. Only some decompression systems used beyond the saturation zone are equipped with accelerated emergency decompression procedures. Frequently, interrupted decompression is available in these systems. Accelerated decompression is performed as a last resort. It may be necessary beyond the saturation zone, when the diving bell is immobilized. In such circumstances, divers face a hazard of hypothermia, and interrupted decompression cannot be performed, for example, when interrupted decompression is not allowed in this diving system, safe decompression cannot be attained in the particular phase of diving, the system is not sufficiently equipped to carry out interrupted decompression, etc.

Different breathing mixtures are used in research. This depends on the characteristics of the particular research. The use of gas mixtures containing heavy noble gases, for example, argon, is one of the most interesting problems. Argon is 40% heavier than nitrogen. It may be used to simulate breathing at greater depths[36] without the necessity of compression to high pressure. This makes research safer as regards emergency situations that may occur. A peculiar phenomenon was recorded in the course research into the use of *Argonit* as a breathing mixture. Breathing with the use *Argonit* causes respiratory oscillations. A decrease in gaseous exchange in pulmonary alveoli caused by breathing a large-density gas has been described earlier. Oxygen gas particles are not able, in some measure, to pass through the bronchial tree to the pulmonary alveoli. Similarly, the molecules of carbon dioxide and the other gases inhaled are mostly not able to flow out of the lungs. This causes a human organism to activate a breathing mechanism that has not been known up to that moment. The chest begins low-frequency oscillations, several or a dozen *Hz*.

Its energy is transferred to gas molecules, and it accentuates their motion within the bronchial tree. It is a peculiar gaseous exchange in the course of breathing aided with a breathing gas having a significant density (Flook, 1992).

4.2 TREATMENT

The historical collection of *treatment tables TT* up to the 1980s can be found in the report by the *US Naval Medical Research Center* (Berghage, Vorosmarti Jr., and Barnard, 1978). One of the methods available in the reviewed literature is the classification agreed upon in NATO. The complete list of the *TT* used by individual countries is presented in the NATO Standards (e.g., NSO, 2020).

4.2.1 INTRODUCTION

The existing treatment tables *TT* can be divided into saturation, subsaturation, and outside saturation zone *TT* (Table 4.4). And the last mentioned *TT* can be divided into those using oxygen or other gas therapy.[37]

TABLE 4.4
Example of Treatment Tables *TT* Apportionment

Type	Example of National Tables	Equivalent US Navy Tables
Short oxygen tables	TABLE V	TT USN 5
	Goodman 1 (BE)	
	OHB (FR)	
	HBO (GE)	
	9SH (NL)	
	60 and 61 (UK)	
Long, shallow oxygen tables	TABLE VI	TT USN 6
	62 (UK)	
	Goodman 2 (BE)	
Long, deep oxygen tables	TABLE VII	TT USN 6A
	6A (mod) (GE, CA)	
	TABLES A and B (FR)	
	SND (GE)	
	63 (UK)	
Saturation or subsaturation tables	TABLE VIII	TT USN 4
	US 7	
	US 8	
	TABLE C (FR)	
	64 and 65 (UK)	
Air tables	TABLE I	1A
	TABLE II, TABLE D (FR)	2A
	52 (UK)	

Source: Berghage, Vorosmarti Jr., and Barnard (1978).

TABLE 4.5
Types of Decompression Sickness *DCS*

type I	*Limbs*. Symptoms: localized pain in joints; skin itching, bends, rashes, and mottling; lymphatic swelling; ascending weakness in the legs, hips, elbows, muscles or skin
type II	*Central nervous syndrome CNSyn*. Symptoms: confusion, anxiety, paralysis, dyspnea, chest pain, breathing difficulty, loss of conscientiousness, lack of alertness, impaired balance problems in keeping upright posture, especially keeping the spine straight
type III	*Inner ear*. Symptoms: impaired hearing, dizziness, ringing in the ears, and tinnitus or nausea; the result of the impact of pressure on keeping balance in the organs in the ear
type IV	*Avascular necrosis*. Symptoms: mechanical bone injury, structural injury, local mineralization; affects mostly long bones

Source: Wienke (2003).

The US Navy procedures constitute the most widespread treatment system[38] (USN, 2016). An interesting extension of air treatment procedures is the set of the air–helium[39] tables developed by the Russians (Военное Издательство, 1968). However, the system which can support the whole range of depths available for divers at present is the system developed by COMEX (Comex Marseille, 1986).

4.2.2 Procedures

There exist different classifications of *decompression sickness DCS*. The classification proposed by Wienke has been adopted here[40] (Table 4.5). The most common cases encompass *type I DCS* and *type II DCS*. They are caused by the formation of gas in free phase in tissues. They can be effectively treated with hyperbaric methods by compressing gas bubbles in free gas phase, as occurred in tissues, to a small volume and creating the possibility for their resorption and elimination by gas exchange through the blood into the respiratory system.

In the traditional operational diving,[41] *type I DCS* is most often caused by the formation of gas in free phase in tissues, especially in blood, adipose, and connective tissue, *whereas type II DCS* forms in nervous tissue. The blocked flow of blood by gas free phase can cause local ischemia, leading to necrosis. The occurrence of gas in free phase in tissues usually causes neurological symptoms.

There exist a few theories relating to the probable pain-inducing mechanism with regard to *type I DCS* (Kłos, 2000). It occurs when bubbles in gas in free phase reach a size big enough for the following:

- Dislocate or irritate nerve endings.
- Block capillaries, leading to ischemia and tissue necrosis, which, while excreting chemical substances, warn the brain about injuries, inducing feeling of pain generated by the brain.
- Activate the biochemical reaction of immunoglobulins production, which also stimulates pain signaling the presence of harmful substances in living cells.

The destructive actions in tissues caused by gas present in free phase are not the only pathological symptoms induced by it. Defensive reactions of an organism induced by gas in free phase do not have to be directly linked to a mechanical impact, but they also have biochemical origins of pain. This theory, called complement activation,[42] seems to be well-documented (Kłos, Konarski, and Olszański, 2004).

type II DCS can be a result of a blocked blood flow to the spinal cord, which causes stimulation or inhibiting higher brain functions to slow down.[43]

type III DCS can be caused by gas in free phase, inert gas counterdiffusion, and increased pressure of bodily fluids in the inner ear. Theoretically, anomalies occurring on the surface of cell membranes can be responsible for *type III DCS*, and gas counterdiffusion caused by occurrence of content gradients of various gases is responsible for these anomalies. One gas can defund in one direction, whereas the other in the opposite direction. Their countercurrent streams cause gas exchange disturbances on the borders of phases, which causes a change in the physical conditions of gas solubility in tissue[44] (Lambertsen and Idicula, 1975). Their mutual competing for presence in the solution can lead to displacing one of them[45] (Lambertsen, 1986). *type III DCS* symptoms can be a result of an osmotic reaction. Bodily fluids tend to balance the concentration of gases contained in it through dislocation, causing dilution of the more concentrated solution. This can entail an increase in the amount of fluids in the inner ear organs, which leads to dysfunction of the labyrinth as a consequence (Doolette and Mitchell, 2003).

It is difficult to determine to what extent *type IV DCS* is caused by wrongly performed decompression. Its formation can be influenced by the free gas phase, ischemia, embolism,[46] and a combination of these factors. Statistically, this sickness is important for professional divers and people working in caissons. Both of these groups are often exposed to high-pressure actions. The same symptoms as in *type IV DCS* are found in racing horses and alcoholics. In this case, this sickness is called aseptic bone necrosis. The mechanisms of formation of *type IV DCS* are considered theoretically, and there is no exhaustive evidence based on medical observation. These symptoms are osteonecrosis that is not induced by infection.[47] If pathological hardening of bone occurs in the central part, it does not affect vitals. If pathological changes occur at the end of a bone, where the head of one of the bones fits into the socket of another one, splinters can form and joints can be damaged. Probably, repeated fast compression and the content of dioxide can have an influence on inducing avascular osteonecrosis. Long exposition to partial oxygen pressure above 60 *kPa* can cause osteonecrosis in people predisposed to developing this sickness. The probable mechanism is associated with blocking blood flow into the marrow and living bone tissue by bubbles in gas in free phase, locating themselves in capillary vessels (Wienke, 2003). A connection has been found of pain located in bone regions with the yellow marrow, which is far less supplied by blood[48] than the red marrow. Connecting this with the generally known good solubility of inert gases in fats, it can be anticipated that the difficulty in transporting the gas from the inside of a bone can be the cause of the sickness symptoms.

type III DCS and *type IV DCS* should not be treated with hyperbaric methods using the recompression treatment. In many cases, the recompression treatment can lead to a worsened condition of the patient (Strauss, 1976). It is important, as *type II DCS* and *type III DCS* are difficult to distinguish from each other.

During saturation dives, *type I DCS* can occur after an excursion to a depth less than the plateau of saturation or during decompression upon the completion of saturation. The symptoms are pain in skeletal muscle areas and joint or joints, especially knee joints. It can be preceded by itching, a rash, or skin spots. Most often, at the beginning, increased stiffness occurs in a knee joint area, hindering movement. Then, within a few hours, the pain steadily grows in joint areas. However, divers should be able to distinguish the pain, which occurs as a result of small injuries or after some physical effort, caused by joint overload or mechanical injuries which occur during work from which the cause may be related to decompression. To this end, it is necessary to thoroughly consider the history of pain formation and its increase. It is little likely that the pain which occurred before the beginning of the decompression was caused by *DCS*. *Type I DCS* formed during an excursion from the plateau of saturation and up to 60 *min* after the excursion over the depth of the plateau of saturation should be regarded as *type II DCS*, because *type I DCS* symptoms are most often a signal of more serious complications.

The treatment of symptoms of diving sickness during saturation differs from that administered after completed decompression following saturation diving and operational diving outside the saturation zone (Kłos, 2014). Usually, the treatment methods in saturation can also be applied as the final *DCS* treatment methods following diving beyond the saturation zone, as in the system developed by Comex (Comex Marseille, 1986). The effectiveness of saturation treatment methods should be higher than the other ones, but they are long-term methods, and that is why treatment using treatment tables can be performed more effectively. Hence, the US Navy system is the most common (USN, 2011).

An almost-complete historical review of the most important trends in the progress recorded with regard to the decompression sickness treatment can be found in many places, but it is easy to find it on the internet, on the website address[49] *www.wikiwand.com/en/Hyperbaric_treatment_schedules*.

It presents a set of tables implemented in the US Navy and used by many NATO countries together with other methods.

4.2.3 US Navy

The set of implemented treatment tables by the US Navy encompasses procedures using only air[50] or air and Nx [51] and procedures using both air/ Nx and oxygen, alternately,[52] as treatment breathing agents (USN, 2016). Procedures utilizing oxygen are more effective and should be administered first. Procedures utilizing air only should be administered when oxygen is not available,[53] in emergency conditions, where the patient does not tolerate oxygen under high pressure[54] and shows symptoms of oxygen toxicity during the treatment or the occurrence of oxygen toxicity is highly probable.[55]

To treat *type I DCS*, the *TT 5 USN* is used (Figure 4.7).

If after $\tau = 10\ min$ of stop at the depth $H = 18\ mH_2O$ the sickness symptoms do not disappear, the treatment has to be continued using the *TT 6 USN* (Figure 4.8).

If after $\tau = 10\ min$ of stop at the depth $H = 18\ mH_2O$ the sickness symptoms disappear, the treatment can be discontinued in accordance with the *TT 5 USN*, or it can be prolonged by maximum two cycles $\tau_{O_2} = 20\ min$ cycle of breathing oxygen and $\tau_{air} = 5\ min$ cycle of breathing air at the depth $H = 9\ mH_2O$. Treatment using *TT 6 USN* can be extended by two $\tau_{O_2} = 20\ min$ cycles of breathing oxygen and

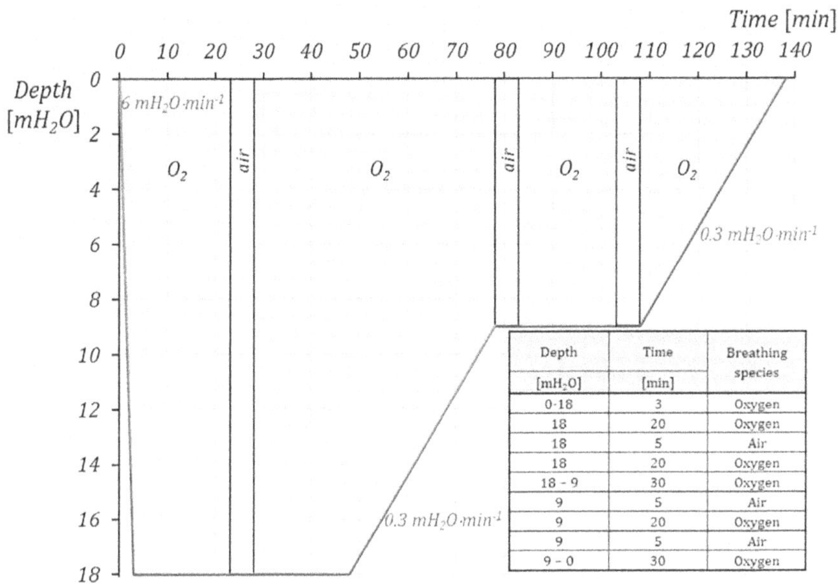

Depth [mH₂O]	Time [min]	Breathing species
0-18	3	Oxygen
18	20	Oxygen
18	5	Air
18	20	Oxygen
18 - 9	30	Oxygen
9	5	Air
9	20	Oxygen
9	5	Air
9 - 0	30	Oxygen

FIGURE 4.7 Treatment table, *TT* 5 *USN*.

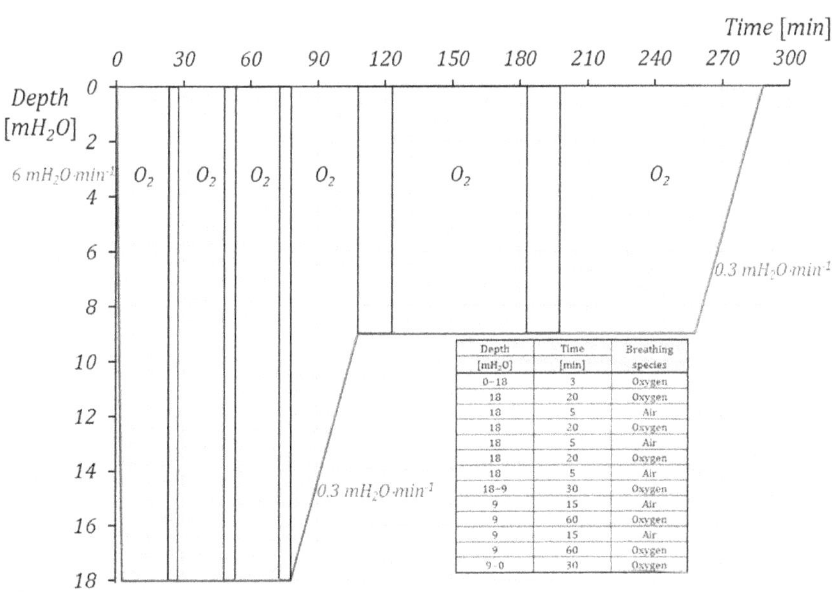

Depth [mH₂O]	Time [min]	Breathing species
0–18	3	Oxygen
18	20	Oxygen
18	5	Air
18	20	Oxygen
18	5	Air
18	20	Oxygen
18	5	Air
18–9	30	Oxygen
9	15	Air
9	60	Oxygen
9	15	Air
9	60	Oxygen
9–0	30	Oxygen

FIGURE 4.8 Treatment table, *TT* 6 *USN*.

$\tau_{air} = 5\ min$ of breathing air at the depth $H = 18\ mH_2O$ and two $\tau_{O_2} = 20\ min$ cycles of breathing oxygen and $\tau_{air} = 5\ min$ breathing at the depth of $H = 9\ mH_2O$. More serious cases, such as *Type II DCS*, are treated following the algorithm presented in Figure 4.9.

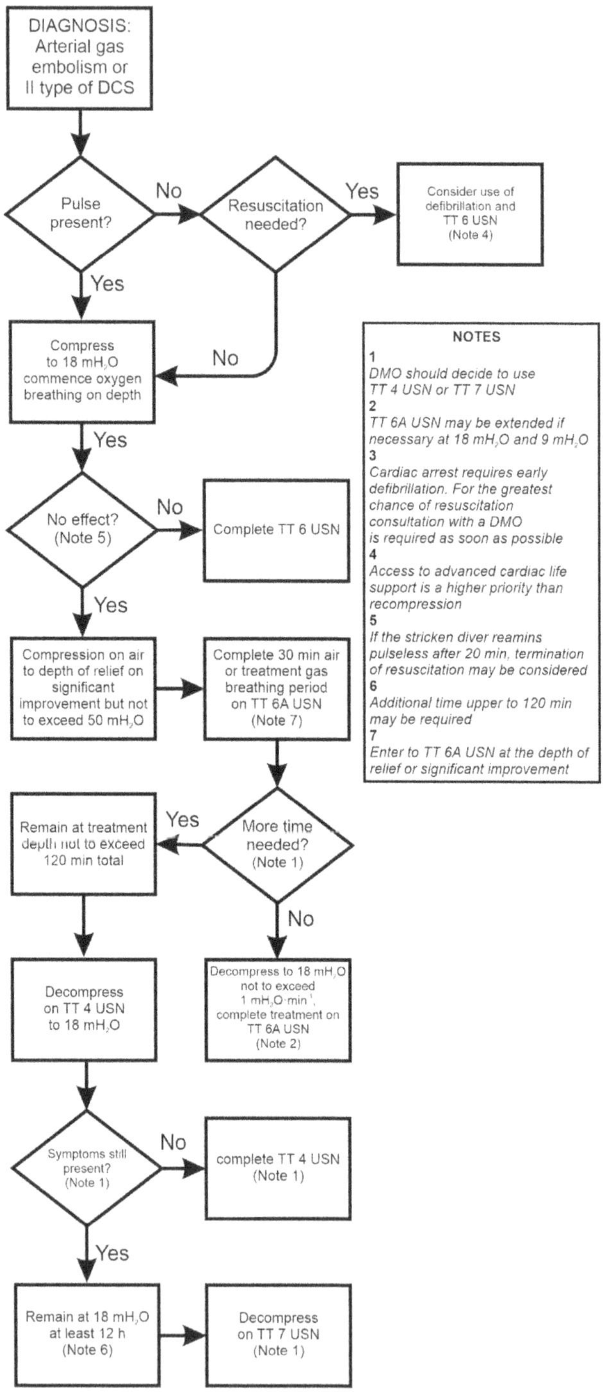

FIGURE 4.9 Algorithm for the treatment of gas embolism, *type II DCS*.

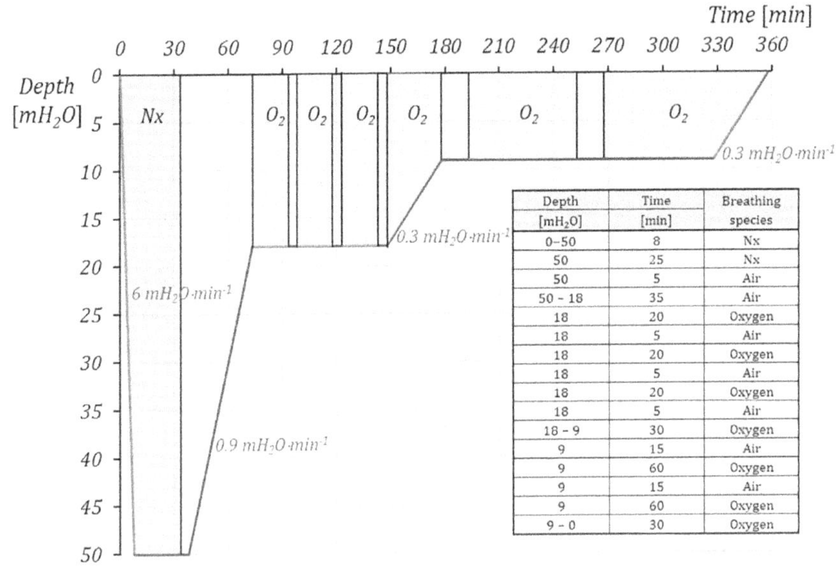

Depth [mH₂O]	Time [min]	Breathing species
0–50	8	Nx
50	25	Nx
50	5	Air
50 - 18	35	Air
18	20	Oxygen
18	5	Air
18	20	Oxygen
18	5	Air
18	20	Oxygen
18	5	Air
18 - 9	30	Oxygen
9	15	Air
9	60	Oxygen
9	15	Air
9	60	Oxygen
9 - 0	30	Oxygen

FIGURE 4.10 Treatment table, *TT* 6*A USN*.

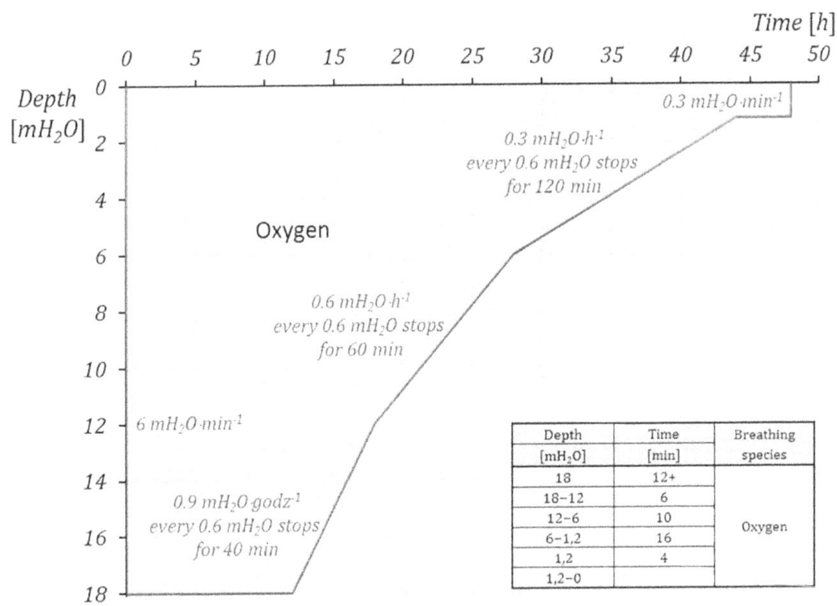

Depth [mH₂O]	Time [min]	Breathing species
18	12+	Oxygen
18–12	6	
12–6	10	
6–1,2	16	
1,2	4	
1,2–0		

FIGURE 4.11 Treatment table, *TT* 7 *USN*.

To treat gas embolism or *type II DCS* treatment tables, *TT* 4 *USN*, *TT* 6*A USN*, and *TT* 7 *USN* are used. They are shown in Figure 4.10 – Figure 4.12, and the procedure in Figure 4.13.

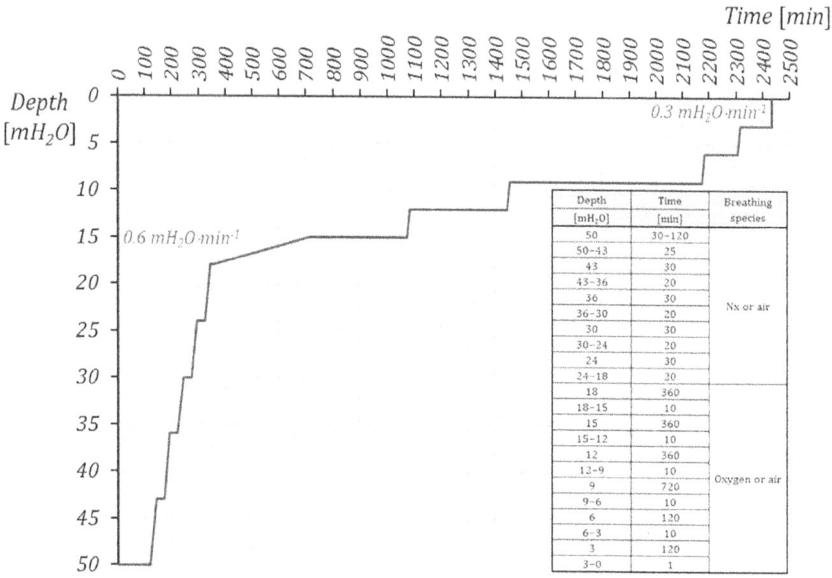

Depth [mH₂O]	Time [min]	Breathing species
50	30–120	Nx or air
50–43	25	
43	30	
43–36	20	
36	30	
36–30	20	
30	30	
30–24	20	
24	30	
24–18	20	
18	360	Oxygen or air
18–15	10	
15	360	
15–12	10	
12	360	
12–9	10	
9	720	
9–6	10	
6	120	
6–3	10	
3	120	
3–0	1	

FIGURE 4.12 Treatment table, *TT* 4 *USN*.

Air treatment table *TT* 1*A USN* is used to treat light symptoms of *DCS*, when the depth of relief does not exceed 20 mH_2O (Figure 4.14).

Tables *TT* 2*A USN* and *TT* 3 *USN* are used interchangeably to treat more serious cases: Figure 4.15 – Figure 4.16. When the depth of relief exceeds 50 mH_2O, the *TT* 4 *USN* should be used (Figure 4.12).

The supplement to the *DCS* treatment system are two *US Navy* tables: *TT* 8 *USN* and *TT* 9 *USN*. The *TT* 8 *USN* is an adaptation of the *TT* 65 *RN* used in the *DCS* treatment system by the Royal Navy and can be used when decompression stations have been omitted or the diver is thrown to the surface in an uncontrollable way. The *TT* 8 *USN* starts with compressing the diver to the depth of relief, but not higher than 68 mH_2O (Figure 4.17).

The *TT* 9 *USN* is used to treat mild *DCS* symptoms following the previous treatment (Figure 4.18).

It is also used to treat cases of carbon monoxide poisoning or poisoning by exhaust fumes.

4.2.4 Russian Air Treatment Tables

An interesting supplement system to the methods for air *DCS* treatment used by the *US Navy* are treatment tables air/*Tx* [56] proposed by *Военно Морский Флот России*, which provide the capability of reaching the pressure of relief up to the depth of 100 mH_2O (Чикер, 1968; Table 4.6–Table 4.7).

Russian treatment tables *TT I BMФ* – *TT III BMФ* use only air/O_2 as the treatment agent. The *TT I BMФ* is dedicated to treating light symptoms of

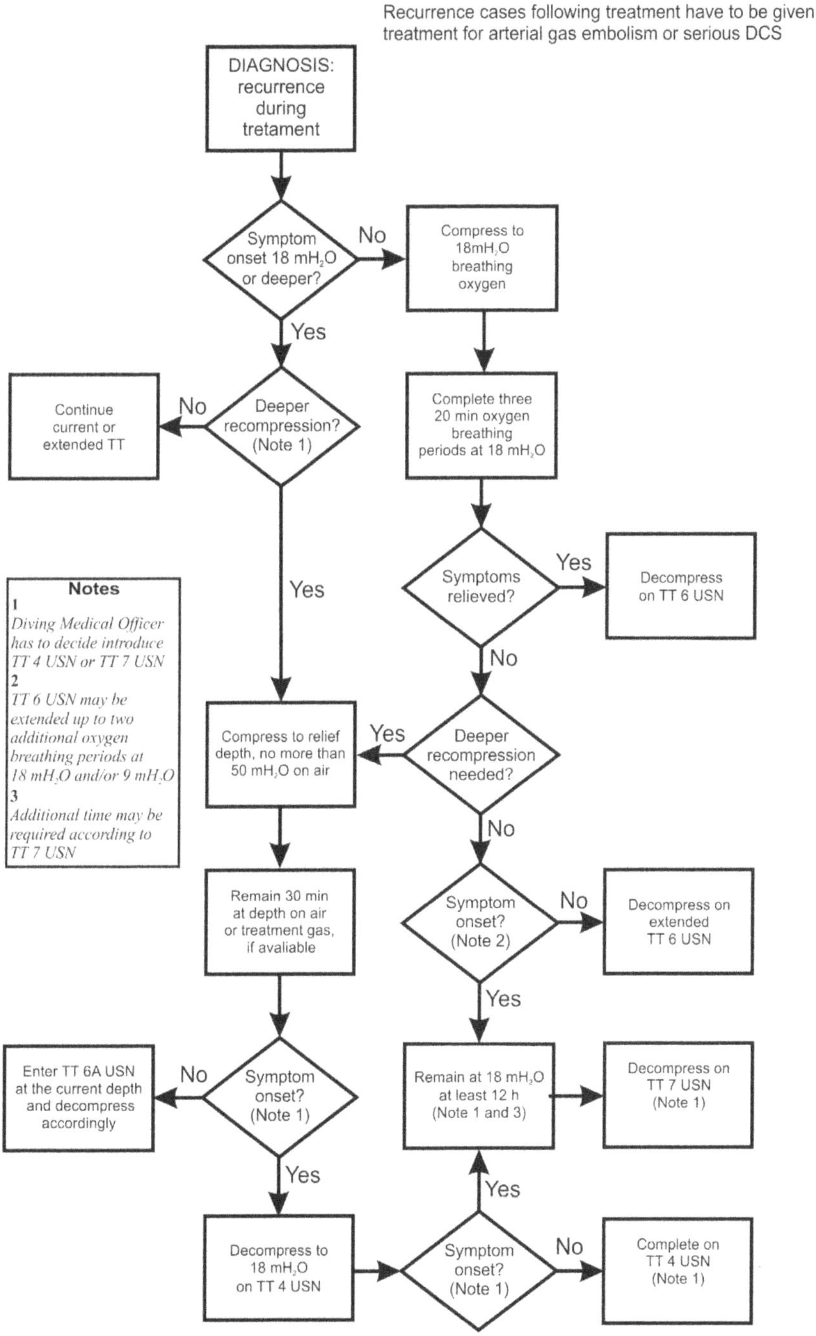

FIGURE 4.13 *DCS* recurrence treatment algorithm.

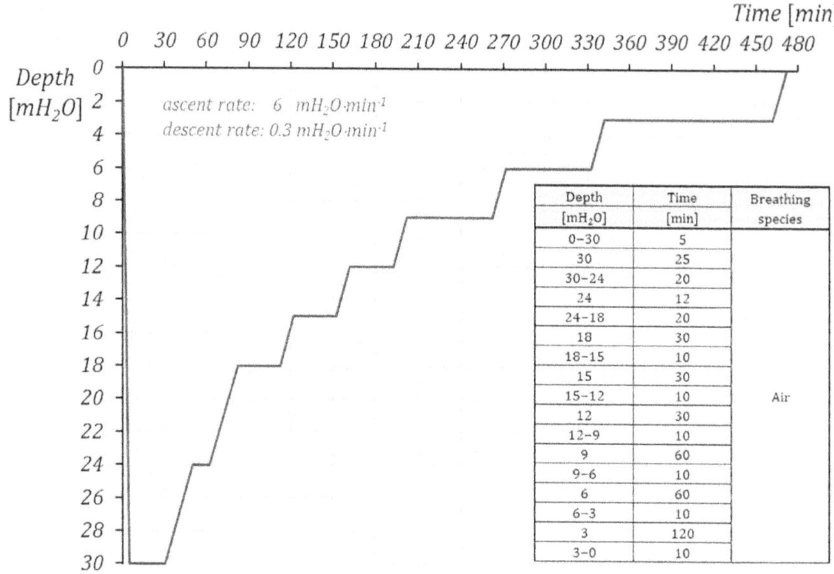

Depth [mH_2O]	Time [min]	Breathing species
0–30	5	
30	25	
30–24	20	
24	12	
24–18	20	
18	30	
18–15	10	
15	30	
15–12	10	Air
12	30	
12–9	10	
9	60	
9–6	10	
6	60	
6–3	10	
3	120	
3–0	10	

FIGURE 4.14 Treatment table, *TT* 1*A USN*.

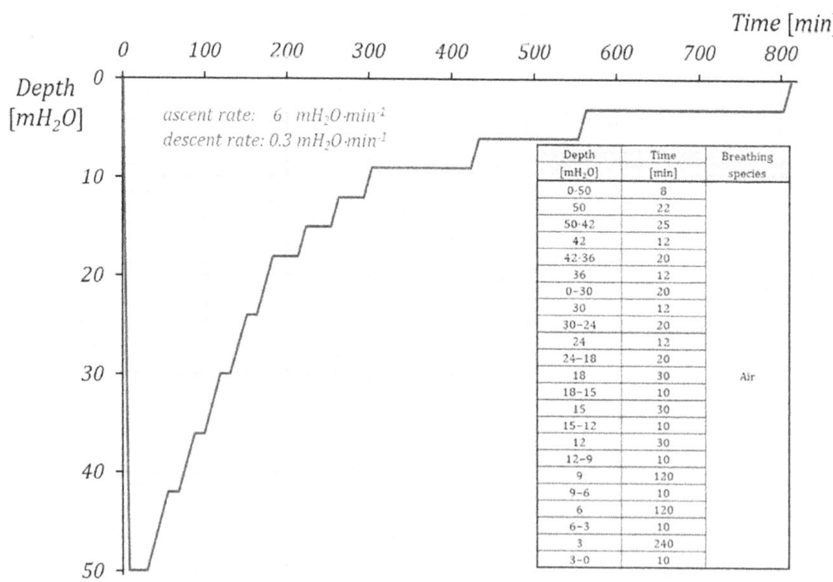

Depth [mH_2O]	Time [min]	Breathing species
0-50	8	
50	22	
50-42	25	
42	12	
42-36	20	
36	12	
0–30	20	
30	12	
30–24	20	
24	12	
24–18	20	
18	30	Air
18–15	10	
15	30	
15–12	10	
12	30	
12–9	10	
9	120	
9–6	10	
6	120	
6–3	10	
3	240	
3–0	10	

FIGURE 4.15 Treatment table, *TT* 2*A USN*.

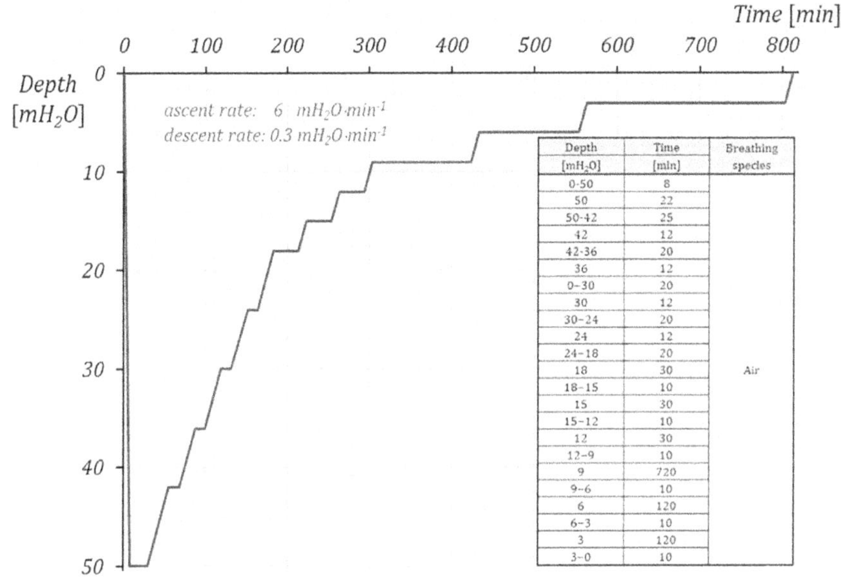

Depth [mH_2O]	Time [min]	Breathing species
0-50	8	Air
50	22	
50-42	25	
42	12	
42-36	20	
36	12	
0-30	20	
30	12	
30-24	20	
24	12	
24-18	20	
18	30	
18-15	10	
15	30	
15-12	10	
12	30	
12-9	10	
9	720	
9-6	10	
6	120	
6-3	10	
3	120	
3-0	10	

FIGURE 4.16 Treatment table, *TT* 3 *USN*.

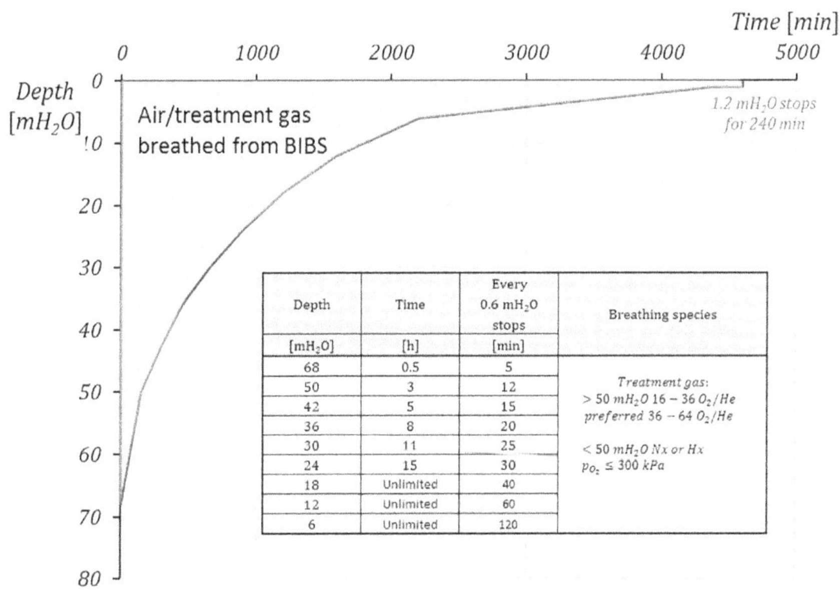

Depth [mH_2O]	Time [h]	Every 0.6 mH_2O stops [min]	Breathing species
68	0.5	5	*Treatment gas:* $> 50\ mH_2O$ 16 – 36 O_2/He preferred 36 – 64 O_2/He; $< 50\ mH_2O$ *Nx or Hx* $p_{O_2} \leq 300\ kPa$
50	3	12	
42	5	15	
36	8	20	
30	11	25	
24	15	30	
18	Unlimited	40	
12	Unlimited	60	
6	Unlimited	120	

FIGURE 4.17 Treatment table, *TT* 8 *USN* / TT65 *RN*.

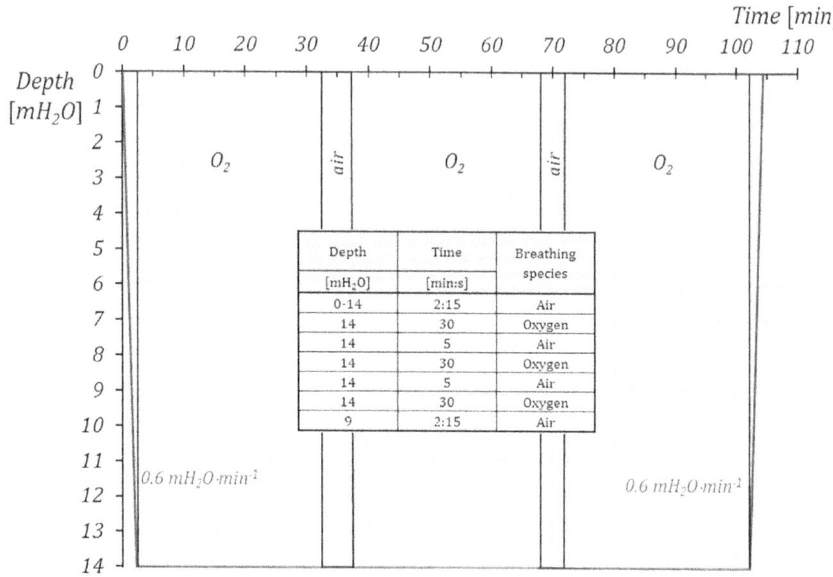

Depth [mH_2O]	Time [min:s]	Breathing species
0-14	2:15	Air
14	30	Oxygen
14	5	Air
14	30	Oxygen
14	5	Air
14	30	Oxygen
9	2:15	Air

FIGURE 4.18 Treatment table, *TT* 9 *USN*.

decompression sickness *type I DCS*, which disappear below the depth of 30 mH_2O during compression. The *TT II BMΦ* is dedicated to treating light symptoms of decompression sickness *type I DCS*, where the depth of relief occurs below the depth of 50 mH_2O during compression. The *TT III BMΦ* is dedicated to treating mild symptoms of decompression sickness *type I DCS*.

Treatment tables *TT IV BMΦ* – *TT V BMΦ* use *Tx* as the treatment agent, because from the depth of 70 mH_2O, pure helium or *Hx*, having the content of helium within $Hx : [18; 20]\%_v O_2 / He$, is used for compression. The *TT IV BMΦ* is used to treat severe forms of decompression sickness *type I DCS* and *TT V BMΦ* to treat neurological *type II DCS* symptoms.

4.2.5 *COMEX*

The treatment tables of the French firm *COMEX* constitute a base for treating the most severe *DCS* forms which can occur during both saturated and outside the saturation zone diving to the maximum operational depth of 450 mH_2O (Comex Marseille, 1986). Light cases are treated out of the saturation zone, but more severe ones require facilities for providing saturation.

Treatment tables $TT\,12_{86}$ *Cx*, $TT\,18_{86}$ *Cx*, and $TT\,30_{86}$ *Cx* are similar to the *US Navy* oxygen treatment tables (USN, 2011). $TT\,12_{86}$ *Cx* is used to treat light *DCS* symptoms, and $TT\,18_{86}$ *Cx* to treat more severe *Type I DCS* (Figure 4.19 and Figure 4.20).

TABLE 4.6
Russian Air/Oxygen Treatment Tables

Recompression Regime	Decompression Regime	Maximum Depth [mH_2O]	Max Time at Max depth (*min*)	Time to 1st Stop (*min*)	1st stop depth ([mH_2O])	Time at 1st Stop (*min*)	Time to 2nd Stop (*min*)	Next Decompression Stops [mH_2O] / Time at Next Decompression Stops (*min*): 30	28	26	24	22	20	18	16	14	12
I	А		60				20										
	Ђ	50	120	20	30	30	14								107	202	245
	В		180				10						101	166	182	202	245
	Г		360			73	2		106	112	121	130	152	166	182	202	245
II	А		60				32							65	182	203	245
	Ђ	70	90	20	50	30	28					39	164	179	196	218	262
	В		120				24			75	131	140	164	179	196	218	262
III	А					60	46				113	152	164	178	196	218	262
	Ђ	100	15	30	70	90	42		63	132	140	152	164	178	196	218	262
	В					120	40	113	123	132	141	152	164	178	196	218	262

Source: Чикер (1968).

TABLE 4.6
Continuation, Russian Air/Oxygen Treatment Tables

Recompression Regime	Decompression Regime	Next Decompression Stops (mH_2O) 10	8	6	4	2	Total time Air	O_2	Totality
		Time at Next Decompression Stops on Air (в) or O_2 (min)					(min)		(h: min)
I	А	30 + 140в	30в + 60	90в + 60	180в + 60	60в + 60	570	270	14:00
		247в	305в	368в	235в	235в	1,460	–	24:20
	Б	280в	60+35в	105в + 60	210в + 60	70в + 60	1,318	240	25:58
			328в	396в	253в	253в	2,128	–	35:28
	В	280в	60+35з	105в + 60	210в + 60	70в + 60	1,656	240	31:36
			328в	396в	253в	253в	2,466	–	40:16
	Г	280в	328в	396в	253в	253в	3,021	–	50:21
II	А	30+150в	35в +60	105в + 60	210в + 60	70в + 60	1,347	270	26:57
		280в	328в	296в	253в	253в	2,287	–	38:07
	Б	300в	60+40в	110в + 60	225в + 60	75в + 60	1,886	240	35:26
			351в	425в	271в	271в	2,754	–	45:54
	В	300в	351в	60+110в	225в + 60	75в + 60	2,500	180	44:40
				425в	271в	271в	3,057	–	50:57
III	А	300в	351в	60+115в	230в + 60	75в + 60	2,490	180	44:30
				425в	271в	271в	3,037	–	50:37
	Б	300в	351в	425в	60+230в	75в + 60	3,048	120	52:48
					271в	271в	3,285	–	54:45
	В	300в	351в	425в	271в	271в	3,487	–	58:07

Source: Чикер (1968).

TABLE 4.6
Continuation, Russian Air/Oxygen Treatment Tables

					Next Decompression Stops (mH_2O)								
					88	**86**	**84**	**82**	**80**	**78**	**76**	**74**	**72**
Recompression Regime	**Decompression Regime**	**Maximum Depth** (mH_2O)	**Max time at Max Depth** (*min*)	**Time to 1st Stop** (*min*)	Time at Next Decompression Stops When Breathing with the *Tx7*% O_2. (*min*)								
IV	A		60	6				2	7	11	11	11	12
	Ђ	100	180	8			5	20	20	25	35	35	45
	B		360	24	90	90	100	90	100	100	110	100	110

Source: Чикер (1968).

TABLE 4.7
Russian Air/ *Tx* Treatment Table Recompression Regime IV

Decompression Regime	Next Decompression Stops (mH_2O)																						
	70	68	66	64	62	60	58	56	54	52	50	48	46	44	42	40	38	36	34	32	30	28	26	24	22	20
	Time at Next Decompression Stops, When Breathing with $Tx10\%\ O_2$ (min)																			Time at Next Decompression Stops When Breathing with Air (min)						
А	8	13	18	18	19	20	29	33	35	46	54	66	79	108	136	136	144	152	159	91	94	98	102	108	117	130
Б	40	40	50	55	65	75	90	95	100	110	115	125	130	140	145	150	155	160	170	95	95	100	105	110	120	135
В	80	80	80	80	90	90	90	100	100	110	120	125	130	140	145	150	160	160	170	180	190	200	220	230	250	120

Source: Военное Издательство (1968).

TABLE 4.7
Continuation, Russian Air/ *Tx* Treatment Table Recompression Regime IV

Decompression Regime	Next Decompression Stops (mH_2O)									Total Time			Totality	
	18	16	14	12	10	8	6	4	2	$Tx7\%\ O_2$	$Tx10\%\ O_2$	air	(h)	(min)
	Time at next decompression stops when breathing with the air (min)													
А	146	162	181	202	227	255	290	334	394	60	1273	2931	71	64
Б	150	170	190	210	240	270	310	360	430	193	2010	3090	88	13
В	140	160	170	210	250	300	370	470	610	914	3470	2800	119	44

Source: Чикер (1968).

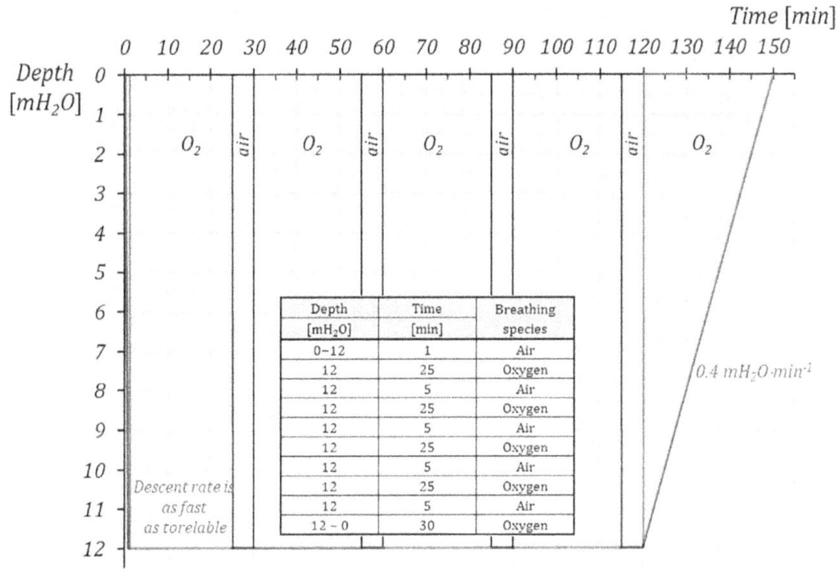

Depth [mH$_2$O]	Time [min]	Breathing species
0–12	1	Air
12	25	Oxygen
12	5	Air
12	25	Oxygen
12	5	Air
12	25	Oxygen
12	5	Air
12	25	Oxygen
12	5	Air
12 – 0	30	Oxygen

FIGURE 4.19 Treatment table $TT\,12_{86}\;Cx$.

Source: Comex Marseille (1986).

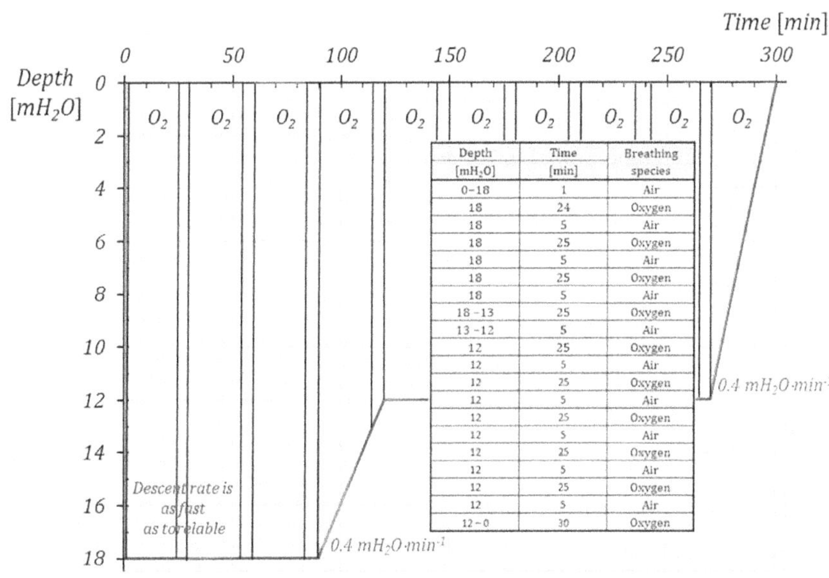

Depth [mH$_2$O]	Time [min]	Breathing species
0–18	1	Air
18	24	Oxygen
18	5	Air
18	25	Oxygen
18	5	Air
18	25	Oxygen
18	5	Air
18 –13	25	Oxygen
13 –12	5	Air
12	25	Oxygen
12	5	Air
12	25	Oxygen
12	5	Air
12	25	Oxygen
12	5	Air
12	25	Oxygen
12	5	Air
12	25	Oxygen
12	5	Air
12 – 0	30	Oxygen

FIGURE 4.20 Treatment table $TT\,18_{86}\;Cx$.

Source: Comex Marseille (1986).

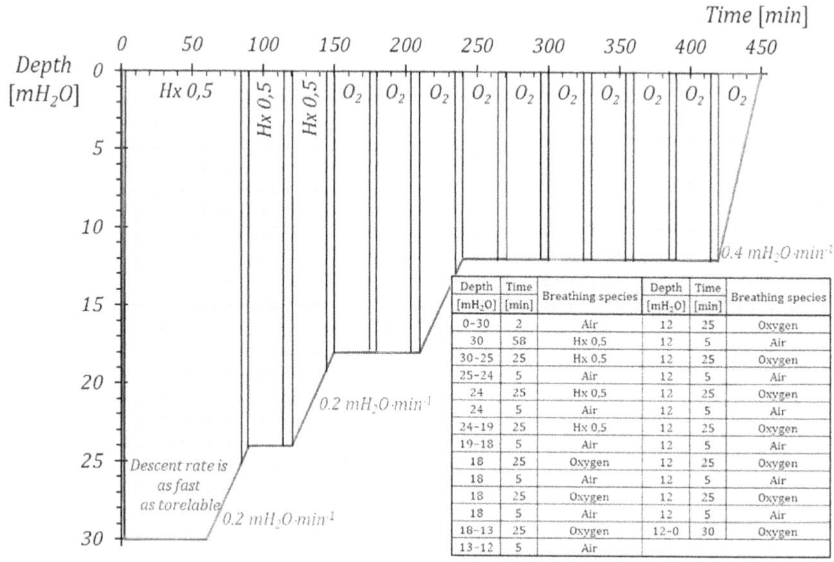

Depth [mH₂O]	Time [min]	Breathing species	Depth [mH₂O]	Time [min]	Breathing species
0-30	2	Air	12	25	Oxygen
30	58	Hx 0,5	12	5	Air
30-25	25	Hx 0,5	12	25	Oxygen
25-24	5	Air	12	5	Air
24	25	Hx 0,5	12	25	Oxygen
24	5	Air	12	5	Air
24-19	25	Hx 0,5	12	25	Oxygen
19-18	5	Air	12	5	Air
18	25	Oxygen	12	25	Oxygen
18	5	Air	12	5	Air
18	25	Oxygen	12	25	Oxygen
18	5	Air	12	5	Air
18-13	25	Oxygen	12-0	30	Oxygen
13-12	5	Air			

FIGURE 4.21 Treatment table $TT\,30_{86}\ Cx$.

Source: Comex Marseille (1986).

$TT\,30_{86}\ Cx$ is used to treat neurological symptoms of *Type II DCS* (Figure 4.21). These tables assume pharmacological support through administering acetylsalicylic acid[57] and, in the case of $TT\,30_{86}\ Cx$, also anti-inflammatory medication, hormone medicine having inflammatory properties, and isotonic infusion liquids.[58]

Treatment tables $TT\,12_{86}\ Cx$ and $TT\,18_{86}\ Cx$ constitute a treatment recompression system for symptoms *Type I DCS* occurring after air or *Hx* diving. If after $\tau = 15\ min$ from compression to $12\ mH_2O$ *DCS* symptoms disappear, the treatment is based on $TT\,12_{86}\ Cx$. If this is not the case, treatment continuation is based on $TT\,18_{86}\ Cx$. Symptoms of *Type II DCS* are treated following $TT\,30_{86}\ Cx$.

When decompression stations have been omitted or the diver has been blown up to the surface during an air dive outside the saturation zone, for *Type I DCS*, $TT\,18_{86}\ Cx$ is used, and for *Type II DCS* symptoms, treatment is based on $TT\,30_{86}\ Cx$. If there exists a need for extending the treatment time in accordance with $TT\,30_{86}\ Cx$, the time of stop at the depth $H = 30\ mH_2O$ can be prolonged to maximum $\tau = 3\ h$.

When *DCS* occurs during air saturation or *Hx* saturation dives, or operational *Hx* diving outside the saturation zone, for *Type I DCS* symptoms, $TT\ B_{86}\ Cx$ is used if the symptoms disappear within 15 *min* with standard decompression; if not, $TT\ SB_{86}\ Cx$ is used with standard decompression also (Table 4.8).

When the condition of the diver worsens, special saturation table $TT\ N_{86}\ Cx$ should be used (Table 4.9).

Saturation table $TT\ N_{86}\ Cx$ is also used to treat *Type II DCS* symptoms following air or *Hx* saturation or operational dives *Hx* outside the saturation zone. When using

TABLE 4.8
Treatment Tables $TT\ B_{86}\ Cx$ and $TT\ SB_{86}\ Cx$

$TT\ B_{86}\ Cx$			$TT\ SB_{86}\ Cx$				
Depth of Treatment (mH_2O)	Hx Bounce Dive and Saturation	Air Saturation	Depth of Accident (mH_2O)	Depth of Treatment (mH_2O)	BIBS Breathing Duration (min)	Hx Bounce Dive and Saturation	Air Saturation
0–12	Oxygen	Oxygen	0–12	+6	120	Oxygen	Oxygen
19–40	$Hx0.50$	$Nx0.50$	13–31	+9	120	$Hx0.50$	$Nx0.50$
41–110	$Hx0.20$		32–98	+12	120	$Hx0.20$	
111–210	$Hx0.10$		99–195	+15	120	$Hx0.10$	
211–360	$Hx0.05$		196–345	+15	120	$Hx0.05$	
361–450	$Hx0.03$		346–450	+15	120	$Hx0.03$	

Source: Comex Marseille (1986).

TABLE 4.9
Treatment Table $TT\ N_{86}\ Cx$

$TT\ N_{86}\ Cx$			
Depth of Accident (mH_2O)	BIBS Breathing Duration (min)	Hx Bounce Dive and Saturation	Air Saturation
20–40	120	$Hx0.50$	$Nx0.50$
41–110	180	$Hx0.20$	
111–210	240	$Hx0.10$	
211–360	300	$Hx0.05$	
361–450	360	$Hx0.03$	

Source: Comex Marseille (1986).

saturation table $TT\ N_{86}\ Cx$ for treating cases of DCS developed during or after Hx dives, both saturated and outside the saturation zone, decompression is conducted on a continuous basis, with the ascent time through individual depth $t = 60\ min \cdot mH_2O^{-1}$ within depth range $H \in [15, 200]\, mH_2O$, keeping the oxygen partial pressure at a level of $p_{O_2} \in [60, 63]\, kPa$ in Hx, creating the atmosphere of the complex. Decompression from a depth $H = 15\ mH_2O$ to the surface is conducted with the ascent time through individual depth $t = 80\ min \cdot mH_2O^{-1}$, keeping the oxygen volume content at level $C_{O_2} \in [23, 24]\%_v$ in *Hx*, creating the atmosphere of the complex. When treating *DCS* cases developed during or after saturation dives based on nitrox *Nx*, for the depth range $H \in [15, 30]\,mH_2O$, decompression is conducted on a continuous basis at the ascent rate of $v = 120\ min \cdot mH_2O^{-1}$, keeping the oxygen partial pressure in *Nx*, making the atmosphere of the complex at the level $p_{O_2} \in [60, 63]\,kPa$. From the depth $H = 15\ mH_2O$ to the surface decompression is conducted at ascent rate $v = 180\ min \cdot mH_2O^{-1}$, keeping the oxygen volume content in *Nx* creating the atmosphere of the complex at level $C_{O_2} \in [23, 24]\%_v$.

As there are no special procedures for cases when *DCS* symptoms appear during decompression in water, as is the case with other systems,[59] when they appear, the decompression is discontinued. The diver is brought to the surface as quickly as possible and further treatment is administered as in the case when a diver is blown up to the surface.

4.2.6 Conclusions

The diving decompression tables commonly used worldwide usually allow for only a low workload of the diver underwater.[60] This is caused by the existence, in hyperbaric conditions, of additional respiratory workload and the action of the water environment itself. For example, divers use suits which require additional effort during work, or because they cannot breathe underwater, divers have to carry a substantial amount of breathing agent, which increases the effort necessary to carry it, etc. Only some of the decompression tables take into account additional effort necessary to perform effective work underwater, determining its maximum border magnitude as the mean load. This limitation is associated not only with performing work in the water environment itself but also with the necessity to later undergo decompression. The decompression process will be disturbed by products of metabolism, which occurs after a high amount of effort is made. For example, acidification of muscles with lactic acid caused by anaerobic changes leads to acidification of blood, and because of this, hemoglobin loses its capacities of carrying oxygen (Kłos, 2012). This hampers gas exchange, and as a result, it disturbs the process of decompression. A detailed analysis of the effect of workload on the panned decompression is very difficult. It is only saturation dives that give divers the possibility to make relatively high physical efforts, as their decompression process is separated from the work period by a suitably long rest period dedicated to stabilization.

Sometimes, there exists a high probability for a diver to get stuck underwater, for example, inside a wreck during penetration. Divers who work in support of a rescue operation, outside the saturation zone, under emergency situations, may, as a result, face improper decompression, or face the hazard of, for example, a blowup to the surface without decompression or emergency-induced omitting of some elements of decompression.

It follows from the carried-out analysis that it is relatively easy to select a treatment recompression system for *DCS* cases which occurred in the course of short-term operational dives[61] outside the subsaturation or full saturation zone. Of the many existing systems, the most common is the system developed and published by the US Navy (USN, 2011). It can be used to support the *DCS* treatment process in divers, providing support to saturation divers in the course of wreck penetration. Such divers can be employed to provide assistance in straightening and winding umbilicals of the saturation diver and underwater vehicles and/or deliver tools, render assistance in transporting casualties, and in other support actions carried out outside the subsaturation or full saturation zone. These methods can also be used to treat light *DCS* occurring after *Nx* saturation diving. The US Navy system does not allow exceeding 70 mH_2O to produce the relief effect in a patient. Extending the US Navy system by the adding air tables BMΦ offers additional treatment capabilities. However, the treatment is conducted in difficult conditions, hard to accept without doubt. Complications will appear when respiratory failure occurs in the casualty, for example, as a result of pulmonary barotrauma,[62] which occurs when the diver is blown up to the surface. In such a case, effective breathing of *Tx*, *Nx*, or oxygen under high pressure can cause airway resistance in the patient so high that it not only will hamper the treatment process but may also prevent him/her from breathing and induce asphyxia.[63] When this is the case, using *Hx* may turn out to be necessary. It should be borne in mind, however, that transfer from a heavy-breathing agent to a lighter one often causes the occurrence of *DCS* symptoms through counterdiffusion, even if this is not accompanied by change in pressure (Hamilton, Adams, Harvey, and Knight, 1982; Harvey, 1977; Hyldegaard and Jensen, 2007; UHMS 22 Workshop, 1979). A comparative analysis of the proposed tables, based on one selected approach to estimation of *DCS* hazard, is worth conducting.

Treatment of *Type I DCS* and *Type II DCS* cases is based, first of all, on the hyperbaric method. The first treatment parameter is the pressure causing compression of gas in free phase precipitated in tissues, allowing it to dissolve in fluids existing in tissues. This causes a decrease in mechanical impact of gas on tissues, unblocking blood flow through capillaries, and the cessation of response reaction of the immunity system identifying gas bubbles as aggressors.

The second parameter is the time needed for homeostasis, resorption of gas in tissues and bodily fluids, and transport of dissolved gas excess to the blood. Then, eliminating the gas excess from the body during the decompression by the blood prefunding the bodily tissues and lung tissue (Figure 4.22).

The factor intensifying the treatment process is the use of oxygen or mixtures enriched with oxygen. Oxygen is rapidly metabolized in the body tissues, leaving free space to transport inert gas by the blood. Sometimes, other drugs are needed in emergency cases and for treatment intensification. The therapeutic recompression process can be supported pharmacologically by blood thinning, maximizing hydration, and/or administering painkillers, anticoagulants,[64] and anti-inflammatory medication.

The systems of treating *DCS* with therapeutic recompression, described here, constitute a vivid example of uses of these briefly presented parameters. But free use of pressure, time, and composition of the breathing agent is often made difficult because of the phenomena such as of oxygen toxicity, counterdiffusion, respiratory failure, etc.

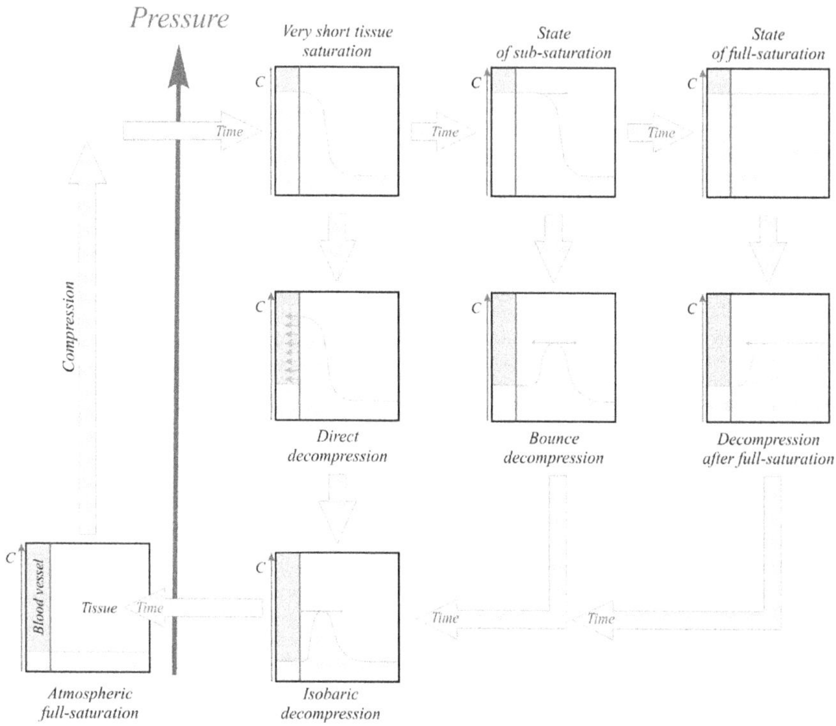

FIGURE 4.22 Main overview for overload of inert gas in body tissues in the hyperbaric environment and its elimination during the decompression process.

Modelling counterdiffusion has, so far, failed to produce a comprehensive theory taking counterdiffusion into account during a decompression process, despite the fact that counterdiffusion cases often occur in diving (Wienke, 2001; Imbert, 1975). The occurrence of *DCS* cases accompanied by counterdiffusion can be divided into two forms: *SICD*[65] and *DTICD*.[66] The *SICD* symptoms occur when the diver is breathing a gas different than the gas surrounding the body (Lambertsen and Idicula, 1975). For example, in the Polish Navy deep diving technology, *Tx* is used as a breathing agent during work performed underwater, and the dry suit is inflated with air. During decompression, the diver substitutes the breathing agent for air during the transfer under pressure and then for decompression oxygen. In all divers, light skin *DCS* symptoms occur, for example, general skin itching. The *DTICD* symptoms accompany sequential changes in the breathing agent, usually during decompression. The sequential changes in the breathing agent from lighter to heavier mixture and from heavier to lighter mixture can cause *DCS*, even if they are not accompanied by changes in depth – *ICD*.[67] A combination of counterdiffusion phenomena through cell membranes and dissolution of gases in tissues can lead to exceeding the dissolution border and the precipitation of gas in free phase in tissues, causing *DCS*, due to the differences in diffusion and solubility coefficients, especially of helium and nitrogen (Karreman and Lambertsen, 1977).

A possible counterdiffusion occurrence has to be taken into account when the US Navy method is used for treatment after an accident during deep diving or saturation diving. When the treatment methods recommended by the French firm COMEX are used, the probability of counterdiffusion occurrence can be minimized by choosing for treatment the breathing agent on which *DCS* symptoms occurred.

4.3 POLLUTION

Pollution of a breathing agent by toxic components which affect the human organism is always undesirable. Generally, it can be assumed that the response of an organism to a toxic component in the form of vapor or gas is dependent on its partial pressure: $p_i \equiv p \cdot x_i$, where p_i is partial pressure of the ith component, p represents total pressure of the gas mixture, and x_i is mole fraction of the ith component. If for the atmospheric pressure p_0 the acceptable content of toxic component is equal to $x_i^{max} = x_i$, for the pressure equivalent to the depth of $10\,mH_2O$, the acceptable content of this component is half lower: $x_i^{max} = 0.5x_i$. If, in the breathing agent, there occur more than one type of pollutants, the law of cumulative pollution action is many times applied as $\sum_i \frac{x_i}{x_i^{max}} \leq 1.3$. Some authors recommend adopting magnitude 1.0 instead of 1.5. The maximum acceptable concentration C_i^{max} of the toxic substances is usually calculated in relation to the atmospheric pressure p_0. Using the definition of *partial pressure*, the maximum pressure p_i^{max} can be calculated for toxic substances: $p_i^{max} = p_0 \cdot \frac{C_i^{max}}{100\%}$. Using the definition of *percentage concentration* $C_i \overset{\text{def}}{\equiv} x_i \cdot 100\%$, the following can be written:

$$\sum_i \frac{C_i}{C_i^{max}} = \sum_i \frac{x_i}{x_i^{max}} = \sum_i \frac{p_i}{p_i^{max}} \leq 1.5 \tag{4.4}$$

where C_i – concentration of the ith polluting component; C_i^{max} – maximum acceptable concentration of the ith polluting component; x_i – mole fraction of the ith polluting component; x_i^{max} – maximum acceptable mole fraction the ith polluting component; p_i – partial pressure of the ith polluting component; and p_i^{max} – maximum acceptable partial pressure of the ith polluting component.

As the partial pressure of the toxic substance i is $p_i = p_0 \cdot \frac{C_i}{100\%}$, dependence (4.4) can be written for hyperbaric conditions: $\frac{p}{p_0} \sum_i \frac{C_i}{C_i^{max}} \leq 1.5$, where p represents total pressure at the depth of diving and p_0 represents atmospheric pressure.

TABLE 4.10
Composition of Dry Air in Volume and Mass Percentage

Gas	N_2	O_2	Ar	CO_2	H_2	Ne	He	Kr	Xe
$C\ (\%_v)$	78.03	20.99	0.93	0.030	0.01	0.018	0.0005	0.001	0.00001
$C\ (\%_m)$	75.47	23.20	28	0.046	0.001	0.0012	0.0001	0.0003	0.0004

Pollutants responsible for polluting breathing agents in an ecologically closed atmosphere can be divided into three groups:

- Pollutants emitted by air or other components of the breathing gas mixture
- Pollutants emitted by technical elements being in contact with a breathing agent[68]
- Pollutants emitted by the human organism

4.3.1 Pollutants Emitted by Gas Components

Typical pollutants emitted by breathing air are carbon dioxide $< 0.05\%_v$, nitrogen oxides $< 0.7 mg \cdot m^{-3}$, hydrocarbons $< 5.0 mg \cdot m^{-3}$, carbon oxide $< 3.0 mg \cdot m^{-3}$, water vapor $[0.01 - 0.1] g \cdot m^{-3}$, etc. The composition of dry air is shown in Table 4.10.

In many cases, technical gases used as components of breathing gas mixtures consist of similar pollutants, but having lover concentration.

4.3.2 Pollutants Emitted by Technical Equipment

Classification societies provide the most common types of pollutants caused by technical equipment and their maximum acceptable concentration magnitudes. The review of the selected pollutants coming from technical equipment is illustrated in Table 4.11 (Kłos, 2000).

Data on the safe concentration of pollutants can be found in some instruction manuals dealing with the issue of survival in special conditions (e.g., Военное Издательство, 1983).

Table 4.12 contains the maximum acceptable concentration of some toxic vapors and gases used in the struggle for survival.

These magnitudes are used in the formula for cumulative action of pollutants. Table 4.13 contains the maximum acceptable concentration confirmed with modern research methods and legally approved limits for concentration of selected toxic compounds in the environment which, in everyday time-limited exposition, do not cause, in a longer period, pathological changes or diseases in the present and the future generations.

TABLE 4.11
Maximum Acceptable Concentration of Some Toxic Substances in Submersible Objects[69]

Chemical Compound		Source of Pollution	Maximum Acceptable Concentration for the Moment of Exposure		
			1*h*	24*h*	90*days*
Acetylene	C_2H_2	Fried dishes	6000 *ppm*	6000 *ppm*	6000 *ppm*
Acrolein	CH_2CHCHO	Fried dishes	—	0.1*ppm*	—
Isopropyl alcohol	C_3H_7OH	Paint diluent	—	—	1 *ppm*
Ammoniac	NH_3	Metabolism	400 *ppm*	50 *ppm*	25 *ppm*
Stibine	SbH_3	Battery gassing	—	0.05*ppm*	0.01*ppm*
Arsine	AsH_3	Battery gassing	—	0.1*ppm*	0.1*ppm*
Benzene	C_6H_6	Diluent	—	100 *ppm*	0 *ppm*
Methyl chloride	CH_3Cl	Diluent	—	—	25 *ppm*
Chloroform	$CHCl_3$	Diluent	—	—	1 *ppm*
Hydrogen chloride	HCl	Freon decomposition	10 *ppm*	4.0*ppm*	0 *ppm*
Nitrogen dioxide	NO_2	Generators	10 *ppm*	0 *ppm*	0.5*ppm*
Sulfur dioxide	SO_2	Sanitation	10 *ppm*	5.0*ppm*	0*ppm*
Carbon dioxide	CO_2	Metabolism	2.5%	0%	0.5%
Ethanol	C_2H_5OH	Diluent	—	—	100*ppm*
Hydrogen fluoride	HF	Freon decomposition	8*ppm*	1.0*ppm*	0.1*ppm*
Formaldehyde	$HCHO$	Ready-made dishes	5*ppm*	5*ppm*	5*ppm*
Phosgene	$COCl_2$	Freon decomposition	0*ppm*	0.1*ppm*	0.05*ppm*
Freon 113	$CClF_2CCl_2F$	Cooling installations	—	—	100*ppm*
Freon 11	CCl_3F	Cooling installations	—	—	100*ppm*
Freon 12	CCl_2F_2	Cooling installations	—	—	100*ppm*
Freon 114	$CClF_2CClF_2$	Cooling installations	—	—	100*ppm*
Methyl-ethyl ketone	$CH_3COC_2H_5$	Diluent	—	—	20*ppm*
Methyl-isobutyl ketone	$CH_3COC_3H_7$	Diluent	—	—	20*ppm*
Xylene	$C_6H_4(CH_3)_2$	Paint diluents	—	—	50*ppm*
Methane	CH_4	Sanitation	3%	3%	3%
Methanol	CH_3OH	Diluent	-	-	10*ppm*
Ozone	O_3	Commutated motors	0*ppm*	0.1*ppm*	0.02*ppm*
Nitrogen oxide	NO	Generators	10*ppm*	0*ppm*	0.5*ppm*
Toluene	$C_6H_5CH_3$	Diluents	—	—	20*ppm*
Trimethylobenzenes	$C_6H_3(CH_3)_3$	Diluents	—	—	3*ppm*
Aromatic hydrocarbons, except benzene		Paint diluents	—	—	10*mgm*^{3}
Aliphatic hydrocarbons, except methane		Paint diluents	—	—	10*mgm*^{3}
Hydrogen	H_2	Battery gassing	1000*ppm*	1000*ppm*	1000*ppm*

Table 4.11 and Table 4.13 present examples of acceptable equivalent values x_{SEV} for some selected pollutants for various exposure periods. The measurement results of the concentration of polluting substances x can be applied to calculate the maximum safe depth for using a breathing agent p_H with regard to the surface equivalent value SEV[70] of toxic substance x_{SEV}:

$$p_H = \frac{x_{SEV}}{x} \cdot p_0 \tag{4.5}$$

where x_{SEV} –surface equivalent value of the pollutant, x– pollutant value in the sample, p_H –total hyperbaric pressure Pa; and p_0 –normal pressure.

4.3.3 Pollutants Emitted as a Result of Metabolism

In order to analyze survival, a model was developed of pollution emitted by a "standard human being" doing only very light work, presented in Table 4.14 (Kłos, 2000).

Use of oxygen and emission of carbon dioxide can be substantially reduced by being in the totally nonactive state (Table 4.15).

The investigations show that in the condition of nonactivity, use of oxygen can be at a level of $8.7 dm^3 hour^{-1} person^{-1} STP$,[71] and carbon dioxide emission at a level of $17.0 dm^3 hour^{-1} person^{-1} STP$ (Kłos, 2008). Higher magnitudes of oxygen use can occur at a level of $30.0 dm^3 hour^{-1} person^{-1} STP$, and carbon dioxide emission at a

TABLE 4.12
Maximum Acceptable Concentration of Some Toxic Vapors and Gases Used in the Struggle for Survival

Time	CO	Nitrogen Oxides	Benzene	Toluene	Xylene	Formaldehyde	Acetaldehyde
(min)					*(mgm³)*		
5	700	60	—	—	—	—	—
10	600	45	90	—	—	—	—
15	400	35	70	270	—	—	—
20	360	30	60	—	—	—	—
30	300	25	50	170	130	8	100
40	240	20	45	—	—	—	—
60	200	15	37	165	123	—	—
120	150	—	25	115	110	—	—
240	100	—	15	80	90	—	—
480	60	10	—	—	79	—	—
1,440	40	—	—	—	60	—	—

Source: Военное Издательство (1983).

level of $25.0dm^3hour^{-1}person^{-1}STP$. Survival is possible as long as partial pressure of oxygen can be maintained above $p_{O_2} \gtrsim 16kPa$, and partial pressure of carbon dioxide up to the level of $p_{CO_2} < 2.5kPa$.

TABLE 4.13
Maximum Acceptable Concentration of Selected Toxic Compounds for the Work Environment

Chemical Compound		Maximum Acceptable Concentration for Exposure (mgm^3)			
		$4h$	$8h$	$24h$	$[200,3000]h$
Stibine	SbH_3	0.5	0.3	0.15	—
Arsine	AsH_3	—	0.1	—	0.003
Benzene	C_6H_6	—	5	—	2
Phosgene	$COCl_2$	—	0.5	—	—
Freon 12	CCl_2F_2	6,000	3,000	—	150
Freon 114	$CClF_2CClF_2$	—	1,000	—	100
Xylene	$C_6H_4(CH_3)_2$	—	50	—	12
Mercury	Hg	—	0.01	—	0.003
Nitrogen oxide	NO	5	5	—	0.5
Carbon oxide	CO	30	20	18	5
Toluene	$C_6H_5CH_3$	—	50	—	8
Hydrocarbons	C_xH_y	—	300	—	35

Source: Военное Издательство (1983).

TABLE 4.14
Pollutants Emitted by a "Standard Human Being" Exerting Slight Effort

Pollutant		Emission of Pollutant over a $24h$ Period
Carbon dioxide	CO_2	$550dm^3$
Methane	CH_4	$3dm^3$
Carbon oxide	CO	$0,7dm^3$
Acetone	$(CH_3)_2CO$	$0,0005dm^3$
Water vapor	H_2O	$2000dm^3$
Heat emission:		
– Latent heat		$5600kJ$
– Other heat		$6300kJ$

Source: Kłos (2000).

TABLE 4.15
Streams of Used Oxygen and Lung Ventilation Depending on Physical Effort

Physical Effort: Intensity	Physical Effort: Example	Stream of Used Oxygen ($dm^3 \cdot min^{-1}$)	Number of Breaths per Minute (min^{-1})	Lung Ventilation ($dm^3 \cdot min^{-1}$)	Border Stream of Used Oxygen
Very light	Lying on bed	0.25	10–20	8–10	To 0.5
	Sitting still	0.30			
	Standing still	0.40			
Light	Walking $3.5 km \cdot h^{-1}$	0.7	20–25	10–20	0.5–1.0
Moderate	Marching $6.5 km \cdot h^{-1}$	1.2	25–30	20–30	1.0–1.5
Hard	Swimming at speed of $3.0 km \cdot h^{-1}$	1.8	30–35	30–50	1.5–2.0
Very hard	Running at speed of $13 km \cdot h^{-1}$	2.0	35–40	50–65	2.0–2.5
Extremely hard	Running uphill	4.0	>40	>65	>2.5

Source: Przylipiak and Torbus (1981).

Short-term conditions under which survival is possible are regarded as those where partial pressure of oxygen is at a level of $p_{O_2} \geq 14 kPa$, and of carbon dioxide not exceeding $p_{CO_2} \leq 5 kPa$. At present, it is claimed that the action of a small concentration of volatile hydrocarbon pollutants of a breathing agent on a human organism usually has a narcotic character.[72] It is high-concentration magnitudes that produce toxicity (Flook, 2002).

4.3.4 Conclusions

Possibilities for survival in hyperbaric facilities generate a multithreaded problem situation. One of the issues discussed briefly in this study is the assessment of the danger generated by toxic pollutants that can occur in a breathing agent. It carries out an initial analysis of breathing agent pollutants which can occur, and their possible action on a human body. Many of them are dependent on time, and they are connected with each other in an interactive manner. These issues are not dealt with in here as they will be the subject of further analyses dedicated to toxicity of the main components of the breathing agent.

4.4 ANALYSIS

Sometimes, specifications of many different measuring instruments approved for the diving operations are presented in literature (e.g., USN, 2008). It is worthy of notice that requirements for measuring instruments are dependent upon conditions of diving operations. For example, in Poland, some attempts were made to reduce the costs of diving operations. The aim of these attempts was to introduce a helium–oxygen–nitrogen gas mixture[73] having the lowest possible helium concentration values as a breathing mixture. Some magnetothermoconductometric oxygen analyzers that are in common use are not applicable under hyperbaric conditions (Kłos, 1989). Similarly, most of the electrolytic cell analyzers are not accurate enough. As it follows from the preceding text, some research should be initiated before an analyzer is used concerning calibration methods, analytic procedures. The aim of this research should be to determine the metrology features of the analyzer. Instrumental analysis is the most important analytical method. The techniques used in instrumental gas analysis enable performing rapid gas analysis, which in turn enables multiple repeating of analysis. That is why statistical verification of the analysis can also be performed. Thus, the calculation of maximal errors in gas analysis performed is easy. It is very difficult to imagine using classical chemical analytical methods in diving operations because they are very complicated and time-consuming. It is said that chemical methods are as accurate as physicochemical instrumental methods. Generally, this view is not substantiated, because in classic chemical methods, it is impossible to take into account all the errors. Despite the fact that instrumental methods are technologically more advanced, they should not always be regarded as better methods. In most instrumental methods, physical/chemical standards are needed to be used for calibration. When these methods are used for accurate, responsible operations, calibration is needed before each analysis. It can be assumed that in quantitative instrumental analysis, almost all measuring instruments need standards-based calibration. It follows from the fact that the experimental results were obtained as relative analytic signals that should be related to the absolute standards. Additionally, the measured values are dependent upon indefinable factors that affect the analytical signal. The effect of these factors can be diminished by the standard sample analysis performed under the same conditions.

4.4.1 Mass Spectroscopy

Mass spectroscopy enables characterizing the mass of ions and ionized molecules based on a path deviation in electrostatic and magnetic fields. Mass spectrometry is one of the most accurate methods used in instrumental analysis. It is a very popular method, employed especially for qualitative organic compound analysis. Mass spectrometry is a destructive method applied for quantitative analysis that needs calibration, for which pure chemical standards are used. The method can be used for gas samples or vaporized substances[74] having a vacuum from $p \in [0.01, 100]\, mPa$. The relative mass spectrum of a gas obtained as a result of analysis is difficult to interpret, but the interpretation process can be supported by computer programs. The mass spectrum is characterized for each chemical compound. The particular stream of ions obtained from the pure substance or mixture is dependent upon several factors, mostly upon an ionizing process procedure. To obtain repeatable results, the spectra of standard/reference substances and mixture being analyzed should be recorded under the same conditions. The solution to a system of linear equations for selected spectral bands gives the percentage composition of the analyzed mixture.

Information obtained from mass spectrum is ambiguous. This is caused by the multiple ionization of molecules. It is observed in a mass spectrum where, except the band of ions of anticipated mass/charge ratio m/e, bands in other places occur: $m/2e$, $m/3e$. Other ions having the same mass/charge ratio m/e can form during the decomposition of different substances. This will lead to band interference. It is another cause of spectrum complexity. Band interference will occur in a spectrum of mixture consisting of pure gases: CO_2, CO, O_2, N_2.

In order to diminish difficulties in the spectrum interpretation, a mass spectrometer is often coupled with a gas chromatograph. The bands of each components separated in the chromatography process are analyzed with a mass spectrometer. The coupling of a gas chromatograph with a mass spectrometer seems to be the most convenient solution applicable for the laboratory analysis of breathing contaminants. The method can also be used for standard mixture analysis used in other chemical methods of analysis. The mass spectroscopy method can be used to analyze all the contaminants selected earlier.

4.4.2 Infrared Spectroscopy

Infrared spectroscopy is an acknowledged and respected method of instrumental analysis used for the qualitative monitoring of organic compounds. The infrared spectra of gaseous molecules not only show significant absorption spectral bands broadening; the method can also be used to measure trace organic contaminants. Carbon dioxide and carbon monoxide give the spectra of good resolution absorption bands partly overlapping.

The spectrum is pure in *the fingerprint region*, that is, in the wave number range of wave number[75] $\bar{\nu}$: $\bar{\nu} = (800, 2000)\,cm^{-1}$. This region includes characteristic/typical absorption spectrum for many characteristic organic compounds. Similarly, as fingerprints enable identifying a person, the characteristic spectrum enables identifying a compound;,however, interpretation of a spectrum is difficult. The spectrum is pure within the approximate range of $\bar{\nu} = 3000\,cm^{-1}$, where absorption spectra of hydrocarbons occur. Organic compounds as metabolic products which can occur in breathing gases are characterized by very high absorption activity within the range of $\bar{\nu} \in (800, 2000)\,cm^{-1}$ and approximately at $\bar{\nu} = 3000\,cm^{-1}$. Organic compounds can be measured together with typical contaminants, such as carbon dioxide and carbon monoxide.

Sometimes, it is very difficult to use the method for quantitative analysis having sufficient accuracy as significant band broadening, band interference in the case of mixtures, impossibility of accurate separation and integration, characteristics curvature, relationship between the signal and concentration,[76] calibration difficulties, and so on occur. Infrared spectroscopy is a widely used instrumental technique, and special equipment is available for use in gas mixture analysis.

Infrared spectroscopy is also used for gas analysis of gas flow. The method cannot be employed to monitor homonuclear and atomic gases, which means that argon content cannot be measured with this method. The other earlier-specified contaminants are possible to measure. Water vapor content is very difficult to measure with absorption IR –spectroscopy because of its very broad absorption bands. A photoacoustics detector is relatively simple to use.

4.4.3 Gas Chromatography

The gas chromatography process is not a measurement analytical method, but the physical separation process relies on the resolution of an analyzed mixture on chromatographic columns. The resolution of a mixture is possible by constituent adsorption on a chromatography column package. Then, a detector is used for assessing the amount of separated substance. The resolution of a mixture into its components is attained by a change in chromatography column working parameters, such as carrier gas velocity, the kind of the carrier gas, measurement temperature, column length, column diameter, package of column, etc. Proper accuracy of chromatography measurements depends on the working parameters of the detector selected. A description of the chromatograph package is given in the manufacturer's leaflet (e.g., Supelco SA, Machery-Nagel GmbH) and chromatographic papers (e.g., *J. Chrom.*; *J. Chrom. Sci.*; *Gas Chrom.*; *Chromatographia*).

The gas chromatography method needs calibration. Gas chromatography is used for determining selected earlier vapors and gases that might contaminate the breathing gas.

4.4.4 Electrochemical Analyzers

For gas determination, electrochemical methods using ion-selected electrodes are often employed. The methods rely on penetrating gas molecules through the permeable or semipermeable membrane of the electrochemical cell. Then, the gas is dissolved in electrolyte, which usually concentrates salt solutions. Ions are formed as a result of the chemical reaction between gas molecules and the electrolyte. Chemical reactions produce a current signal that is electronically transformed into an analytical signal.

The methods need calibration. However, the measurement accuracy of ion-selected electrodes is less than that of the chromatographic and spectroscopic methods; ion-selected electrodes are widely used because measuring simplicity can be applied in the field condition. Ion-selected electrodes are used for determining the gas content with regard to the following: CO_2, CO, NH_3, NO_x, CH_4, and other hydrocarbons' vapor. Hydrocarbon analyzers are characterized by low selectivity with respect to the compounds being monitored; therefore, the measurement can be disturbed by the other constituents of the gaseous atmosphere. The selection of proper electrodes is difficult; therefore, it must be professionally done by analytical chemists for each analytical problem.

4.4.5 Chemical Methods

Of the chemical methods, detector tubes are the most often used for determining the contaminants of the breathing gas (e.g., Drägerwerk, 2011; and some internet versions; UNIPHOS, 2022; Honeywell, 2022). The principle of detector tube measurement relies on color reaction between the constituent and proper chemical compound based on silica gel package placed inside the transparent tube. The course of the chemical reaction is the same as the gas flow direction.

The measurement of the reaction front shift per unit of the flowing gas enables determining the concentration of a chemical compound. There are several methods to force the gas to flow: bellows hand pump of a known volume,[77] pressure-reducing devices with rotameter gas flow control, and different kinds of pumps of the constant flow or diffusion process. Detector tubes are blanked-off tubes at the ends. Usually, detector tubes are made of glass, and their ends are blanked off. To make gas flow possible, it is necessary to break off the tube ends before the measurement starts. Chemical reagents are precipitated on the solid support[78] or placed inside the internal ampoules. The ampoules should be broken before or after adsorption. The solid support should be saturated with the chemical reagent. The ampoules are crushed with the punch or by breaking the internal tube.[79] Detector tubes are scaled. Special measurement and calculation instructions are attached. The specification of chemical reactions and chemical compounds that should disturb reaction[80] is a very important part of instruction. When a detection tube is used the first time, it is necessary to analyze the possibility of the specific detector tube use.

Detector tubes are developed for selected chemical substances, different measurement ranges, and analytical reactions. Analytical chemists should be responsible for selecting the proper detector tube. As there are different analytical situations, the selection of detector tubes and the procedures for its use[81] are essential. The detector tubes methods are widely used because of their simplicity. The methods are used as threshold indicators for rapid approximate evaluation of contaminant content.

4.4.6 Specialized Analyzers

Some chemical compounds have special properties that enable developing special gas analyzers. Special gas analyzers are disturbance-resistant. This enables measuring compounds characterized by special properties with high measuring accuracy without interference by the other mixture constituents.

However, before such an analyzer is used, it is necessary to check it for the possibility of measurement disturbances. The magnetothermoconductometric method used in paramagnetic oxygen analyzers may be an example of its successful use for making nitrox *Nx* and heliox *Hx* mixtures; however, in some cases, the method can be unreliable. When some analyzers' constructions were applied for *Tx* of variable nitrogen/oxygen ratio mixture analysis, the aforementioned method was found unreliable (Kłos, 2000).

A new optical oxygen sensor for gas measurements is based on luminescence quenching of the sensor dye. The dye is excited with red light, and the properties of the resultant luminescence are measured in the near-infrared region. The presence of molecular oxygen quenches the luminescence, changing its intensity and lifetime fully reversibly. This principle is very robust. It shows virtually no interferences by other gases.

Measurement methods are very often applicable to specific measurement cases. For this reason, specialized measuring devices are developed. As a rule, universal laboratory measurement methods cannot be used in the field. Gas chromatographs used for analyzing combustion gases, IR spectrometers used for carbon dioxide emission, or specialized thermoconductometers used for hydrogen emission are typical specialized analyzers.

The organic contamination content is usually determined and recalculated with respect to organic carbon or selected hydrocarbons – most often, it is benzene or toluene. One of the specialized analyzers is a gas chromatograph GC equipped with the flame ionization detector FID. Sometimes, FID is used independently, separated from GC. Water vapor, carbon monoxide, and carbon dioxide cause disturbances in organic carbon analyses.

An interesting solution is a nitric oxide analyzer which makes use of chemiluminescence of metastable NO_2^* particle formed in reaction of nitric oxide NO with ozone O_3. Under normal conditions, nitric oxide NO is unstable and is oxidized by oxygen from air to NO_2 and N_2O_3. In order to use chemiluminescence, the reduction of stable nitric oxides to NO is needed. The reduction is performed in a $NO_2 - NO$ converter. Then, nitric monoxide NO reacts in the reacting chamber with ozone formed in the generator. Metastable molecules of NO_2^*, during relaxation to NO_2, emit energy by means of electromagnetic waves of characteristic frequency. Radiation intensity is an analytical signal.

The method is resistant to disturbances caused by intermolecular collisions, that is, collisions of the other vapor and gases[82] with metastable molecules of NO_2^* and relaxation free of radiation. The effects can be properly compensated by the use of the standard mixtures, whose quantitative and qualitative composition approximate to the composition of the sample being analyzed.

In many analytical processes, the content of the contaminant can be determined despite the low sensitivity of the measuring method. This can be attained by using concentrators. The concentrators enable attaining the average value of contaminant concentration at a chosen, specified time. The contaminant content can be determined following desorption[83] that occurs after a sufficiently determined contaminant adsorption time. The concentration of the desorbed chemical compound or compound mixture is sufficiently high, which makes it possible to use fewer sensitivity analyzers.

4.4.7 Gage R&R

In the following, presented is a discussion focused on the methods used to assess the reliability of measurement systems without using reference materials – $Gage\, R\,\&\, R$[84] (Allen, 2006; Automotive Industry Task Force, 2010). $Gage\, R\,\&\, R$ methods are an attempt to use stochastic methods[85] to search for sources of variability in the results of measurements carried out by measurement systems, later affecting their reliability, which should match the CTQ[86] requirements.

In the literature on $Gage\, R\,\&\, R$ methods, two trends can be distinguished. The first is utilitarian studies (Burdick and Graybill, 1992), whereas the second includes exhaustive statistical analyses (BIML, 2000; Burdick, Borror, and Montgomery, 2005). There are few works on $Gage\, R\,\&\, R$ methods that relate to the reliability of the measurement process by comparison with reference materials, which are rather the domain of work in the field of metrology (Jakus and Kłos, 2007).

An assessment will be presented of the measurement system reliability with *Gage R & R* methods utilizing etalons, which are standard reference materials for correct values $\hat{\mu}$ for various physicochemical quantities X.

The correct value $\hat{\mu}$ should be understood as a quantitative estimation of the value reproduced by etalon, burdened with negligible systematic errors[87] ε_s from the perspective of the defined applications. This means that although the reproduced correct value $\hat{\mu}$ is burdened with residual systematic errors ε_s, in particular applications,[88] it may be seen as a sufficient[89] estimation of the true value[90] μ.

4.4.8 Measurement System

In assessing a system of measurement,[91] an evaluation of the reliability of both the measuring instrument and the equipment, as well as the *standard operating procedure*, or *SOP* (Allen, 2006; Automotive Industry Task Force, 2010), is highly important.[92] Despite significant progress made in the field, systems of measurement usually use the same physiochemical phenomena. Unless sensors are changed, the use of modernized electronic support systems in the measuring process will only have little impact on the limitations the measurement methods have been affected by so far.

Knowledge of the physiochemical phenomena underlying the operation of the sensors used is crucial in assessing the possibilities of increasing the reliability of systems of measurement. Traditional metrological knowledge[93] continues to be valid and constantly updated and broadened.

Accurate measurements require relatively frequent calibration of measurement systems with etalons.[94] Moreover, calibration[95] may be a prospective method of assessing the capacity of a measurement process[96] finding its main use in legal metrology,[97] in calibration of measuring instruments,[98] in measures applied in responsible measurements,[99] etc.

There are a few types of methods using etalons: comparisons performed with a single standard, comparisons with two standards, the calibration curve method, etc. (Kłos, 1990).

Methods utilizing standard reference materials are not always profitable[100] or applicable, as is the case with retrospective methods.[101] A *post-factum* checkup of the correctness of operation of a system of measurement by comparison with standards should not be used to make assumptions on the previous performance quality of the system.

The calibration process is carried out with the use of etalons having particular physiochemical quantity X, which are characterized by an assigned and recognized[102] value for that physiochemical quantity calculated with uncertainty acceptable for concrete applications[103] – the correct value $\hat{\mu}$.

Calibration based on etalons having correct values $\hat{\mu}$ results in the minimization of systematic errors ε_s, thus making it possible to maintain[104] CTQ[105] requirements Critical to Quality[106] for the measurement system. *Gage R & R* methodologies which do not use etalons cannot offer the same possibilities of assessing various sources of systematic errors.

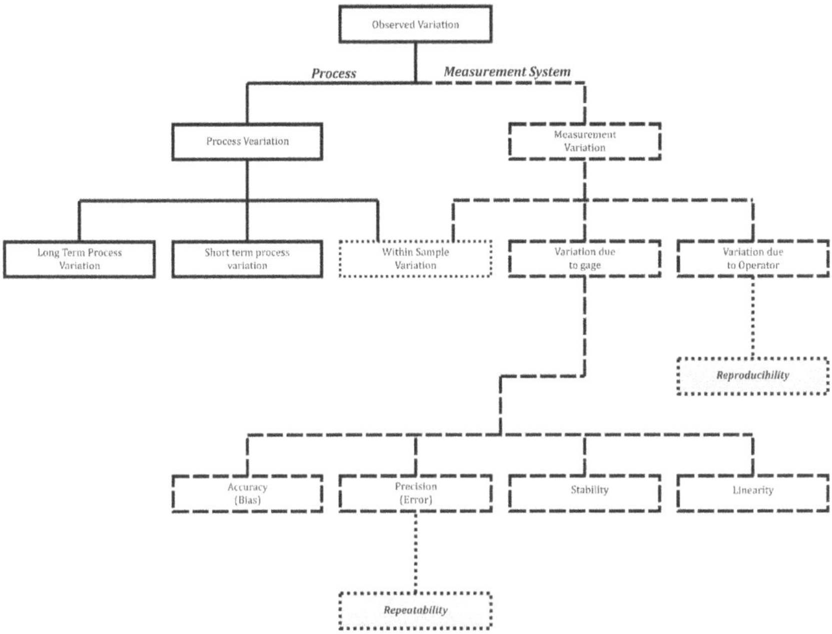

FIGURE 4.23 Types of the observed process variability and overlapping variability of the measurement system itself.

Source: Montgomery (2009).

The preparation of systems of measurement without the use of etalons guarantees only a minimization of the deviation related to reproducibility and repeatability of measurements (Figure 2.23).

Deviations related to scale linearity[107] or stability in time[108] are determined by means of metrological tests with the use of etalons and, most commonly, are to be found in the test inspection certificate of a system of metrological measurement.

Their determination is usually possible only with the use of etalons. Compensation of various types of deviations is most commonly attained following good laboratory practice of *SOP*, which may be found in manufacturers' recommendations, metrological manuals, or scientific studies (Kłos, 1990, 2007b).

In assessing a measurement system, we need to consider the reproducibility and repeatability related to its operation when various allowable modifications in the measurement system[109] have been made, including changes in operating modes (Figure 2.23).

4.4.9 Reproducibility and Repeatability

Reliability assessment of a measurement system, both with and without the use of etalons,[110] can be carried out on the basis of uncertainty assessment related to reproducibility[111] and repeatability[112] of measurements. Reproducibility and repeatability may be estimated by conducting an analysis of historical measurement data[113] – the retrospective method.

In such a case, *reproducibility* characterizes the potential capabilities of a measurement system,[114] whereas *repeatability* characterizes the practically achievable process capability related to the use of a particular instrument, *SOP*, and *GLP*.[115]

Reproducibility-related variability is observed in the case of measurements of the same sample performed by different analysts and/or modifications of the measurement system. *Repeatability*, on the other hand, is an observed changeability in the measurement results of the same sample, when the measurements are performed by the same analyst, using the same measuring instrument, and at not-too-long time intervals[116] (Figure 4.23).

4.4.10 Methods of Assessment of a Measurement System

The need for the development of methods of reliability assessment[117] of a measurement process is mainly generated by large-scale and mass production industries (Burdick and Graybill, 1992). The most important methods include R&R methodologies used to assess process capabilities[118] (Allen, 2006).

Methods used in metrology relate to the use of reference of measurement results to a chain of etalons having correct values. Thanks to this, they can be used to estimate the propagation of nearly all constituents of residual systematic uncertainty[119] expressed by systematic errors.[120] *Systematic error* ε_s is a difference between the value of the measurand obtained from the measurements $X \leftarrow x$ and the correct value[121] $\hat{\mu}$ adopted for comparison, most commonly expressed as a difference between the mean value from the conducted measurements $\overline{x}$ and the correct value $\hat{\mu}$: $\varepsilon_s = \overline{x} - \hat{\mu}$.

Often, *reproducibility* is taken for systematic uncertainty. It is required to distinguish between repeatability determined for an etalon having the correct value $\hat{\mu}$, which should not be taken for reproducibility, connected with $l-$fold measurement repetition $X \leftarrow x_l$ on the same sample X having unknown or only roughly estimated correct value $\sim \hat{\mu}$.

The use of methods of mathematical statistics allowed the working out of evaluation methods[122] for measurement systems on the basis of measurement of the same samples from a stabilized or low-frequency process[123] by estimation and separation of the variability of a measurement process and measurand[124] variability[125] and thus determining reproducibility and repeatability of the measurement system.

$R\&R$ analysis is a simplified measurement system assessment method as compared with the methods used in metrology[126] (Kłos, 1990; Allen, 2006; Jakus and Kłos, 2007). They can be used to evaluate a measurement system in the case of absence of standards or when their application is not possible (Kłos, 1990). However, such an evaluation no longer makes it possible to estimate all sources of residual systematic errors ε_s in a measurement system (Table 4.16).

4.4.11 General Assessment of Measurement System Reliability

The general approach to the assessment of measurement system reliability with the use of R&R procedures with and without etalons involves determining the proportion

TABLE 4.16
Methods Suggested for Use in Measurement System Reliability Assessment Depending on Measurement Type and Availability of a Standard Correct Value $\hat{\mu}$

Access to Standard Correct Value	Type of Measurement	
	Destructive	Nondestructive
Available	Comparison with standard correct value	Comparison with standard correct value
Unavailable	*Crossed R&R* method*	*Nested R&R* method**

*Method does not make it possible to estimate all constituents of a systematic error of the measurement system ε_s and only enables estimation of uncertainty related to reproducibility and repeatability.

**Method does not make it possible to estimate variability related to reproducibility and repeatability and only enables estimation of measurement uncertainty as a whole, for which reason it could be formally omitted.

Source: On the basis of[127] Allen (2006).

of the obtained measurement precision to the scope of tolerance *PTR*,[128] defined as (BIML, 2000; Allen, 2006):

$$PTR = k \cdot \frac{s}{USL - LSL} < 0.1 \Big| s = \sqrt{\frac{\sum_i (\bar{x}_i - \hat{\mu}_i)^2}{n-1}} \tag{4.6}$$

where *PTR* is the ratio of the obtained measurement precision to tolerance, representing quality requirements *CTQ* for the measurement system; $k-$ is the coverage factor, in which most commonly two values are adopted, $k \in \{5.15, 6\}$; $s-$ is standard deviation from the correct value, USL, LSL, respectively, the upper and lower specification limit, which is a *CTQ* requirement for the measurement system; $i-$ is the number of standards; $\bar{x}_i -$ is the mean value from l measurements for $i-$ th standard; $\hat{\mu}_i -$ is the correct value for $i-$th standard; $n-$ is the total number of replications consisting of all combinations of the number of system modifications m, the number of standards i, and replications l for each standard i: $n = \sum_i l_i \cdot m_i$.

Coverage factor k is a number expressed in the units of the observed standard deviation $[k] = s$ related to expanded uncertainty u, estimated as $u = k \cdot U$, where U stands for *standard uncertainty*, estimated on the basis of variance of measurement results obtained by a measurement system. *Expanded uncertainty* defines the confidence interval $\Delta(P) \equiv \pm u$ at the level of confidence P for measurement results. The confidence interval constructed this way is a typical method of estimating uncertainty for a replicated measurement (Allen, 2006).

When measuring reference values reproduced by etalon i, the deviations $\Delta_i(P)$ are estimated from observable deviations of the mean values $\bar{x}_i$ from measurements x_i[129] from the correct value $\hat{\mu}_i$ reproduced by etalon. If the variability caused by systematic errors ε_s is significantly smaller $\varepsilon_s \ll \varepsilon_r$ than the variability due to random errors ε_r, then the deviations $\Delta_i = \overline{x}_i - \mu_i$ of the mean $\overline{x}_i$ z $n_i = l_i \cdot m_i$ from replications related to repetitions l and system modifications m in measurement implementation $X \leftarrow x_i$ for the standard i from its correct value $\hat{\mu}_i$ should have a normal distribution, since by a significant majority, they represent variability connected only with random error propagation ε_r. In order for this distribution to have the central value at point $k = 0$, the variabilities making constant contribution from systematic errors ε_s causing the center to shift from value $k = 0$ in the form of an adjustment (see lower, Figure 4.28).

Coverage factor k is selected according to the preferred method. In metrology, the field representing the probability of occurrence of a variance $x_i - \mu$ constituting $P(x_i - \mu) \cong 99\%$ of the overall field below the *Gaussian curve*, representing a certain event, is commonly adopted as satisfactory for the class of laboratory measurements. In such an approach, the coverage factor is at the level of $k \cong 5.15 \cdot s$. For technical measurements, confidence interval is adopted at the level of $P(x_i - \mu) \cong 95\%$ with coverage factor k at the level of $k \cong 3.92 \cdot s$ (Figure 4.24).

Factor $k = 6 \cdot s$ is compliant with the approach referred to as $Six - Sigma$ (Figure 4.25).

It is possible to assume that the confidence intervals construed this way, sometimes called coverage intervals of measurement capability of the measurement system, constitute significantly weak reliability for such a system. The greater the interval around the observed divergence in the measurement method, the easier it is to differentiate and classify the measured properties with its use.

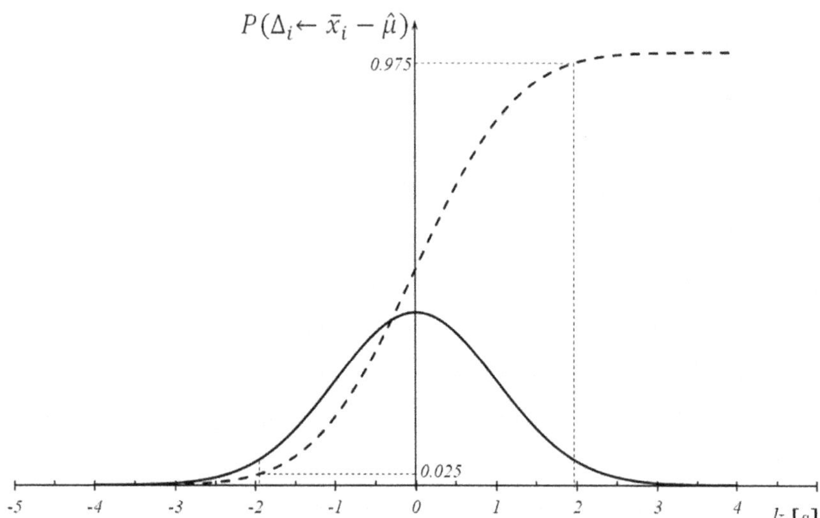

FIGURE 4.24 Theoretical coverage interval for deviations $\Delta_i = \overline{x}_i - \mu_i$ of the mean $\overline{x}_i$ from $n_i = l_i \cdot m_i$ from replications connected with repetitions l and system modifications m in the measurement implementation for the standard i at the technical accuracy level corresponding to probability at the level of ca. $P(x_i - \mu) \cong 95\%$ of overall probability constituting a symmetrical interval for values $k \in [-1.96; 1.96] \rightarrow -1.96 \cdot s \leq \hat{\mu} \leq 1.96 \cdot s$.

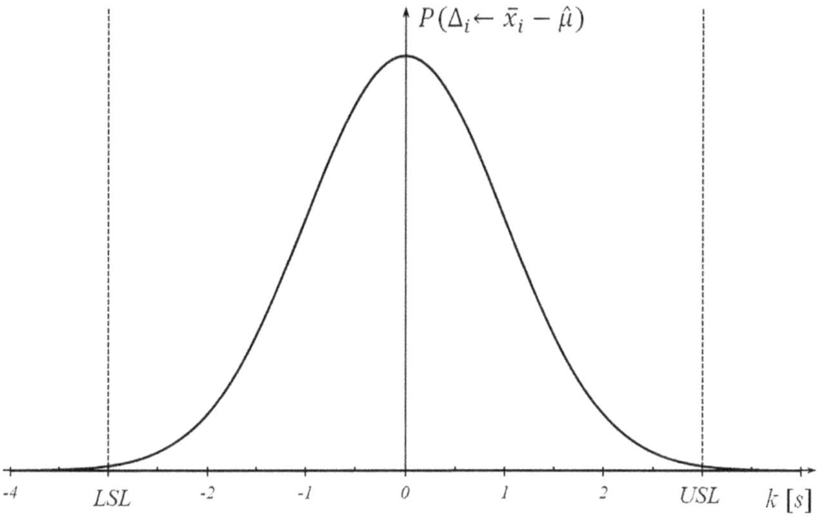

FIGURE 4.25 Theoretical coverage interval for deviations $\Delta_i = \overline{x}_i - \mu_i$ of the mean $\overline{x}_i$ from $n_i = l_i \cdot m_i$ replications connected with repetitions l and system modifications m in measurement implementation for the standard i as measurement coverage interval for sextuple-value $k = 6 \cdot s$ of the observed standard deviation s constituting ca. 99.73% of the probability of variability occurrence Δ_i: $P(x_i - \mu) \cong 99.73\%$.

Experience has shown, however, that in stabilized processes, it is still possible to observe fluctuations connected with the occurrence of systematic errors ε_s causing maximum shift in the reliability of measurement results from the mean $\overline{x}$ by the adopted maximum value of $\varepsilon_s^{max} = 1.5 \cdot s$. With such an assumption, the mean value may deviate by the value ε_s^{max} toward both directions $\overline{x} \pm 1.5 \cdot s$. Leaving tolerance limits $[LSL, USL]$ at the same level as shown in Figure 4.25 will cause that with the maximum value of a systematic error ε_s^{max} in the part being beyond one of the tolerance limits $[LSL, USL]$, probability will be reduced by: $P(x_i - \mu) \cong 6.67\%$ (Figure 4.26).

In the experimental determination of the ratio of the obtained measurement precision to tolerance *PTR*, it is needed to head toward an even coverage with standards i of the required measurement scope[130] $[USL, LSL]$. The measurement of each standard should be performed through its multiple replication $l > 5$, with several possible modifications[131] in the measurement system $m > 2$, including work of different operators. The thus-designed plan of action should be randomized.[132]

Should a measurement system not require operation, it is necessary to group measurements and implement them by applying significant time intervals between the groups in allowable yet different conditions, with various charging levels of the integral battery pack, with the use of different available software versions, allowable system modifications, different times of day, locations,[133] etc., also with consideration of the randomization process for such modifications.

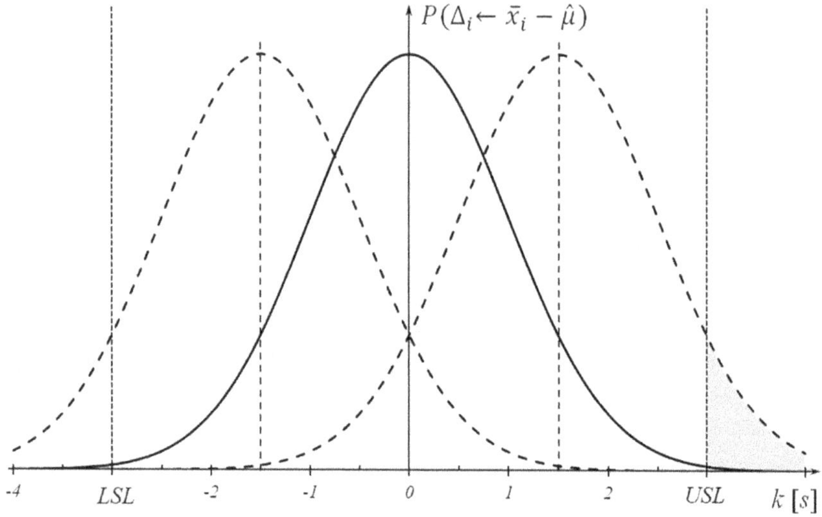

FIGURE 4.26 Illustration of probability related to exceeding upper tolerance limit *USL* caused by the adopted maximum fluctuations of a systematic error at the level of $\varepsilon_s^{max} = \pm 1.5 \cdot \sigma$.

Relation (4.6) reflects the generally adopted assumption that the measurement system should enable differentiation of at least ten different values for the predicted scope of the expected measurand variability, as one of the quality requirements *CTQ* regarding the measurement process.

Figure 4.27 presents an example assessment of the value of ratio *PTR* of the reached measurement precision to tolerance in measurement performance for two standards and the application of the $Six - Sigma$ approach.

The observed etalon divergence was used to define the coverage factor for tolerance interval $[LSL, USL]$.

Nearly half of the field below the Gaussian curve is found within the distance from the center of ca. $k \cong 3 \cdot s$. Therefore, the mean for the two standards will be depicted by the combination of the halves of distribution curves, one for each standard, respectively. The thus-obtained mean precision interval of a measuring method constituted the measure to determine the scope of tolerance. The calculations carried out in Figure 4.27 show that the tolerance interval was covered over eight times. In concord with relation (4.6), however, the system does not meet the adopted quality requirements *CTQ* for the adopted measurand variability range.

Relation (4.6) depicts the familiar principles of mnemotechnic stipulating that the measuring precision should be at least, by an order of magnitude, larger than the anticipated measurand variability range, for which the measurements need to be conducted. Hence, the value of the ratio of the reached measurement precision to tolerance PTR from relation (4.6) should be lower than $PTR < 0.1$.

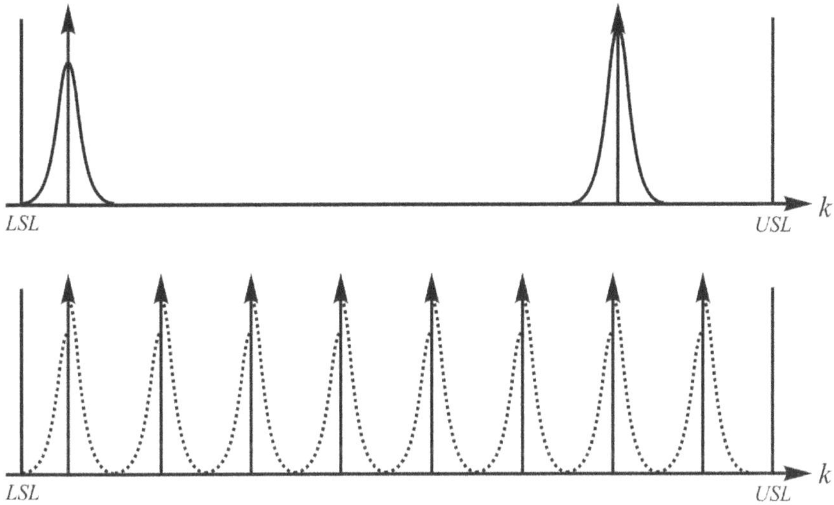

FIGURE 4.27 Representation of an assessment of precision to tolerance ratio *PTR* for the measurement system (*bottom figure*), with coverage factor $k = 6 \cdot s$ and the tolerance scope $\lfloor LSL, USL \rfloor$ for which the measurements were performed, with two standards (*top figure*). However, the conducted calculations show only approximately octuple coverage of the tolerance interval $\lfloor LSL, USL \rfloor$; hence, the system failed to meet the requirements of relation (4.6).

4.4.12 Measurement System Assessment Based on a Standard

Metrology commonly utilizes estimation of an absolute error Δ in the measurement result in the form of an arithmetic mean $\overline{x}$ constituting the estimator[134] of the true value μ from the performed n measurements: $\Delta = k \cdot \frac{s}{\sqrt{n}} = k \cdot u \Rightarrow \frac{\Delta}{u} = k$, where s is a standard deviation calculated on the basis of the results obtained in the measuring cycle, whereas u is a standard error. The choice of values k for technical measurements is illustrated in Figure 4.24.

Generally, deviations $\Delta = \overline{x} - \alpha$ in the mean $\overline{x}$ from l measurement replications $X \leftarrow x_l$ from true value μ should have normal distribution, as they should represent only random errors. Coverage factor k is selected on the basis of this distribution. However, often what is available are only results obtained for a small sample of $n < 30$, and in such a case, the value of the coverage factor k selected on the basis of $t-$ student distribution. For sample multiplicity $n \geq 30$, it is assumed that differences between the $t-$ student and normal distribution may be omitted in the majority of problem situations with factor k adopted on the basis of normal distribution. In technical measurements and with normal distribution, the most commonly adopted value of the coverage factor of $k = 2$.

There is the mnemotechnic principle stipulating that the result[135] $\overline{x}$ of a measurement cycle[136] should be at least, by an order of magnitude, larger than its uncertainty, expressed with the use of absolute error $\Delta : \frac{\overline{x}}{\Delta} \gg 10$.

Making an assumption that a measurement is a value of weighted absolute error $\overline{\Delta}$ from measurements i standards calculated as mean $\overline{\Delta} = \frac{\Delta}{n}$ from the values of absolute differences $\Delta_i = \left|\overline{x}_i - \hat{\mu}_i\right|$ of the mean $\overline{x}_i$ of l measurement replications $X \leftarrow x_n$ for all modifications in the measurement system n, analogously to the preceding representation, the absolute error for this value $\overline{\Delta}$ will be the product $k \cdot u$, where u is standard uncertainty $u = \frac{1}{\sqrt{n}} \cdot s$.

Since it was previously mentioned that the reasonable minimum value of the coverage factor k is $k = 2$, then going back to the principle stating that the measured value should be, by an order of magnitude, greater than the absolute error for this value, we may write down the following: $\forall_{n \geq 30 \Rightarrow k=2} \frac{\overline{\Delta}}{k \cdot u} \gg 10 \Rightarrow \frac{\overline{\Delta}}{u} \gg 5$. This makes it possible to estimate the value of measurement system reliability assessment index for the method based on the use of etalons, as a parameter constituting the ratio $\frac{\overline{\Delta}}{\overline{s}}$ of mean absolute error $\overline{\Delta}$ to the estimated mean value of its standard deviation $\overline{s}$ (Jakus and Kłos, 2007):

$$\frac{\overline{\Delta}}{u} \gg 5 \left| \overline{\Delta} = \frac{\sum_i \left|\overline{x}_i - \hat{\mu}_i\right|}{n}; u = \frac{1}{\sqrt{n}} \cdot s = \frac{1}{\sqrt{n}} \cdot \sqrt{\frac{\sum_i (\overline{x}_i - \hat{\mu}_i)^2}{n-1}} \right. \tag{4.7}$$

where $\overline{\Delta}$ – standard deviation from correct values $\hat{\mu}_i$ for i standards; u – standard uncertainty calculated for all standards i.

In essence, principle (4.7) constitutes a variance analysis carried out in the same manner as estimation of measurement uncertainty, where the measurement results are absolute values of the deviation $\left|x_{n,i} - \hat{\mu}_i\right|$ of measurement results $x_{n,i}$ from the correct value of the measurand $\hat{\mu}_i$.

The averaging of deviations is justified in this case, as the variables accumulate only the allowable differences in the modifications of the measurement which may be neglected from the point of view of quality requirements *CTQ* for the measurement process (Figure 4.28).

Such a choice of variables should, in the majority of cases, ensure their normal distribution. If, in the course of an obligatory checkup of deviation distribution $\left|x_{n,i} - \hat{\mu}_i\right|$, it proves to be disturbed, we will be bound to suspect the presence of an excessive number of systematic errors.[137]

The deterministic disturbance caused by the presence of systematic errors should be first identified and minimized before carrying out another analysis on the reliability of a measurement system. Seeking the sources of systematic errors is only possible with etalons reproducing correct values $\hat{\mu}_i$ for the measurands.

4.4.13 Conclusions

Relation (4.6) indicates that a measurement process is capable of covering the range of the expected measurand variations, whereas inequality (4.7) implies fulfilment of

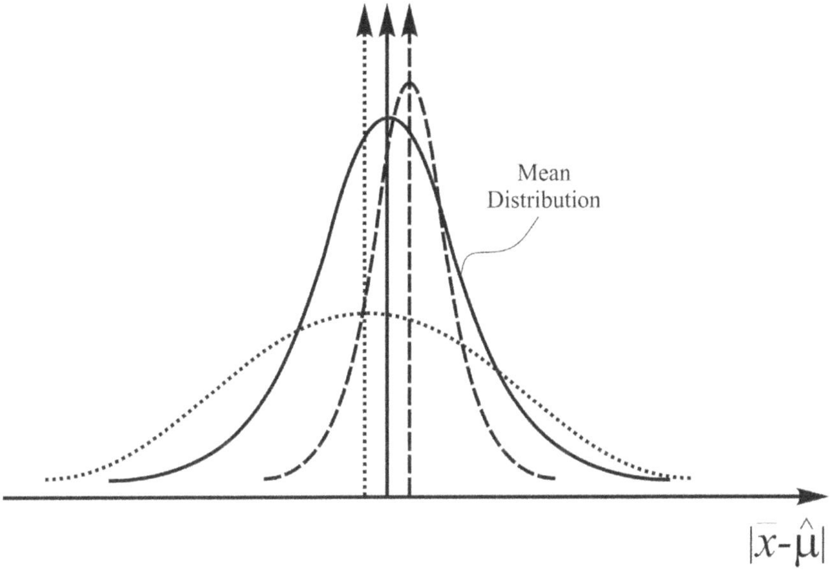

FIGURE 4.28 Deviations $\left|\bar{x}_i - \hat{\mu}_i\right|$ for $i = 2$ standards of correct values $\hat{\mu}_i$ and the distribution of their mean value $\frac{1}{n}\cdot\sum_i\left|\bar{x}_i - \hat{\mu}_i\right|$.

quality requirements for each measurement. For a measurement system which meets the *CTQ* requirements, conditions (4.6) and (4.7) should be fulfilled simultaneously.

Fulfilling only one condition (4.6) expressed as the precision-to-tolerance ratio *PTR*, which may fall below the value $PTR < 0.1$ and thus meet the quality requirements, *CTQ* related to the ability to distinguish the measured value is insufficient.

Together with the fulfilment of condition (4.6), the value of ratio (4.7) of mean absolute error $\overline{\Delta}$ to its standard uncertainty u may fall below the value $\frac{\overline{\Delta}}{u} < 5$, and the system will not guarantee the required precision with approved reliability. Similarly, meeting condition (4.7) will ensure measurement precision; however, at the same time, the measurement process may not ensure coverage of the required interval and thus fail to fulfil condition[138] (4.6).

Information concerning the reliability of a measurement system should be provided by the manufacturer when requested for appropriate declaration. Often, values declared by the manufacturer may be amended by applying proper operational procedures *SOP*, as usually manufacturers provide, in their declarations, average values selected according to the worst-case analysis method. In the course of determining the true metrological values of a measurement system, it is necessary to carry out metrological tests.

These tend to be costly, not only in terms of money spent on carrying the tests themselves, but also in that in majority of cases, they require investments in the form of purchasing a measurement system. Companies rarely agree to lend measurement systems for conducting metrological tests. For this reason, this stage of work is often omitted, which leads to expensive disappointment when a measurement system is being operated.

The proposed indicators (4.6)–(4.7) relating to the fulfilment of *CTQ* requirements of a measurement system are useful, albeit their application is rather conceptually complicated. A natural, self-explanatory method in the assessment of measurement systems involves establishing confidence intervals for the fractions of observed divergence, estimated with the variance analysis (Burdick, Borror, and Montgomery, 2005). Nonetheless, this approach is rarely used in the analysis of measurement systems or statistical inference (Pyzdek, 2003).

This is probably due to the necessity to consciously adopt the level of the power of inference involving the awareness of relatively large, impossible-to-accept values of errors of the second type for small samples. If metrological tests reveal unacceptable value of errors of the second type, they need to be repeated on a larger population, thus significantly increasing the costs and causing researchers to their own, as well as those of sponsors, the management, etc., experience of dissatisfaction. Often, due to a lack of knowledge, or for their peace of mind, they are omitted or undisclosed by researchers. In designing a life support system, such cynical omissions or lack of sufficient knowledge may have very dramatic repercussions.

4.5 PREPARATION

The safety of divers in the hyperbaric environment[139] directly depends upon the quality[140] of the breathing gas used. Multicomponent gas mixtures used as a breathing gas can be prepared using different methods. To prepare a gas mixture, one of the following methods can be used: gravimetric, volumetric, dynamic, and pressure transfer methods.

4.5.1 The Gravimetric Method

This method relies on the successive addition of the component gas to the tank. The mass of each gas component is determined by weighing the gasholder before and after filling it. It enables determining the content of each gas component. Despite its high accuracy, the method was not in common use in diving due to the small yield of gas mixture. The gas for *UBAs* can be produced with this method, and its price is not much higher when compared to pure gases. Special scales for private users are already available on the market.

This method is commonly employed for preparing standard gas mixtures used for the verification of gas analyzer indications.

4.5.2 The Static Volumetric Method

The gasholder of known volume is filled up with the gas components at the atmospheric pressure and ambient temperature. Determination of the mixture composition is based on knowing the volume of each gas component. The method is not in common use in diving due to the small yield of the gas mixture.

The static volumetric method is used for preparing small yields of standard gas mixtures. The dynamic method is a certain modification of the static volumetric method.

4.5.3 The Dynamic Method

It relies on the simultaneous mixing of two or more streams of gases. The procedure is used to prepare the mixtures for immediate use or when the mixture prepared is to be compressed into a gasholder (Bøe and Hartung, 1983).

Nitrox/other breathing gas mixture blenders are already available on the market, not only for commercial, but also for personal use, like the Dräger Nitrox unit panel (Figure 4.29).

4.5.4 The Partial Pressure Method

The possibility of preparing large amounts of mixture using simple equipment and procedure makes the method useful in the field. The method involves mixing the component gases under pressure in cylinders. After mixing, the gas must be carefully analyzed, and the percentage adjustment of the component gases is essential. The adjustment has to be preceded by calculating the magnitude of the pressure increase with pure component gases. To determine this pressure, the equation of state is applied. Based on the relevant assumptions and simplifications, different analytical expressions of the equation of state can be used. The most important analytical forms of the equation of state are presented in Table 4.17.

All equations presented in Table 4.17 are special cases of the virial equation, whose accuracy is dependent upon the expansion degree, similarly as for series expansion Taylor's series.

The principle of the pressure transfer used for gas mixing is shown in Figure 4.30.

FIGURE 4.29 Dräger Nitrox unit panel.

Source: Author's own photos.

TABLE 4.17
Analytical Expressions of the Equation of State

Formula Name	Mathematical Formula
Clapeyron	$p \cdot V = n \cdot R \cdot T$
van der Waals	$\left(p + \frac{a}{V^2}\right) \cdot (V - b) = n \cdot R \cdot T$
Beattie–Bridgeman	$p \cdot \upsilon^2 = R \cdot T \cdot \left[\upsilon + B_0 \cdot \left(1 - \frac{b}{\upsilon}\right)\right] \cdot \left(1 - \frac{c}{\upsilon \cdot T^3}\right) - A_0 \cdot \left(1 - \frac{a}{\upsilon}\right)$
The virial equation	$p \cdot V = n \cdot R \cdot T \cdot \left[1 + B(T) \cdot p + C(T) \cdot p^2 + ..\right]$

Note: $R = 8314{,}7 J \cdot mol^{-1} \cdot K^{-1}$ a, b, c, A_0, B_0 – the constants; $B(T), C(T)$ – the virial coefficients; T = the absolute temperature; V = the gas volume; υ – molar volume.

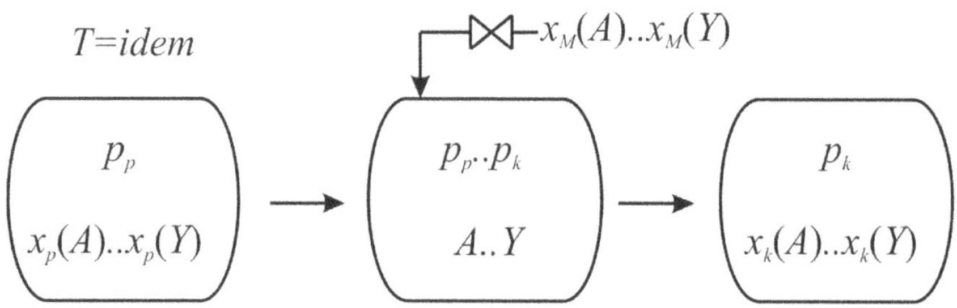

FIGURE 4.30 Preparation of a breathing mixture in a gasholder at a constant temperature.

Calculations are based on the equation of state for an ideal gas. More accurate computations are not necessary due to difficulties in keeping the constant temperature in the course of mixing.

Initially, the gasholder is filled with the gas mixture containing pure gases $A..Y$, the molar fraction of each component at pressure p_p being $x_p(A)..x_p(Y)$. The filling procedure involves entering a mixture of gases $A..Y$ into the gasholder, the mixture causing the pressure to increase up to p_k.

The composition of a complementary gas mixture is given as $x_M(A)..x_M(Y)$. Finally, the gas pressure is p_k, and molar fractions of gases are $x_k(A)..x_k(Y)$. The amount of the gas at the initial and the final state is calculated from the ideal gas equation and is given as:

$$\sum_{i=A}^{Y} n_p(i) = S_p = \frac{p_p \cdot V}{R \cdot T} \tag{4.8}$$

where $n_p(i)$ – initial mole amount of the i – th gas; S_p – initial mole amount of gas; p_p – initial pressure; V – volume of the gasholder; R – universal gas constant; and T – absolute temperature.

Similarly, the final sum of gas moles contained in the gasholder can be written as:

$$\sum_{i=A}^{Y} n_k(i) = S_k = \frac{p_k \cdot V}{R \cdot T} \tag{4.9}$$

where $n_k(i)$ – final mole amount of the i – th gas; S_k – final mole amount of gas; and p_p – final pressure.

The amount Δn of the gas supplied to the gasholder is calculated from equations (4.8), (4.9), and is given as:

$$\Delta n = S_k - S_p = \frac{(p_k - p_p) \cdot V}{R \cdot T} \tag{4.10}$$

where Δn – the amount of the gas supplied to the gasholder.

The amount of the j gas component at the final state, after the container has been filled with the gas mixture of composition $x_M(A)..x_M(Y)$, is given as:

$$n_k(j) = x_p(j) \cdot S_p + x_M(j) \cdot \Delta n \tag{4.11}$$

The mole fraction of the gas j in the mixture in the final state can be obtained by dividing equation (4.11) by equation (4.9) and is written as:

$$\forall_{T=idem}\ x_k(j) = \frac{p_p \cdot [x_p(j) - x_M(j)]}{p_k} + x_M(j) \tag{4.12}$$

Equation (4.12) makes it possible to calculate the partial pressure of the j gas in the gasholder at the final state:

$$\forall_{T=idem}\ p_k(j) = p_p \cdot [x_p(j) - x_M(j)] + p_k\ x_M(j) \tag{4.13}$$

Equations (4.12) and (4.13) are the basic equations describing the process of obtaining arbitrary gas mixtures from pure gases and other mixtures. In case of preparation, when the gas mixtures are prepared exclusively from pure gases, the calculation method will be simpler. In preparing gas mixtures from pure gas components by pressure transfer equation, method (4.13) is given as: $\forall_{T=idem}\ \Delta p \equiv p_k - p_p = p_k \cdot x_k(j) - p_p \cdot x_p(j)$.

To make the preparation procedure more efficient, a special calculation blank was prepared. Table 4.18 shows the blank that allows preparing the gas mixtures involving oxygen and helium – *Hx*.

The blank can be used when in the gasholder there are some residues of not utilized *Hx* of the known oxygen percentage. Most deep diving operations involve *Hx* , in which the contribution of oxygen is less than $20\%_v$. With regard to fire safety, it is necessary to keep the percentage of oxygen less than $25\%_v$ during mixture preparation ("Oxygen safety in the production of enriched air nitrox breathing mixtures," 1992). The blank presented as Table 4.18 is especially dedicated for preparing *Hx* if the oxygen percentage is less than $20\%_v$.

TABLE 4.18
Calculating the Blank for the Isothermal Preparation of Helium–Oxygen Mixtures from Pure Gases with the Pressure Transfer Method

Correct calculations guarantee that at any moment of gas mixture preparation, the oxygen content will not exceed $20\%_v O_2$.

The presented method of preparation of gas mixtures guarantees several percent accuracy at the constant temperature when the oxygen content is less than $15\%_v O_2$ and the final pressure is $15 MPa$.

1. The calculation data.

Pressure of Hx remaining in the gasholder: $p_1 =$	MPa
Molar fraction of oxygen in remaining Hx: $x_1 =$	$mol \cdot mol^{-1}$
Required final pressure of oxygen $p_2 =$	MPa
Required final molar fraction of oxygen in Hx $x_2 =$	$mol \cdot mol^{-1}$

2. The introductory calculations

$$\Delta p = p_2 \cdot x_2 - p_1 \cdot x_1 = \quad \cdot \quad - \quad \cdot \quad = \quad MPa$$

$$p_x = 5 \cdot x_2 \cdot p_2 - \Delta p = 5 \cdot \quad \cdot \quad \cdot \quad - \quad = \quad MPa$$

If $p_x > p_1$, then go to p. 3; if $p_x \leq p_1$, then go to p. 4.

If $\Delta p 0$, it is necessary to increase the final pressure, as in this case preparation of the breathing mixture is not possible.

3. Calculation continued.

$$p_y = p_x + \Delta p = \quad + \quad = \quad MPa$$

Process of obtaining heliox:

a. Increase the pressure with He from $p_1 =$	MPa to	$p_x =$	MPa
b. Increase the pressure with O_2 from p_x up to the pressure		$p_y =$	MPa
c. Fill the gasholder with He up to		$p_2 =$	MPa

4. Calculation continued.

$$p_z = p_1 + \Delta p = \quad + \quad = \quad MPa$$

a. Increase the pressure with O_2 from $p_1 =$	MPa to	$p_z =$	MPa
b. Fill the gasholder with He up to		$p_2 =$	MPa

5. The verifying calculations.

If S calculated from the following equation is equal to 0 to 0.01, the calculations are correct.

For p. 3: $$S = \frac{p_y - p_x + (p_1 \cdot x_1)}{p_y} - 0.2 =$$

For p. 4: $$S = p_2 \cdot x_2 + p_1 \cdot (1 - x_1) - p_z =$$

When gas mixtures are prepared for shallow dives, where the oxygen content in the gas mixture can exceed $20\%_v$, the gas mixtures should be prepared in gasholders authorized for oxygen conditions (see further). When the percentage of oxygen exceeds $20\%_v$, the blank should be modified. A part of the blank concerns verifying calculations. Mistakes in the mixing are very expensive, because the preparation of the mixtures is very laborious and the high-purity gas components are costly. The quality assurance of the breathing gas is essential for the diver.

The gaseous mixtures contain oxygen as the main biological component and inert gases: nitrogen, helium, hydrogen, argon, neon, etc. There are different breathing gas mixtures, and they are specified in Table 4.18. Binary mixtures are diving gases used most often. When there is a lack of own laboratory, during the preparation of complex mixtures,[141] the gas is not indirectly analyzed in a chemical laboratory because of the costs and inconvenient procedures. Now, inexpensive high-accuracy oxygen analyses are available on the market. That is why oxygen content can be analyzed with a portable high-accuracy oxygen meter (Kłos, 1990, 2000).

Each developed diving base should be equipped with a portable oxygen analyzer, for example, used for monitoring the habitat atmosphere. As opposed to oxygen, the analysis of inert gases is difficult and sometimes impossible to be done in the field due to a lack of cheap high-accuracy gas analyzers. Therefore, the content of the inert gases should be measured in a special physicochemical laboratory. If the breathing mixtures are prepared in the diving base not equipped with its own laboratory, preparation of the complex gas mixtures is difficult due to a lack of reliable measurements of each component content. For this reason, a special procedure of gas mixture preparation should be employed. The preparation procedure will be presented based on an example of trimix *Tx* preparation.

Oxygen and nitrogen are the gases that mix quite well,[142] and there is no stratification of gases. The first stage at *Tx* preparation is to prepare the basic *Nx* having the same oxygen and nitrogen proportions as in *Tx* which is to be prepared. The oxygen content will be determined with the oxygen analyzer. For example, in order to prepare *Tx* having the following composition of $17\%_v O_2$, $50\%_v He$, and $33\%_v N_2$, the oxygen/nitrogen ratio is the same as the oxygen/nitrogen ratio in the composition $34\%_v O_2$ and $66\%_v O_2 - 0.34Nx$. Therefore, the first stage at gas mixture preparation is to prepare the basic *Nx* containing $34\%_v O_2$. The basic nitrox composition can be checked only with an oxygen analyzer.

Tx is prepared on the basis of helium and precisely prepared basic *Nx*. During preparation of *Tx*, only oxygen content needs to be measured, because the ratio of oxygen and nitrogen has already been established in the basic *Nx*; helium is a complementary gas.

After *Tx* has been prepared and its composition established, the gas sample can be analyzed in a laboratory to confirm its composition. Upon request, a toxicology test of the prepared gas mixture is also available in the laboratory. The following organic compounds are determined in the gas sample (USN, 1993):

- Highly volatile compounds, for example, freons (11, 12, 114, 113)
- Medium-volatility hydrocarbons, for example, benzene, toluene, xylene, etc.
- Highly polar chemicals, for example, alcohols

The aforementioned compounds are the only representatives of the compound groups that can possibly occur in a gas mixture. The composition of breathing mixtures is usually determined with gas chromatography. Gas chromatograph, or another alternative gas analyzer, should be able to determine freon 113 to $0.5 ppm$ in the gas sample. The relative measurement error of freon 113 concentration must be less than 5%. The mixture containing $C \in [5,10]\, ppm$ of freon 113 or toluene can be used as a standard gas. Standard gases should have a certificate stating that their relative manufacturing error is less than 2%. Usually, the contaminant content is related to the standard gas that contains freon 113 or toluene. All chromatograph peaks for which the relative contaminant content (that is, in relation to freon 113 or toluene) exceeds $1\, ppm$ should be identified. For the purpose of correct interpretation, $1\, ppm$ or more concentration of impurities has to be recorded, which means that the signal has to be minimum-intensive, like for the sample of $1\, ppm$ freon 113.

The use of concentrators is permissible to attain a greater concentration of the contaminant. *Concentrators* are chemicals which make adsorption and desorption of contaminants possible, for example, by applying a thermal technique.

Testing breathing gas mixtures involves determining the content of the permanent gases: oxygen, argon, nitrogen, helium, carbon dioxide, carbon monoxide, and the contaminants mentioned earlier (Section 4.3). To overcome the difficulties related to carbon monoxide analysis, the content of carbon monoxide can be measured together with carbon dioxide[143] using a flame ionized detector FID. The measurement system should be able to determine $5\, ppm$ of carbon monoxide CO. In order to analyze the permanent gases, a gas chromatograph equipped with a cathetometer detector[144] should be used. The maximum systematic relative error of permanent gases should not exceed 1%.

The standard gas mixtures used for gas analyzer testing should be prepared with the relative error less than 1% before they are used for analyzing the permanent gases. The results of the gas chromatography analysis of contaminants should include thus:

- The interpretation method of chromatography peaks
- The height of the peaks or area below chromatography peaks related to freon 113, where only the chromatography peaks having intensity higher than $0.1 ppm$ related to freon 113 are identified
- The determination of the measurement errors

Additionally, the results of the analysis of permanent gases should be described. The required accuracy is as follows:

- Oxygen, nitrogen, helium – relative error not exceeding $0.2\%_v$
- Carbon dioxide – absolute error less than $1\, ppm$
- If carbon monoxide content in the mixture exceeds $5\, ppm$, carbon monoxide CO content to be determined, with absolute error not exceeding $1\, ppm$

The results of the trace gas contamination analysis are useful for determining the maximum permissible depth of diving with regard to surface equivalent value *SEV* of the toxic contaminant. The *SEV* is defined as:

$$C_{SEV} = C \cdot \frac{p_H}{p_0} \tag{4.14}$$

where C_{SEV} –the surface equivalent value of contaminant; C –content of contaminants in gas sample; p_H –the total pressure at the maximum permissible diving depth; and p_0 –barometric pressure.

The surface equivalent value of the chosen contaminants at different exposures is given in Table 4.12 and Table 4.13. The measured contaminant content enables determining the maximum permissible depth pressure p_H of diving that can be reached.

Detailed methods are applied to analyze the main compounds and the contaminant the author has described earlier (Kłos, 1990, 2000).

It is convenient to manufacture gas mixtures in horizontal cylinders or typical oxygen transportation cylinders. As it follows from the experiments, the cylinders refilled with a gas mixture need some time for homogenization. This process proceeds slowly for a gas mixture in transportation cylinders. Homogenization of gas mixture may be accelerated by overturning the cylinder or heating its bottom and cooling its top. A substantial amount of helium accumulation was recorded in the space below the cylinder valve, particularly when the cylinder was stored vertically. A significant amount of helium was accumulated after it had been stored for $[0.5,1]\,month$ (Kłos, 1990).

Each cylinder or rack[145] used for manufacturing or storing gas mixtures should be color coded and used only for these purposes. Compressed gas cylinders must have durable markings on their surface, such as a serial number placed on the shoulder of a cylinder. This number can be additionally painted below the shut-off valve. Gas mixtures could be supplied in cylinders marked with inscriptions: ATTENTION, BREATHING MIXTURE. It is very important that appropriate caution is exercised to ensure the safe storage, handling, and use of compressed gas cylinders. Often, the user exchanges gas cylinders with the shipping company.

According to regulations, gas cylinders should contain some gas residues to avoid contamination of the gas cylinder. Gas residues are not tested before the cylinder is filled up by a gas company. The remains of the breathing mixture can contaminate the pure gas after the cylinder has been filled up. Gases like, for example, helium are sometimes used as protective atmosphere, for example, to protect catalysts against oxidizers. Oxygen contamination from a breathing mixture containing a pure gas may lead to catalyst damage and huge financial losses, as was the case in one of the refineries. Helium is also used as a carrier gas in gas analyzers, for example, chromatography. Contaminated helium can cause some problems in conducting correct instrumental, physicochemical analyses.

The composition of the gas mixture contained in the gas cylinder, the number of the cylinder, and the mixture production date should be provided in the material safety data sheet, as well as the laboratory standard procedure sheet and markings on the gas cylinder.[146] The oxygen concentration percentage C should not exceed $C < 25\%_v$ at any moment of the mixture production process. In the case of higher oxygen concentration, special cylinders made to meet oxygen standards should be used. It is recommended that gas mixtures be prepared from duly certified gas components.

In manufacturing breathing gases containing nitrogen, two trends are observed. One of them recommends using pure component gases which have special purity certificates, preventing the mixtures from being polluted in the course of their preparation. The deficiency of the method is the time necessary for the gases to homogenize; however, stratification usually occurs. A mistake may be made if there is not enough analytical checking of the mixing process – the most important is checking oxygen concentration in the gas composition, which is essential for breathing systems.

The second method recommended involves using air as natural *Nx* component of the breathing gas which contains nitrogen, despite some contamination of permanent gases, for example, argon. That method minimizes the risk of making a gross error. A hazard to the air quality can occur, especially when the compressors are lubricated with mineral oils. For this procedure, the quality of the air filters and the concentration values of the natural contaminants in the air are of utmost importance. The gas must be analyzed at least twice before it is used. If any differences are recorded, a third gas analysis is needed.

The use of the ideal gas model in the pressure transfer method is successful within the total pressure p range of $p \in [10,12]\,MPa$. Above this limit, errors resulting from assuming the ideal gas model are significant. The approximate pressure effect of an isothermal mixing can be determined as the difference of the pressures calculated from the Clapeyron and Beattie–Bridgeman equations. The pressure effect of oxygen–helium mixing at the constant temperature is presented in Figure 4.31.

At a pressure up to $p > 12\,MPa$ and low oxygen content in helium or low content of helium in oxygen, the errors are comparable to these resulting from inevitable variability of temperature at mixing. To eliminate these errors, a two-stage mixing of the gases is recommended. The first stage involves preparing the mixture having approximate composition. The cylinders refilled with the gas should be analyzed after they are allowed some time to stand still for homogenization of the mixture. At

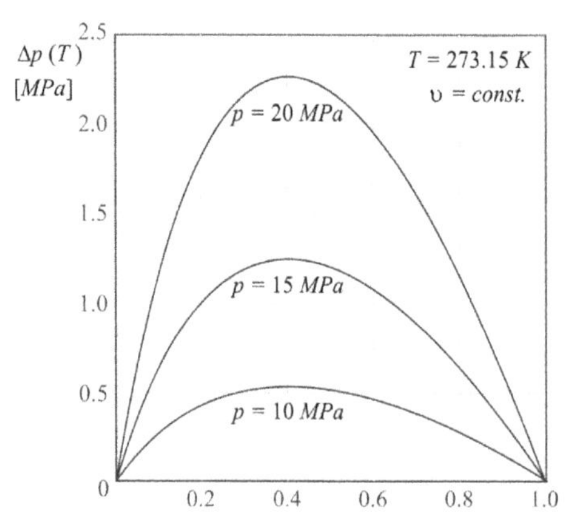

FIGURE 4.31 The pressure effect of an isothermal mixing of oxygen and helium.

the second stage of mixing (correction of composition), adjustment of the component gases is performed. When the refill station is equipped with appropriate computer programs, the calculations may be performed using the Beattie–Bridgeman equation of state. It is particularly recommended when the mixtures can be prepared very slowly or when the gasholder is equipped with a thermostat. This procedure eliminates the temperature variations at gas compression.

4.5.5 Fire Risk

All breathing gas mixtures must contain oxygen. However, oxygen is a dangerous gas because of its fire hazard. To reduce the possibility of fire when an enriched atmosphere is being prepared and operated, the regulations concerning fire precautions must be observed. Oxygen is not a flammable gas itself; however, it is a strong oxidizer[147] that takes part in strongly exothermic oxidation reactions. The courses of oxidation reactions can differ. They can be slow, rapid, or explosive. *Corrosion* is a process of slow oxidation of metals and other materials in reaction with an oxidant, such as oxygen. The same phenomenon occurs in other materials, like rubber, for example, used in diving apparatuses, compressors, etc. Very often, slow oxidation leads to the destruction of neoprene gaskets type O–ring. The rapid oxidation reaction is called *combustion*. *Explosion* is defined as a very rapid oxidation process.

Problems concerning fire safety during the preparation of an oxygen-enriched gas mixture are referred to as *technical oxygen safety*. There are several methods for dealing with this problem. Two of them will be briefly discussed here. The first problem is concerned with the concept of similarity of the typical safety rules applied to compressed oxygen operational use. The second one is concerned with the concept of absolute flammability of all materials ("Oxygen safety in the production of enriched air nitrox breathing mixtures," 1992).

The first concept is based on the fact that compressed oxygen is widely used in industry, aviation, medicine, etc. Suitable oxygen operational safety rules, special constructions, use of operating materials, and skills of properly trained personnel will contribute to safely using compressed oxygen. According to this concept, most of the oxygen operational safety rules can be applicable to operating gas mixtures containing oxygen.

The concept of absolute flammability is based on the assumption that all materials can be flammable under special conditions. There are three fundamental prerequisites for fire: flammable material, oxidant, and sufficiently high temperature to initiate reaction. This "fire triangle" becomes complicated under hyperbaric conditions (ASTM, 1997; NFPA, 1999; NOAA, 2017)

As pressure increases, the amount of oxygen molecules in a volume unit increases as well. For this reason, an ignition source is not necessary to initiate oxidation reaction; however, self-ignition can occur. This phenomenon can occur under pressure at temperature conditions lower than that at a normobaric condition. However, there are some materials being oxygen-resistant in a hyperbaric environment. The oxygen-resistant materials should be taken into account *during SCR* design. *Oxidation-resistant materials* are defined as materials for use under oxygen conditions.

4.5.6 Ignition Sources

To prevent fire in compressed breathing gas systems, it is necessary to use materials authorized for use under oxygen conditions and maintain temperature below the self-ignition point. In practice, maintaining the temperature at such a low level is difficult. This is because of the high and rapid local compression in hyperbaric systems.

Compression that occurs in breathing systems is similar to that in compression-ignition engines, for example, a gaseous collector operating at the pressure $20\,MPa$ can be supplied with gas having a temperature higher than 850°C. The temperature significantly exceeds the auto-ignition point of used lubricants. Flow-resistant places are other sources of high temperature. In cases where stream flow is rapidly broken in places where there are valves, elbows, chokes, etc., the compressed gas energy is transformed into heat. Also, friction that occurs during high-velocity gas flow causes the compressed gas energy to transform into heat.

Transformation of different kinds of energy into heat can cause a local increase in temperature. Ball valves are not recommended for use in oxygen systems as they may open unexpectedly or close the outflow of the breathing gas under high pressure. In such systems, the gas flow should be controlled with poppet valves, reducers of pressure and gas flow controller, or valves/reducers fitted to control the fundamental main reducer. Such operating procedures, control systems, and devices having special construction for the preparation of gas mixtures reduce shifting the gas temperature inside the breathing gas systems to a minimum.

As far as fire hazard is concerned, hydrocarbons are the most dangerous contaminants in breathing gas systems. The lubricants used in compressors are the main hydrocarbon source. Usually, the lubricants are cumulated in the gas system. When the temperature rises above $t \in [50,230]$, most lubricants characterized by low corrosion rate recommended by the manufacturers of compressors may self-ignite at oxygen presence. The fire-hazardous hydrocarbon concentration under high pressure has been determined experimentally. It is approximately $200\,mg \cdot m^{-2}$ of the internal tube area. The amount of these contaminants is directly proportional to the amount of air being compressed and depends upon the quality of the fitted filters. Other flammable contaminants can originate from lubricants used in valves, reducers, seals, etc. Generally, all materials applied should be characterized by *oxygen purity*. Oxygen purity is also required in breathing gas systems where the oxygen content C is higher than $C > 25\%_v$. To reduce the ignition risk, it is necessary to replace mineral greases used for lubricating valves and seals with those compatible to the oxygen environment.

Now, most of *UBAs* are made of high-quality materials; however, sometimes the materials are not authorized for the *oxygen conditions*.

Oxygen purity is recommended for common use, as well as for constructions authorized for the oxygen conditions. To prepare the system for the oxygen conditions, it has to be purified. Purification involves decreasing the flammable contaminant content. The maximal flammable contaminant level is determined for the designed allowable difficult conditions.

It is difficult to determine the permissible concentration of flammable[148] materials for different oxygen content values in a gas mixture; therefore, all the systems should be prepared like for pure oxygen operations. The standards for contaminant content, the methods of removal, and their effectiveness are experimentally determined by specialized companies; however, sometimes the results are kept secret.

However, there are cases when experimental results are published ("Oxygen safety in the production of enriched air nitrox breathing mixtures," 1992). Cleaning technologies offered by the manufacturer guarantee proper functioning of the system under the oxygen conditions. At present, *UBAs* prepared, by manufactures, for the oxygen conditions are available on the market. These *UBAs* are especially marked, and in some of them, different kinds of threaded joints are used to avoid casual contamination, for example, diving air from oil-lubricated compressors. This is the basic safety rule during an operation with artificial breathing gas. Steel or electrolytic zinc–coated steel is the most recommended material for gas cylinders suitable for the oxygen conditions. Aluminum is not recommended because of its combustion. In the oxygen atmosphere, oxidation reaction is initiated. It is a well-known aluminothermy reaction[149] used for iron casting, for welding, or in thermite munition/grenades. Steel ignition point temperature in the oxygen atmosphere is very high. The aluminum combustion process is more intensive than that of steel. Additionally, chlorinated-fluoresced lubricants – commonly used in the oxygen technique – react with aluminum. Some manufacturers do not recommend lubricating the elements of their products made of aluminum, magnesium, or their alloys. However, when construction is required to have nonmagnetic property, especially purified aluminum is applied.

The use of proper wear-and-tear materials is a separate problem in operating the systems authorized for the oxygen conditions. Lubricants are the most important group among such items. As was mentioned earlier, greases based on chlorinated-fluoresced hydrocarbons cause corrosion of elastomers. At present, the lubricants produced on the basis of perfluorinated polyether oil are the best greases for use in the oxygen atmosphere.

This is particularly important in breathing regulators, the first stage of reduction of which can supply additional devices, for example, an additional second stage with a mouthpiece,[150] automatic deflector of a buoyancy vest, additional valve for a suit with adjustable buoyancy compensator, etc. Intensive, high-pressure gas outflow causes gas expansion, which leads to a significant decrease in temperature.[151] Cooling of the gas can cause solidification of the lubricant, which, in turn, can lead to abnormality or impairment in the operation of the demand regulator. Another advantage of lubricants is low consumption and relatively low price. A comparison of the most often used lubricants in the oxygen atmosphere is given in Table 4.19.

4.5.7 Cleaning Procedures

The systems working on oxygen-enriched atmosphere[152] should be cleaned from flammable impurities. Oils, lubricants, paper, hair, rags, wood pieces, carbon dust, solvents, welding slag, rust, sand, etc. are the most often met impurities. When they are very difficult[153] or impossible to remove, it is necessary to oxidize them in the oxygen

TABLE 4.19
The Properties of the Most Often Used Lubricants in the Oxygen Atmosphere

Type of Lubricant	Recommended Range of Working Temperatures (°C)	Autoignition Point (°C)	Mass Decrement During 24*h* Heating (%)
Cristo-Lube MCG-111	−70–204	–	1.7 (204 °C)
Krytox GPL-205	−30–210	–	4 (99 °C for 30*h*)
Krytox 240 AZ	−55–150	>427	70 (204 °C)
Fomblin OT 20	−70–100	–	7 (66 °C)
Halocarbon 25–5S	−18–177	>427	–
Halocarbon X90–10M	−40–93	>427	–

Note: Autoignition point has been determined according to ASTM G72–96 (ASTM, 1997).

atmosphere so that the systems can be accepted for further use. In addition, in order to avoid explosion, periodic inspections or continuous monitoring of the concentration of contaminants in the gaseous system should be performed. Monitoring the temperature in gaseous collectors with thermometric paints produces good results.

To select the proper cleaning procedure, several factors should be analyzed. The most important are as follows:

- Kind of mechanical impurities
- Place and contamination degree
- Access to the surfaces being cleaned
- Rate and effectiveness of the cleaning agent applied to the particular kind of impurity skills of the personnel using the cleaning agent in the case of poisonous or caustic agents
- Costs of the cleaning method, etc.

Selection of the method should be preceded by an accurate visual inspection and specifying the cleaning requirements. The most important cleaning methods are as follows:

- Mechanical cleaning, for example, removal of gaseous contaminants by blowing them through, surface grinding, sandblasting, wire nail brush cleaning, etc.
- Hot water and steam cleaning supported with detergents
- Organic solvent cleaning; also solvent vapor and ultrasonic cleaning
- Caustic alkalis cleaning
- Acid cleaning, etc.

Steam cleaning is performed most often with steam generators using injected detergents and the steam produced under pressure. The method is applied when mechanic impurities, such as oils, etc., are loosely packed. Steam should be free of oil contaminants. Detergent should be properly selected to the impurities and the material being cleaned.

Steam removes contaminants through their plasticization, abrasion, heating, melting, dispergation, and emulsification in water formed in steam condensation. The residual detergent should be flushed away after the surface has been cleaned. Then the surfaces should be dried, for example, in a stream of oil-free air or nitrogen. If strong alkaline reaction detergent has been used, washing should be repeated using neutral-reaction detergent – $pH \cong 7$.

Detergents have to be used during the hot water cleaning process. Dispergation of hot water cleaning is applied when high temperature is not needed for contamination removal. To ensure good contact with the surface being cleaned, the proper detergent should be selected. Detergent removal from the cleaned surface is of utmost importance. Most detergents have good water solubility. Detergent removal relies on accurate flushing away of the cleaned surface with hot or cold water. If the strong alkaline reaction detergent has been used, repeated washing should be applied using neutral reaction detergent – $pH \cong 7$.

Alkaline cleaning is used for heavy and very adhesive adherent impurities. The particular matter is removed using strong alkaline solutions. This method offers high removal effectiveness. The method enables removing most contaminants, especially petroleum contaminants; however, it should not be used to clean materials that react with the alkalis, for example, aluminum alloys.

The majority of alkalis used for cleaning are well soluble in water and do not cause any fire hazard or any significant toxic gaseous contamination; however, they are very caustic. The alkaline solutions are used for sprinkling, immersion, or washing. Caustic bath can consists of additives, for example, corrosion inhibitors. In order to avoid corrosion, the remains of the alkaline solution should be from the surface being cleaned by flushing it thoroughly with water. The temperature of the flushing water is within the range of 40–80°C.

To protect the cleaned surfaces against corrosion, some materials need to be further chemically treated. After the surface has been cleaned and flushed, it has to be dried. The alkalis used should be utilized. Because alkalis are caustic and toxic, the personnel should be well-trained and provided with skin-protective equipment.

The aim of acid cleaning is to remove metal oxides and other impurities, soluble in the acid solutions, from surfaces. Most often, cleaning is performed at room temperature. The selection of the proper acid is dependent upon the material that has to be cleaned and the kind of impurity. The characteristics of some of the most popular acid baths used for cleaning metal surfaces are presented in Table 4.20.

Corrosion inhibitors, brighteners, etc. can be components of an acid bath. Acid-resistant tanks, circulation pumps, and fittings are applied in acid cleaning systems. Small details can be cleaned by immersion in an acid bath. Large surfaces may be cleaned by washing them using acid solutions. In the course of acid cleaning, toxic gases and vapors are often emitted. For this reason, acid cleaning should be performed using special stands equipped with suitable ventilation systems. Because of caustic and toxic properties, the personnel should be well trained and provided with skin-protective equipment. After the surface has been cleaned and flushed, it is dried. The liquid wastes from the cleaning should be utilized. After the cleaning has been completed, most often the surface is washed with hot water.

TABLE 4.20
The Basic Rules Applicable to the Selection of Acids Used for Cleaning Metal Surfaces

Phosphoric acid	Phosphoric acid can be used for cleaning all metals. It removes oxides and loose rust, light oil impurities, and the remains of welding and solder fluxes. After cleaning has been completed, protective baths (passivation) are not needed.
Hydrochloric acid	It is the basic agent used for cleaning steel and steel alloys. It removes rust, boiler scale†, forge scale, oxidizing layers. Chromium, cadmium, and electrolytic zinc coatings or plating can be removed using hydrochloric acid. It is mostly used as mixture with nitric acid‡. The remaining chloride ions very often cause pitting corrosion. For this reason, after metals have been cleaned, they should be thoroughly washed, and the surface should be protected by passing.
Chromic acid (chromic acid cleaning mixture) and nitric acid	These acids are recommended for cleaning surfaces made of metal and alloys, aluminum, copper, etc. Chromic acid having high dilution is used for passivation, brightening, and removal of black patina* after alkaline cleaning. The acids used for cleaning have concentration of $C \in [5, 50]\%$.

TABLE 4.21
Some Properties of Selected Organic Solvents Used for Impurity Removal

Solvent	Chemical Formula	Boiling Point (°C)	Remarks
1,1,1-trichloroethane	$C_2H_3Cl_3$	73.89	Flammable when concentration C in air is of $C \in [7.5,15]\%$ at temperature of $t = 25°C$
Methyl chloride	CH_2Cl_2	39.78	Flammable when concentration in air is 12–22% at temperature of 25°C
Perchloroethylene	C_2Cl_4	121.22	
Freon 11 (trichlorofluoromethane)	CCl_3F	23.78	
Freon 113 (trichlorofluoromethane)	$C_2(Cl_2F)_2$	47.56	
Trichloroethylene (tri)	C_2HCl_3	87.00	Flammable when concentration in air is 8–10.5% at temperature of 25°C, flash point 32.2°C

Organic solvent cleaning is most often used for organic impurity removal. The process is performed in an ultrasonic washing stand where the oscillation energy is applied for mechanic disintegration and the removal of contaminants. The properties of some organic solvents are presented in Table 4.21.

Because of its high toxicity, tetrachloride is not recommended for cleaning. Trichloroethylene is used only in some cases.[154] When low-boiling organic solvents are used for cleaning, they release vapors, which do the cleaning.

Most organic solvents destroy plastics, rubber, neoprene, etc. The plastics which are organic solvent–resistant include polyethylene, polypropylene, Teflon, etc.

Cleaning the particular matter can be performed by washing, immersion in solvents, or solvent vapor treatment. Industrial cleaning methods need special cleaning chambers equipped with solvent emission detectors.

Organic solvent vapors are toxic. Most of them produce fire and explosion hazard. Before a solvent is used, it is necessary to make sure it does not destroy the surface, a detail to be cleaned. The color of a solvent may be compared before and after the cleaning. Sometimes, special instruments have to be used, which, for example, may compare light transmission (light absorption) through. Infrared spectroscopy can be applied as well. As for the other analytic methods, chromatography can be used. The difference between the solvent and the contaminant boiling points can be applied to evaluate the method. Investigation of dry residue after solvent evaporation enables evaluating the effectiveness of the cleaning method. The contaminated organic solvent used can be regenerated or used for preliminary cleaning.

The liquid wastes should be utilized. The cleaned details should be dried and vented to remove vapor produced by solvents. Venting is most often performed with oil-free air or nitrogen. Sometimes, after organic solvent cleaning has been completed, additionally, steam or hot water cleaning is performed. However, as a rule, it is not necessary.

Mechanical cleaning can be done by sandblasting, using wire nail brushes, grinding, peening, etc. Mechanical cleaning is a preliminary cleaning. After the mechanical cleaning process has been completed, dust should be removed from the cleaned surfaces using oil-free air or nitrogen venting.

The surfaces that have been cleaned and are intended for use in the oxygen conditions should be periodically inspected. The simple methods that can be used (some of them have been partly described earlier) are as follows:

- Daylight inspection – impurity particles not exceeding 50 μm can be detected.
- Ultraviolet[155] inspection – fluorescent substance can be detected, for example, oil contaminants.
- Inspection done by cleaning with a white rag, not leaving the fibers,[156] and ultraviolet radiation detection of impurities.
- Inspection done by applying a sprayed-water test. Water is observed sprayed on the surface for $5s$. If water droplets form, it means the surface is greasy.
- Inspection done by extraction/washout of contaminants using an organic solvent, and their analysis with, for example, the spectroscopy or chromatography methods.

The permissible level of contaminants in the oxygen conditions should not exceed $500 mg{\cdot}m^{-2}$. The permissible amount of single contaminant particles should not exceed 215 particles $\cdot m^{-2}$, and their dimensions should be within the range of $[500,1000]\,\mu m$ up to $2000\,\mu m$ for elongated particles (CGA, 1985).

4.5.8 Compressors

Oil-lubricated compressors deliver the required quality of diving air; however, sometimes it does not meet the established purity standards for the oxygen conditions. For example, in accordance with Compressed Gas Association Inc. established purity standards, the maximum permissible content of condensed hydrocarbon contaminant in the diving air should not exceed $5 mg \cdot m^{-3}$ at NTP conditions.

The standard was established on the basis of observation of the human respiratory system's defense mechanism. Hydrocarbon concentration below the mentioned value is regarded as nontoxic (Flook, 2002). In respect of fire hazard, for concentration of a few parts per million *ppm*, 1 ppm is regarded as a fire hazard ("Oxygen safety in the production of enriched air nitrox breathing mixtures," 1992). It is evident after recalculation that the permissible content of exposure to benzene is twice than that permissible with regard to the fire hazard value.

The best solution is to use lubricant-free compressors. Because the costs of such compressors are high, most often, modified systems employing lubricated oil compressors are used. The systems are protected against noncontrolled emissions of oil contaminants by monitoring the organic carbon content, for example, by means of FID (Kłos, 2000). With respect to the lubrication system used, compressors can be divided as follows: lubricated with petrochemical oil, agents authorized for oxygen conditions, oil-free compressors, lubricant-free compressors.

Majority of the high-pressure compressors designed for diving applications are lubricated with petrochemical oil. In such compressors, oil is in direct contact, on the cylinder/piston surface, with the air being compressed. Oil, atmospheric water vapor, contaminants such as carbon monoxide and carbon dioxide,[157] and/or mechanical impurities must be removed from the compressed air in filters.

Polish Montejus SP-2 systems authorized for the oxygen conditions are lubricated with a glycerol–water mixture. As follows from their names, the aforementioned devices are not designed to compress gas but to force gas to flow through pipes. Practically, it is recommended to deliver the gas initially compressed to the minimal pressure p of $p \in [2,4] MPa$. Montejuses are used for preparing gas mixtures by means of the pressure method. The gas being forced through should be dried after compression.

Oil-free compressors are a combination of the compressors made of self-lubricating materials and compressors with oil-lubricated mechanical details that are not in contact with the gas being compressed. The compressed air obtained in such systems needs to be purified, that is, moisture, particles of erased seals, piston sealing rings, etc. must be removed – most often, there are Teflon or composite plastics particles. In this case, contaminants such as carbon dioxide from the atmosphere can be removed by filtration or, like carbon monoxide, by first, catalytic oxidization, and then removal by air filtration.

Lubrication-free compressors are wholly made of self-lubricated materials. This enables eliminating the risk of contaminating the system with petrochemical components. The air compressed in such systems is to be purified at inflow to and outflow of the compressor similarly to the air compressed in oil-free compressors.

4.5.9 Conclusions

The safety of divers in the hyperbaric environment directly depends upon the components and pollution of the breathing gas. Multicomponent gas mixtures used as a breathing gas can be prepared using different methods, for example, gravimetric, volumetric, dynamic, and pressure transfer.

The probability of fire occurrence in high-pressure gaseous systems containing oxygen cannot be eliminated. It can only be minimized. The following measures can contribute to minimizing the probability of fire occurrence:

- Using materials authorized for the oxygen conditions to build diving systems and *UBAs*
- Building the systems and *UBAs* according to the design rules developed and authorized for the oxygen conditions
- Using wear-and-tear materials authorized for the oxygen conditions, including seals, lubricants
- Making sure that the personnel operating the gaseous systems is aware of fire hazards and well-trained for work with oxygen
- Exercising proper engineering, laboratory, and medical supervision when artificial breathing gas mixtures are being used
- Making sure that equipment approved for the oxygen conditions is especially marked and operated in compliance with its intended purpose

Oxygen is used in a wide range of partial pressure (Table 2.4). Aviation and space medicine deal with *hypoxia*. Problems inherent in hypoxia are also important for survival in high mountains. For divers, this phenomenon is important in altitude diving. Acclimation before diving and decompression problems in these conditions are related not only to the lower ambient pressure but also to the reduced oxygen concentration C_{O_2}. Sometimes, the oxygen content of C_{O_2} is intentionally reduced, as in the case of type *INERGEN* fire-extinguishing-mixtures[158] type (Table 2.4).

Oxidative stress[159] of O_2 accompanies not only diving activities but also the very nature of the impact of O_2 on the human body (Bartosz, 2008). There is a postulated theory which states that O_2 contained in the air and its long-term carcinogenic impact are responsible to a large extent for the effects of aging.[160] Undoubtedly, *hyperoxia* and *hypoxia* are linked to a direct threat to health and life.

The pathophysiology of the oxygen toxic effect is understood here as a branch of science concerned with the study of changes and disturbances in body cells, organs, and systems following an exposure to hyperbaric oxygen.

4.6 OXYGEN TOXICITY

Oxygen is necessary to maintain homeostasis[161] in the human body, but under hyperbaric conditions, it causes toxicity, affecting the central nervous system, lung tissues, and other tissues, as well as other side effects.

4.6.1 Central Nervous System Toxicity

Oxygen is necessary to maintain homeostasis[162] in the human body, but under hyperbaric conditions, it causes toxicity, affecting the central nervous system, lung tissues, and other tissues, as well as other side effects.

According to some sources, in the recent years, no significant advances in the knowledge of central nervous system oxygen toxicity *CNSyn*[163] have been made (Bitterman, 2004). Many researchers still continue to refer to the classical studies carried out during the Second World War by Keneth Donald, which have been updated and published with the new findings (Donald, 1992). The most commonly used scenarios of exposure to oxygen under combat conditions were developed in the 1920s (Butler and Thalmann, 1984, 1986a, 1986b).

The phenomenon of *CNSyn* is difficult to study because of the complex interactions of many factors, which include age, gender, individual predisposition, or current mental and physical condition (Donald, 1992). *Toxicity* is a phenomenon affecting many organs important to maintain homeostasis. Frequently, the theory of biochemical systems is used to explain the phenomenon of oxygen toxicity, assuming the adverse effects generated by the free radicals[164] and other metabolites which manifest as *CNSyn* symptoms (Torbati, Church, Keller and Pryor, 1992; Bartosz, 2008).

During oxygen exposures, hydrogen peroxide H_2O_2 was observed in human blood, and its effect was tested on various brain areas (Bitterman, 2004).

It is suggested that exposure to oxygen has an effect on various neuroreceptors,[165] for example, *GABA*[166] receptors, and that higher oxidized metabolic forms have an effect on enzymes[167] important for the human body, like, for example, acetylcholine[168] (Bitterman, 2004). Some of the methods used for reducing the risk of *CNSyn* mention air breaks that can be applied during oxygen decompression or hyperbaric treatment *HBOT*[169] (Clark and Thom, 2003).

Upon exposure to oxygen, an initial narrowing of cerebral blood vessels was recorded. This results in decreased blood flow to the cerebral cortex[170] and is followed by expansion of blood vessels. The moment at which this relaxation occurs seems to be the *CNSyn* threshold point, beyond which the *CNSyn* symptoms develop until an attack of convulsions.[171] Substances that reduce blood flow through the cerebral cortex can inhibit the development of *CNSyn* symptoms (Bitterman, 2004).

The *CNSyn* phenomenon is not presented here because it has been briefly presented earlier using, as an example, the general biochemical theory of oxygen toxicity (Kłos, 2023).

With a significant increase in oxygen tension, the production of toxic compounds rises, and the biochemical protection system is not able to deactivate them, resulting in biochemical and physiological changes in the functioning of the body. They manifest themselves as *CNSyn*[172] symptoms. No cases of occurrence of these symptoms immediately after exposure of the body to oxygen have been recorded.[173] As a result of poisoning, some *CNSyn* symptoms were observed preceding an attack of convulsions. They were as follows: restlessness, pale face, trembling lips and eyelids, nausea, cramps, dizziness, lack of coordination, visual and auditory hallucinations, narrowing of the field of vision,[174] or speech disorders (Table 4.22).

TABLE 4.22
Symptoms and Signs of Oxygen Poisoning

Degree of Risk	Symptoms	Number of Cases
	Nausea	75
	Restlessness, dyspnea, insomnia, depression	12
	Headache	5
	Numbness, burning sensation	13
	Dizziness	63
	Cramps	335
	Auditory disorder	7
	Vision disorder	17
	Loss of consciousness, speech disorder	16
	Convulsions	91
Total		634

Source: Harabin, Survanshi, and Homer (1994).

These symptoms are rarely noticeable before the onset of seizures[175] (Brubakk and Neuman, 2003; Harabin and Survanshi, 1993).

Generalized convulsions occur suddenly. The attack begins with the tonic phase, usually lasting $30\,s$, during which the diver loses consciousness and stops breathing. Then there is the clonic phase of uncoordinated movements of the entire body. The whole attack usually lasts about $2\,min$. After a long period of breathing oxygen in a hyperbaric chamber, where oxygen can be replaced with air, the allowable period of apnea, without causing harm to the poisoned diver, is about $t \in [5-8]\,min$[176] (Clark and Thom, 2003).

Factors which increase cerebral blood flow, such as immersion, hypothermia, workload, increase of CO_2 concentration, etc., increase the sensitivity to *CNSyn* symptoms. Carbon dioxide CO_2 can be present in the inhaled breathing medium or come from the so-called dead space.[177] Through the CO_2 tension receptors, the body increases the intensity of ventilation. The increase in ventilation rate is also accompanied by increased density of inhaled breathing medium, increased breathing resistance, etc.

In normobaric conditions, the excess of CO_2 could be more effectively eliminated than under hyperbaric conditions. Generally, under increased pressure, CO_2 accumulates in the body. Initial hyperventilation leads to *hypocapnia*,[178] resulting in, observed during exposure to oxygen, decrease of respiratory action. The defense mechanism that causes blood vessels in the brain to narrow leads to increased concentration of CO_2 in cerebral vessels in relation to peripheral vessels. Thus, CO_2 receptors in the initial phase cannot increase ventilation. This defense mechanism, however, has its drawbacks. The increased CO_2 concentration in cerebral vessels increases the concentration of hydronium ions H_3O^+, which results in hemoglobin rapidly losing oxygen in blood and increasing its tension in plasma.[179] Because of this, the brain tissue is exposed to higher oxygen tension π_{O_2}. It seems, however, that the body tries to compensate these effects with the enzymes which deactivate the free radicals produced.

During work, CO_2 emissions from tissues to peripheral blood may result in the following: increase in ventilation; vasoconstriction at the periphery, which raises blood pressure; and expansion of cerebral blood vessels. This may increase the cerebral blood flow, potentially causing increase in O_2 stream flowing through the brain.

4.6.2 Oxygen Exposures

The Polish Navy has maintained, until recently,[180] the rules for fixing exposure time and maximum allowable oxygen partial pressure during a dive (Kłos, 2023). They corresponded to the applicable provisions of the US Navy in the 1960s (Kenny, 1973). The manual released by the *Ministry of National Defence* in 1981 distinguishes between routine and exceptional exposures, where the boundary between them runs at a pressure of $0.175MPa$. Currently, during dives performed with closed circulation apparatuses, where oxygen is used as a breathing medium, the Polish Navy accepts the recommendations by the US Navy (USN, 2016).

It is understood that O_2 shows no toxic effects on the central nervous system when partial pressure p_{O_2} is equal to or less than $p_{O_2} \leq 0.1MPa$[181] (Betts, 1992). During combat dives, oxygen partial pressure p_{O_2} often goes beyond this value.[182]

In the 1970s, the US Navy changed the standards for oxygen exposure due to the risk of appearance of *CNSyn* symptoms. The changes covered reduction in the partial pressure limits to $p_{O_2} \leq 0.20MPa$, division of the exposure type into standard and exceptional, and reduction in allowable time limits for each of the two exposure types (USN, 2008). In the 1990s, the US Navy changed the allowable oxygen exposure times for $CCR - SCUBA$[183] to allow for more flexible diving operations (USN, 2008). In 1991, NOAA[184] changed its rules concerning the partial pressure limits of oxygen during *nitrox Nx* exposures[185] (NOAA, 2001). The amendment covered a longer time of stay under the allowable partial pressures of oxygen in *Nx* and the determination of the maximum exposure time during the $24h$ period – (Table 4.23).

In Poland, the US Navy approach based on survival analysis was validated, and it is used for predicting *CNSyn* hazard (Figure 2.32; Kłos, 2023).

During deep dives, breathing resistance increases due to an increase in density of breathing medium. The breathing resistance may be accompanied by a rise in accumulation of CO_2 in the diver's body for the reasons described previously. With the increasing distance from the water surface, the possibility of providing assistance or self-rescue to a diver/divers who has/have developed oxygen poisoning becomes more complicated. Therefore, the maximum permitted partial oxygen pressure for deep dives should be reduced.[186]

Similarly, when diving in caves or wrecks, because of limited access to the surface, the rules for deep dives should be followed. The distinction between the standard exposure and the exceptional exposure is related to the conditions of a particular

TABLE 4.23
Exposure Time and Partial Pressure Limits of Oxygen in the *Nx* Accepted by *NOAA*

Standard Exposures				
Oxygen Partial Pressure	**Exposure Time Limits**		**Maximum Exposure Time in 24*h***	
(*MPa*)	(*min*)	(*h*)	(*min*)	(*h*)
0.16	45	0.75	150	2.5
0.15	120	2.0	180	3.0
0.14	150	2.5	180	3.0
0.13	180	3.0	210	3.5
0.12	210	3.5	240	4.0
0.11	240	4.0	270	4.5
0.10	300	5.0	300	5.0
0.09	360	6.0	360	6.0
0.08	450	7.5	450	7.5
0.07	570	9.5	570	9.5
0.06	720	12.0	720	12.0
Exceptional Exposure				
0.20	30	0.50		
0.19	45	0.75		
0.18	60	1.00		
0.17	75	1.25		
0.16	120	2.0		
0.15	150	2.5		
0.14	180	3.0		
0.13	240	4.0		

Note: If in one of the dives the time of exposure has been reached, or the limit exceeded, the diver must rest for at least 2*h* on the surface before the next exposure. Also, if in one or more dives over a period of 24*h* the maximum exposure time assumed for 24*h* has been reached or exceeded, the diver must rest for at least 12*h* on the surface before the next exposure (Rutkowski, 1990).

Source: NOAA (2017).

dive. The exceptional exposure can be allowed only for the purpose of saving a human life or other important random events.

4.6.3 Pulmonary Toxicity

Oxygen is toxic to the respiratory system. Oxygen toxicity was first observed during prolonged[187] breathing of pure oxygen at atmospheric pressure – and it was called either the *Lorrain Smith effect*, after the name of the discoverer, or oxygen pulmonary

TABLE 4.24
The Frequently Used Oxygen Partial Pressure Ranges

Oxygen Partial Pressure (*MPa*)	Specification
0.010	Extinguishing gases type *INERGEN* – it is safe to breathe in limited time a mixture of nitrogen, argon, carbon dioxide, and oxygen in which oxygen partial pressure falls down to the level of $0.008MPa$ under condition that partial pressure of CO_2 will be $0.005MPa$ (Fire research, Test, Development, and Education Centre, 1993).
0.012	Lower limit of safety due to hypoxia.
0.016	First symptoms of hypoxia.
0.021	Normal oxygen partial pressure in atmospheric air.
0.035–0.040	Typical saturation exposures.
0.050	Maximum oxygen partial pressure during saturation dives and the beginning of the lung oxygen toxicity.[190]
0.10	Pure oxygen breathing on surface.
0.16	The most commonly accepted upper limit of safety for nitrox dives out of the saturation zone[191] (Hamilton, 1989; NOAA, 2001).
0.20	Treatment table *CX30*[191] worked out by *COMEX* in 1986 (Comex Marseille, 1986).
0.24	Suggestion to use nitrox $Nx0.4$ at a pressure of $0.6MPa$ in the treatment of diving diseases (Rutkowski, 1990).
0.25	The upper limit of permitted combat dives with oxygen[192] (Butler and Thalmann, 1986b).
0.28	$20\,min$ test of oxygen tolerance (NOAA, 2001) and oxygen treatment tables.[193]
0.30	Suggestion to use nitrox $Nx0.5$ at a pressure of $0.6MPa$ in the treatment of diving diseases[194] (Rutkowski, 1990).

toxicity (Shilling, 1981). The symptoms of pulmonary toxicity are very similar to those of *pneumonia*.[188] During dives outside the saturation dive protocol, this effect plays a minor role as compared with the *Paul Bert* effect,[189] but it is recommended that it should be monitored during oxygen dives. This effect is important during saturation dives and prolonged hyperbaric treatment, so it is not discussed here – it was widely described earlier (Kłos, 2023).

4.6.4 Somatic Toxicity and Other Threats

Oxygen is used in a wide range of its partial pressures (Table 4.24).

Aviation and space medicine deal with *hypoxia*. Problems with hypoxia are also important for survival in the high mountains. For divers, this phenomenon is important when diving in high-altitude waters. Acclimation before diving and decompression problems in these conditions are caused not only by the lower ambient pressure but also by reduced oxygen concentration C_{O_2}. Sometimes, the oxygen content of

C_{O_2} is intentionally reduced, as in the case of type *INERGEN* fire-extinguishing mixtures[195] (Table 4.24).

The carcinogenic effect of O_2 accompanies not only diving activities but also the very nature of the impact of O_2 on the human body (Bartosz, 2008). There is a postulated theory which states that O_2 contained in the air and its long-term carcinogenic impact are responsible, to a large extent, for the effects of aging.[196]

Undoubtedly, hyperoxia and hypoxia are linked to a direct threat of loss of health and life.

The chronic effects of breathing under increased oxygen partial pressure p_{O_2} are called general O_2 toxicity or oxygen somatic toxicity.[197] One of the described effects is a reversible reduction in the amount of hemoglobin and the number of red blood cells in the blood of saturation divers similar to the increase in their number after a long-term acclimation to hypoxia. These effects in healthy people are reversible but must be taken into account when planning dives or rest after the dive.

Other, more dangerous effects are *paresthesia*[198] and avascular necrosis of bone. However, their direct relationship to diving may be difficult to clearly demonstrate (Hamilton, 1989).

An unpleasant effect observed after oxygen dives may be the collapse of the tympanic membrane resulting from chemisorption of O_2 from the middle ear cavity.

During change in dive depth, the diver is forced to equalize pressure in the middle ear cavity through the *eustachian tube*. Multiple changes of depth in the course of diving can cause a significant increase in O_2 concentration in the gas space of the middle ear. Even with minor symptoms of cold, the *eustachian tube*, the channel, has a reduced patency, and even trained divers[199] are then forced to perform strenuous *Valsalva blowout*.[200] The oxygen locked in the ear canal undergoes diffusion through the membrane of the oval window of the inner ear, where it is absorbed and consumed,[201] resulting in a reduction of pressure in the middle ear and an increase in the outside force exerted on the eardrum.

If this phenomenon happens during a rest period, the diver will wake up with a headache resulting from the strain of the tympanic membrane. Typically, there occurs a congestion of tympanum, that is, effusions into external and middle ear and increased release of earwax in the outer ear. It is accompanied by irritating weakening of hearing sensitivity. Usually, there is no perforation of the tympanic membrane. This may occur as a result of earlier injuries, which reduce the membrane's flexibility and create scarring.

Another dangerous effect can be an excessive accumulation of dissolved oxygen in *perylimph*,[202] which may be accompanied by its movement, resulting in symptoms similar to those of neurological disorders, which occur during a decompression sickness (Farmer and Thomas, 1976). There are recorded cases of irritation of the ear labyrinth causing its increased activity during oxygen decompression or during hyperbaric oxygen treatment. This effect is associated with moving of the *lymph* due to differences in tension of gases dissolved in it. Such cases have been rare during diving operations in Poland. However, there are credible descriptions of such problems in the international literature (Strauss, 1976).

Constriction of cerebral blood vessels as physiological effects associated with exposure of brain tissue to high partial pressure of oxygen can be especially dangerous for the eyesight.

During hyperbaric oxygen therapeutic procedures[203] *HBOT*, recorded were cases of constriction of carotid artery feeding the retina *CRAO*,[204] leading up to full blockage[205] and, in effect, to blindness. Quick restoration of circulation resulted in restored vision (Anderson, Saltzman and Barbee, 1965).

It has also been noticed that a disturbing percentage of newborn babies, after a long stay in incubators with atmosphere enriched with O_2, irreversibly lost their eyesight (Bartosz, 2008). For this reason, in today's incubators, clean air is used.

Visual disturbances associated with exposures to high oxygen partial pressure have been studied since the beginning of scientific investigations on diving (Donald, 1947). It has been repeatedly confirmed that oxygen under high pressure has an adverse effect on eyesight (Clark and Thom, 2003).

There is a wide range of literature dealing with the phenomenon of oxygen blackout related to oxygen hypoxia during breath-holding dives. The issue of *hypoxia* occurring while returning to the surface during such dives, however, will not be discussed here. In relation to oxygen combat dives, the term *oxygen blackout* is usually understood as loss of consciousness by a diver after switching from breathing oxygen to breathing atmospheric air after the combat mission. One of its reasons may be the already-mentioned physiological defense reaction involving cerebrovascular constriction, which results in reduced cerebral blood flow. It is usually accompanied by peripheral vasodilatation, which can lead to *hypothermia* induced by the cooling effect of the aquatic environment.

In hyperbaric conditions, in peripheral blood, oxygen tension occurs at a higher level. Hence, oxygen receptors in the glomerulus will decrease the respiratory rate and reduce the blood flow. Rapid switching to breathing air, which contains less O_2 and which is usually combined with the need to get the job done,[206] before the relaxation of cerebral vessels and before increasing cerebral blood flow may cause the phenomenon of *hypoxia*. In addition, the pressure reduction related to ascent is accompanied by a decrease in the partial pressure of O_2, which reduces respiratory stimulation as in the case of hyperventilation. If this effect occurs in water, the diver runs the risk of choking with water or drowning, but in most cases, it only leads to temporary confusion and loss of concentration.

There are hypotheses that suggest taking into account the toxic effect of increased CO_2 tension in the cerebral vessels, but in light of the experiments *in vivo*, it appears that the effect of the hypercapnia on oxygen blackout is limited. This is because the recorded increase in CO_2 tension is, in normal conditions, tolerated asymptomatically (Lamberdsen, Kough, Cooper, Emmel, Loeschcke, and Schmidt, 1953). Although increasing the tension of CO_2 in the cerebral vessels significantly affects the drop of O_2 in hemoglobin, the effect is not significant in the case of oxygen blackout but, rather, in the case of the described-earlier O_2 toxic effect on brain tissue. It seems that the mechanism of oxygen blackout during the oxygen mission may be associated rather with decreased CO_2 tension in peripheral vessels, causing a lack of sufficient stimulation for the respiratory center.[207]

Animal studies have shown that the occurrence of decompression sickness symptoms *DCS*[208] may be caused by compression while using pure oxygen, followed by rapid decompression (Donald, 1955; Bennett and Elliott, 1993). However, the course of the so-induced bends was milder (Donald, 1992). *DCS* induced in such a way usually does not require hyperbaric treatment, since it undergoes spontaneous compensation in time (Vann, 1989). Although the phenomenon of oxygen bends is possible, it does not pose a significant risk during a normal diving operation with the use of O_2 as breathing medium. It can have a negative impact on the air transport in divers after exposure, for example, during a recovery of special group/section.

4.6.5 Oxygen Tolerance Test

The procedure oxygen tolerance test *OTT* is used in the armed forces. The description covers only the most important information on the purpose and method used to conduct the test (Kłos, 2023).

In the armed forces, when qualifying candidates for diving with artificial breathing medium, it is recommended that they undergo selection tests in a decompression chamber: a pressure test and an oxygen tolerance test or both tests simultaneously. A single oxygen tolerance test should be repeated with an interval of at least a week, and pressure tests are recommended to be run as often as possible, especially when Navy divers are to be advanced in ranks and before the periodic medical examination (DMW, 2007).

Discussions whether to use *OTT* test have been held since the beginning of systematic research into the use of O_2 as breathing medium (Clark and Thom, 2003). The fundamental problem in obtaining a conclusive *OTT* result is the considerable and significant variability in both individuals[209] and groups[210] (Donald, 1947). The same studies found, however, that divers exposed to oxygen poisoning in a decompression chamber and submerged in water showed higher resistance to *CNSyn* while submerged in water. This proves that the test performed in a chamber provides useful information (Donald, 1992). Initially, *OTT* has been used in several countries, but at the beginning of the century, it was only the armed forces of Germany and Poland that used it for initial screening.

In 2000, the *US Navy* conducted an analysis of the results of 6,250 oxygen tolerance tests which involved candidates for combat divers in 1976–2000. During this period, only six candidates were rejected; hence, the benefits-to-cost ratio was found to be negative, and it was recommended that the screening tests in the *US Navy*[211] be cancelled (Walters, Gould, Bachrach and Butler, 2000). The most important reason, however, for the recommendation to discontinue the screening was the introduction by the *US Navy* of a new procedure for purging the respiratory volume of the *oxy–CCR SCUBA* with oxygen. As a result of this action, the diver practically breathes $74\%_v O_2 / N_2$, which, when enriched with O_2 during a dive, should not exceed $85\%_v O_2 / N_2$ (Harabin, Survanshi, and Homer, 1994; Walters, Gould, Bachrach, and Butler, 2000).

A candidate for the *OTT* should have an updated medical certificate and, prior to the start of *OTT*, should be familiarized with the way the test will be conducted and the rules of behavior inside the hyperbaric chamber.

Before entering the chamber, divers undergoing *OTT* should answer some questions regarding their general well-being and should be examined by a physician. The scope of examination can be extended, but not by elements which could significantly overload the diver on the day of such an examination.[212]

The hyperbaric chamber for running *OTT*s should be prepared appropriately for such tests. The candidate goes through an *OTT* accompanied by an attendant, who closely monitors his/her behavior and who runs the commissioned tests (Kłos, 2023).

The tested candidate and attendant enter the chamber, which is then pressurized up to the equivalent of a depth of $18\,mH_2O$. During the compression process, they breathe the chamber air. After reaching the depth of $18\,mH_2O$, the tested diver should be again briefly instructed in the use of the oxygen inhaler. Then he puts on the inhaler mask, fits it closely, and starts breathing O_2 for $30\,min$ at rest in the recumbence. The period of time from the start of compression to the start of breathing O_2 should be no longer than $15\,min$. Breathing takes place in the low-resistance open system with the discharge of exhaled medium out of the chamber.[213]

During the *OTT*, an attendant, who is breathing the chamber air, monitors the candidate and, at $5\,min$ intervals, carries out measurement of breathing rate, heart rate, and blood pressure of the diver undergoing the *OTT*. The attendant should, all the time, monitor the tested diver for any symptoms. If such symptoms are noticed, the subject should make the diver stop breathing oxygen from the inhaler and, as soon as possible, make him breathe the chamber air.

During the *OTT*, the hyperbaric chamber should be ventilated, preventing concentration of O_2 inside the chamber from rising above $C_{O_2} < 25\%_v$. CO_2 partial pressure of the chamber air should be less than $p_{CO_2} < 1\,kPa$.

After $30\,min$ of O_2 breathing, the tested diver removes the mask and, for 2 minutes, breathes the chamber air pressurized to the equivalent of $18\,mH_2O$. Then decompression takes place. The recommended decompression speed should not exceed $10\,mH_2O \cdot min^{-1}$ because of the presence of the attendant. If, in accordance with the Polish air decompression table, decompression stops are not required for the attendant, it is recommended that a $1\,min$ stop be made at a depth of $3\,mH_2O$, which is considered to be a safety stop.

The first test in a series of *OTTs* positively completed should be repeated after a minimum of a weeklong break. Only the positive results of the second *OTT* provide a basis for considering the successful completion of it. Such a proceeding is related to the mentioned-earlier variability of individuals in O_2 tolerance under hyperbaric conditions (Donald, 1992). Frequently, the *CNSyn* symptoms detected during the first *OTT* are not confirmed in the second *OTT* (Butler and Knafelc, 1986).

If, during the *OTT*,[214] the diver convulses, the test is terminated, deemed failed, and is never repeated. The failed diver can only be authorized for dives with air as breathing medium.

When, during the *OTT*, the candidate experiences symptoms of malaise, his/her breathing rate drops below four breaths per minute, or, other than seizures, symptoms of *CNSyn* occur, the *OTT* should be interrupted and repeated after of minimum gap of one week. Recurrence of symptoms or signs of *CNSyn* is subject to the same rules as those applicable for cases of seizure occurrence in the tested diver. If the diver during the repeated *OTT* shows no *CNSyn* symptoms, this is regarded as successful completion of the first stage of *OTT*, and the second test is performed after a minimum gap of one week. If the third exposure goes without comment, the diver passes *OTT*. If, however, during the third *OTT* the diver shows even the slightest sign of *CNSyn*, he/she should be treated as if he has experienced a seizure.

If the first test is successful and during the second one malaise symptoms or symptoms of *CNSyn* other than seizures occur, the *OTT* may be repeated twice at intervals of at least one week. If, while repeating *OTT*, symptoms of *CNSyn* occur, the diver should be treated as if he has had seizures. After each *OTT*, the diver should remain in the vicinity of the decompression chamber for around 1*h*.

Due to shortages of qualified physicians specialized in diving medicine, there is an ongoing discussion concerning regulations applied to diving military units in view of the necessity to lower the previously used standard for financial reasons. Given national economic problems and practices in other countries, it seems that resigning from *OTTs* is one of the first permissible steps in this regard.

OTT would have been a long time ago abandoned if there was no common conviction shared by the *Polish Navy* divers about its usefulness and the rationale behind its implementation. Owing to the positive *OTT* results, the divers have built a sense of security and confidence in their own abilities. A negative result was not used to judge them as unfit for diving; it only imposed limits on their powers to take risks. Additional theoretical and practical training which accompanied *OTTs* consolidated the need for self-control and strengthened the knowledge of *CNSyn*. It must therefore be concluded that the psychological effect was more important than the effects of screening.

It seems that the psychological effect of *OTT* is completely overlooked in the undergoing discussions. Its importance is confirmed by the continued employment of psychologists in military units despite the fact that divers are now especially selected professional servicemen. When it comes to military divers, the psychological effects are marginalized, although a significant reduction in the population of active military divers is recorded, which is balancing on the edge of the capability to maintain combat readiness. Probably, there are many reasons for this state of affairs. They have one common denominator, that is, decrease in the ethos of service, understood here as the standards and values that make up the character of a diving professional, which, in turn, defines their identity and uniqueness. Not without significance is the effect of lowering the standards of medical service, which also reduces the attractiveness of service in a diving unit. The higher incidence rate of *CNSyn* symptoms recorded in *NATO* navies prompted some countries to consider reintroducing regulations[215] that might stop the growing concerns of divers and their defection to other types of service.

Because of the need to rationalize the costs, the expenditures incurred by the diving service must be reduced. However, this should not deprive divers of maintaining their feeling of security. It seems that introducing such a pragmatic[216] way of tackling the problems mentioned can only be achieved through organizational changes.

Young and fit candidates, following thorough medical examination during the recruitment, can undergo the selection tests during mandatory training, while the medical staff can be used as *GP*[217] medical diving supervisors and members of recruitment commission. This will allow implementing comprehensive care[218] and will contribute to enhancing the service ethos and strengthening the sense of security among divers. It will also allow for the more efficient use of physicians serving in diving units, who also provide a range of psychological services.

4.6.6 Conclusions

Oxygen exhibits toxicity to the human body. However, during oxygen dives, and in most combat dives, it is its neurological form – *CNSyn* – which plays the most important role. This chapter deals with the biochemical view on *CNSyn* associated with the formation of free radicals and forms at a higher oxidation state that can potentially be toxic to the human body. It also includes some considerations concerning the biochemical protection mechanisms against these harmful products, which protect the diver's body from *CNSyn* only to a limited extent.

NOTES

1 Mix-diving.
2 Hyperbaric chamber.
3 The diver was subjected only to compressed gas, whereas during normal diving, the diver is additionally subjected to the water environment.
4 Enriched air.
5 In this time, *USA* was the only one producer of helium.
6 *Hydrox*.
7 Diving in a wet pot of hyperbaric complexes, where the pressure exerted on the water surface in the swimming pool simulated the pressure at the assumed depth of diving.
8 Reached a depth of $222\,mH_2O$.
9 Short-duration diving, bounce diving.
10 More than 120 mH_2O.
11 As is shown in Figure 4.3.
12 Keeping constant overpressure in relation to the ambient pressure.
13 Nozzle metering.
14 Opposite to an oxygen nozzle.
15 Removed by chemical process.
16 Transfer under pressure *TUP*.
17 Usually, short decompression at a depth is recommended.
18 Surface decompression.
19 For example, diving bell.
20 Not sufficiently ventilated space.
21 Rescue operation.
22 Central nervous syndrome.
23 Nitrogen narcosis; the depth ecstasy.
24 Narcotic.
25 Or oxygen-free energy conversion.
26 Metabolic.
27 Such as decrease in gas density and gas toxicity, faster decompression, etc.
28 Hydrogen and helium thermal conductivity is seven times greater than that of nitrogen, and thermal conductivity of neon is twice that of nitrogen.
29 Use of thermal insulation.
30 Active thermal protection.
31 For example, procedure for changing gas mixture composition in case of fire in the hyperbaric complex.
32 For example, tank crew.

33 In this, case oxygen.
34 Gas mixtures used in compression phase.
35 For example, when divers have to leave the hyperbaric complex.
36 Greater gas density.
37 Air, air–helium, helium–oxygen *Hx* tables.
38 Understood as a method for moving between the treatment procedures in the course of treatment decompression or when the next sickness symptoms occur after the completion of treatment.
39 Formally *Trimix*, but it is the same as in air nitrogen–oxygen ratio – it could be marked as $air - Tx$.
40 Here is presented other than preferred in *NATO*.
41 Outside the saturation zone.
42 *Complement* is a system of plasma proteins contributing to defensive reaction that occurs on the surface of pathogens and generates active components with various effector functions.
43 *The central nervous system CNS* does not tolerate becoming deprived of the flow of information; that is why its activity dies out quickly, causing *Type II DCS*.
44 *Natural solubility* is understood here as the condition in which gases occur independently.
45 Mechanism similar to the salting-out effect in water solutions; addition of sodium chloride to another solution to reduce the solubility of the latter.
46 Gas embolism.
47 Hence the term *aseptic*.
48 Prefunded—*perfunde* is a verb derived from *perfusion*, which means the passage of fluid through the circulatory system or lymphatic system to an organ or a tissue.
49 Available June 2024.
50 *TT* 1*A USN*, *TT* 2*A USN*, and 3 *TT USN*.
51 *TT* 7*USN*; nitrogen–oxygen mixture is called *nitrox* and is marked *Nx*.
52 *TT* 4 *USN*, *TT* 5 *USN*, *TT* 6 *USN*, *TT* 6*A USN*, *TT* 9 *USN*.
53 For example, as a result of built-in breathing system malfunction, exhaustion of stocks, etc.
54 Develops oxygen toxicity (Kłos, 2007a).
55 Considered when the diver was, just before, exposed to high partial pressure of oxygen, and in this case, occurrence of symptoms indicating central nervous system oxygen toxicity or pulmonary oxygen toxicity is highly probable, for example, when the dose of oxygen toxicity was clearly exceeded during the dive (Kłos, 2000, 2012).
56 Nitrogen–helium–oxygen mixture, called *trimix*.
57 *Aspirin* (Germany), *aspegic* (France), *polopiryna* (Poland) itd.
58 Ringer's solution, or balanced salt solution (physiological pH and isotonic salt concentration).
59 For example, the US Navy diving manual (2008).
60 Almost without doing any extra work, except diving.
61 Except for severe cases, when the diver was thrown to the surface.
62 Pulmonary barotrauma.
63 Insufficient amount of oxygen in the body, leading to loss of consciousness.
64 Anticoagulants are medicines that reduce the ability of the blood to clot.
65 Superficial isobaric counterdiffusion.
66 Deep tissue isobaric counterdiffusion.
67 Isobaric counterdiffusion.
68 For example, volatile components of paints, volatile components of thermal insulation, maintenance preparations, etc.
69 Normal conditions.
70 Surface equivalent value.
71 Standard temperature and pressure – STP.
72 Anesthetic.
73 Trimix *Tx*.
74 Extremely small sample.

75 *Wave number*, as used in spectroscopy and most chemistry fields, is defined as the number of wavelengths per unit distance, typically in centimeters: cm^{-1}.
76 Deviation from the Beer–Lambert law.
77 By proper number of sucks performed.
78 Reaction base.
79 In this case, the internal tube is divided and connected tightly with flexible transparent tape.
80 Cross-sensitivity.
81 For example, application of introductory or protective detector tubes.
82 The effect of CO_2 has been tested.
83 For example, thermal desorption.
84 Gage repeatability and reproducibility.
85 That is, random; this term is used for the purpose of differentiation from the term "statistical," which is used not only in the mathematical statistics.
86 Critical to quality.
87 Systematic errors are minimized to the accepted level by introducing alterations or using special measurement methods.
88 If, in a given problem situation, the assumed accuracy level can be accepted by the purpose of the undertaken action.
89 Acceptable in particular conditions.
90 Correct value is an estimator of true value.
91 Current understanding of a measurement system points to a set providing in a simple way easily interpretable results, without complicated operating procedures requiring the presence of highly qualified operators. In the past, a measuring instrument was a device designed to carry out special types of measurements. At present, due to substantial progress in the available methods and the electronization of measurements, it is difficult to encounter a measuring device that would not be a complicated electromechanical system.
92 A measuring instrument is a device designed to perform independent measurements or in combination with one or many auxiliary devices (BIML, 2000).
93 The achievements of metrology are still an indispensable element in the knowledge of an analyst of measurement systems – quite often, negligence of this fact may lead to serious problems.
94 A standard unit of measurement (etalon) is a measuring instrument, reference material, or measurement system designed to define, realize, preserve, or reproduce units of measurement or one of many values of a certain quantity and acting as a reference (BIML, 2000).
95 Calibration is a set of defined operations carried out in defined conditions, relationship between the values of a measured quantity indicated by a measuring instrument or measurement system, or values represented by standard measurement or reference material and proper values of the quantity realized by standard units of measurement.
96 The capacity of a measurement process should be understood as the level of meeting CTQ requirements.
97 For instance, type test constituting a set of activities aimed at determining whether a measuring instrument of a certain type meets the requirements for approval in accordance with the act on measurement units. Legalization comprises a number of activities including checkup, confirmation, and certification with a proof of legalization that a measuring instrument meets the requirements stipulated in the act on measurement units. Authentication resulting from calibration which may be confirmed with a document known as a calibration certificate or protocol.
98 Universal measurement systems require adjustment and calibration to the conditions of measurement performance, as depending on the choice of working parameters, they may be utilized for different types of measurements – for example, a gas chromatograph, depending on column selection, its working parameters, detector type, and SOP, may be used in qualitative or quantitative measurements of various liquid and gaseous mixes.

99 Sometimes it is necessary to aim at precise estimation of the current reliability of measurements, for example, in security technologies, health protection, work conditions, legal metrology, etc.
100 Sometimes its use is impossible, not for technical reasons, but rather due to the absence of an economic justification.
101 Sometimes it is difficult to apply calibration on an operating production line, as it may require interrupting a production process, shutting off or performing a switchover to a parallel instrument, which quite often is an element of an automation system – of course, properly designed measurement systems need to allow calibration; however, it may be unjustified or problematic at a given time.
102 For instance, on the basis of a certificate issued by a recognized company.
103 The best, commonly available in the market.
104 Checking.
105 *Systematic errors* have been compensated to such a degree that they are an order of magnitude smaller than random errors; *gross errors* are subject of statistical analysis of outlying point.
106 *Systematic errors* have been compensated to such a degree that they are an order of magnitude smaller than random errors, *gross errors* are subject of statistical analysis of outlying points.
107 Dependence of the response of a system of measurement O in the function of the value measured, for example, the standard $W : O = f(W)$.
108 Stability of the response O of a system of measurement in the function of time $\tau : O = f(\tau)$.
109 For instance, it may be possible to use two types of software to operate the measurement system.
110 However, in such a case, it will be impossible to estimate all the constituents of a systematic error in the measurement method.
111 *Reproducibility* means the level of compliance of measurement results of the same measured value, performed in varying conditions (BIML, 2000).
112 *Repeatability* means the level of compliance of measurement results of the same measured value, performed in the same conditions (BIML, 2000).
113 The choice of the time segment for performing such estimates is crucial, as it has a significant effect on the values of reproducibility and repeatability – a shorter time usually means better values of the evaluated parameters.
114 Measuring capacity.
115 Good laboratory practice.
116 Repeating of measurement.
117 Also frequently applied is determination of the capability of a measurement process which is concerned with the confirmation of fulfilment of quality requirements *CTQ* of a measurement process for particular applications, whereas reliability assessment underpins the formal confirmation of such capacity.
118 Reliability and reproducibility.
119 Traditionally, and correctly, it is taught that systematic errors should be minimized to an acceptable level, thus perhaps implying the possibility of their complete elimination, whereas in fact, in many cases, it is impossible to achieve (Kłos, 1990, 2007b).
120 *Systematic error* constitutes the absolute value of deviation, whereas uncertainty is understood as an extension of this value with the boundaries of accuracy of its determination.
121 Sometimes, a greater value of a systematic error is adopted, taking into account also various modifications in the measurement system, such as various software versions, adjustment methods, operational performance, etc., most commonly in the form of manufacturer's declaration regarding maximum systematic error for a given measurement system.
122 Estimation of the level of fulfilment of *CTQ* requirements by a measurement system.

123 Natural fluctuations in the measurand.
124 Measured physical quantity, property, state, etc.
125 With the use of a standard correct value.
126 Lack of complete compliance with the quoted source.
127 For example, a production process.
128 Being a realization of random variable $X \leftarrow x_i$.
129 *USL* – upper specification limit; *LSL* – lower specification limit.
130 Settings.
131 Also called *PTR* – precision-to-tolerance ratio.
132 Being a realization of random variable $X \leftarrow x_i$.
133 Random order of realization of various sample combinations and measurement system settings.
134 Function used to estimate parameter distribution value.
135 Usually provided in the form of arithmetic mean $\bar{x}$.
136 Understood as a result of a cycle of research activities leading to the estimation of the final result in the form of research conclusions.
137 Presence of a deterministic factor disturbing analysis.
138 Frequently, the scope of a given measurement system fails to cover the required.
139 Habitat, diving bell, atmosphere of the *UBA*, and so on.
140 Accuracy of preparation, content of contaminants, and so on.
141 Other than binary mixtures.
142 As opposed to mixing those gases with helium.
143 After methanation.
144 Thermal conductivity detector, TCD.
145 Connected-together cylinder set.
146 When labels are used, they must be waterproof or placed in a waterproof socket.
147 For example, like chlorine.
148 Oxidable.
149 Reaction temperature is very high.
150 The so-called octopus.
151 Joule – Thomson effect.
152 Oxygen partial pressure exceeds 20 *kPa*.
153 In the case of limited access.
154 Its toxicity is higher than that of 1,1,1-trichloroethane.
155 Wavelength range $[2500{,}3700]Å$.
156 For example, cotton-made.
157 Oil burnout.
158 It is worth mentioning that *hypoxia* and *hypercapnia* partially cancel each other out in the atmosphere of *INERGEN*.
159 *Oxidative stress*, defined as a disturbance in the balance between the production of reactive oxygen species as free radicals and antioxidant defenses.
160 For example, the carcinogenic effects of oxygen occurring during long, multiple oxygen exposures.
161 The ability of a living organism to retain constant equilibrium of, for example, blood composition or temperature, by proper coordination and regulation of life processes.
162 The ability of a living organism to retain constant equilibrium of, for example, blood composition or temperature, by proper coordination and regulation of life processes.
163 Central nervous system oxygen toxicity is called the *Paul Bert effect*, or the acronym *CNS* (*central nervous syndrome*) is used; for the purposes of this chapter, the acronym *CNSyn* will be used to distinguish it from the acronym *CNS* (*central nervous system*).
164 Group of atoms, generally short-lived, which have unpaired valence electrons – free valencies.

165 Nervous receptor.
166 Two types of *GABA* receptors binding *γ-amino butyric acid*:
- Receptor $GABA_A$ regulates the influx of chloride ions into the cell, hindering the formation of action potentials responsible for providing information in the nervous system.
- Receptor $GABA_B$ regulates the flow of potassium ions and calcium to neutralize the effect of chloride ions and regulate the release of neurotransmitters.

167 In majority of proteins, macromolecular chemical compounds regulating the processes of life.
168 Neurotransmitter converting an electric signal to a chemical signal at the synapse, which plays a fundamental role in the rapid conduction of nerve signals.
169 Hyperbaric oxygen therapy; sometimes used as *HBO* acronym.
170 Observed as the defensive response to increase in partial pressure of oxygen.
171 Tremor.
172 Of them, convulsions are most often mentioned.
173 Indirectly, is a proof of the mechanism described earlier.
174 Called tunnel vision.
175 However, it is advisable to familiarize divers with them; sometimes trained divers can quickly recognize the beginnings of poisoning, thus preventing severe *CNSyn* symptoms (Harabin, Survanshi, and Homer, 1994).
176 Because of good oxygenation of the body.
177 For example, from parts of diving apparatus with limited gas exchange.
178 *Hypocapnia*, also *hypocarbia*, is a state of reduced, below the norm, partial pressure of CO_2 in the blood.
179 Together with decrease in the pH value, the capacity of hemoglobin to bind oxygen is reduced – as in the *Bohr effect* described earlier.
180 Diving provisions were withdrawn due to a legal defect associated with their introduction (DMW, 1984).
181 *NOAA* tables adopted this threshold as $0.6\ MPa$.
182 An additional complication is the fact that the oxygen partial pressure during the dive may change in a fairly wide range.
183 Independent closed circuit diving apparatus with oxygen as breathing medium: *closed circuit rebreather* (*CCR*)–*self-contained breathing apparatus* (*SCUBA*).
184 *National Oceanic and Atmospheric Administration.*
185 These changes were preceded by more than ten years of medical research on dives with nitrox mixtures as breathing medium and *Repex* program (Hamilton, 1989; Hamilton, Kenyon, Peyerson, Butler, and Beers, 1988; Hamilton, Kenyon, and Peterson, 1988).
186 Often, in this type of operations, the stress factor that can significantly increase the cerebral blood flow similarly to the other aforementioned factors is neglected.
187 Above $24\ h$.
188 Symptoms include dry cough, increase in breathing resistance, problems with taking full breaths, etc.
189 *Paul Bert* is considered the first to have described *CNSyn* (Shilling, 1981).
190 *Lorrain Smith* effect.
191 The most commonly accepted safety limit for dives using diving apparatus.
192 Short exposures.
193 Optimal conditions for "washing out" the body from nitrogen.
194 Effective reductions in the size of gas bubbles at the decompression disease.
195 It is worth to mention that *hypoxia* and *hypercapnia* partially cancel each other out in the atmosphere of *INERGEN*
196 For example, the carcinogenic effects of oxygen occurring during long, multiple oxygen exposure were described (Bartosz, 2008).
197 *Somatic* – meaning concerning body: bodily, physical.

198 Distortion of tactile sensation involving the wrong location of stimuli, and distortion of their experience as tingling, numbness, prickling sensation, etc.
199 Who do not have problems with equalizing the pressure through swallowing saliva or movement of the lower jaw.
200 Blowing out breathing medium from lungs to nose with closed mouth and squeezed nostrils.
201 Physical absorption combined with chemical reaction is *chemisorption*.
202 The fluid filling the bony labyrinth of the inner ear.
203 Therapeutic procedure often used for general and local tissue oxygenation against anaerobes, after respiration poisoning, to secure the grafts, frostbites, burns, post-radiation damages, osteomyelitis, safeguarding the slow-healing wounds, sudden deafness, etc.
204 Central retinal artery occlusion – $CRAO$.
205 Blockages often concern the main artery branches supplying the retina and can be seen as partial problem with visual acuity; similar symptoms manifest many cardiovascular diseases, including systemic hypertension, inflammation of the arteries, or diseases such as syphilis.
206 For example, associated with the need to remain on the surface of the water, going to the shore, leaving the area, counteraction against waves.
207 Any lack of stimulation of the nervous system causes perturbations of conduction of signals, which result in the exclusion of control, leading to cessation of breathing.
208 Decompression sickness.
209 Symptoms of *CNSyn* were induced in one diver once or twice a week for a period of $90\,days$ and significant differences in tolerance to increased partial oxygen pressure as a breathing medium were observed (Donald, 1947).
210 Symptoms of *CNSyn* were induced in 36 divers, and significant differences in the tolerance time to oxygen partial pressure at 373 *kPa* were observed (Donald, 1947).
211 One has to be aware that to serve in special forces, candidates with special physical qualities, are healthy, and are young are selected. Hence, it should be assumed that the threat to the wider population is bigger than that observed during the tests.
212 For example, physical performance tests.
213 Breathing work $W < 0.3 J \cdot dm^{-3}$ and the maximum instantaneous inhalation resistance $\Delta p < 2.5 kPa$.
214 The first or the second.
215 Underwater Diving Working Group, NATO Standardization Agency Meeting, Istanbul, 2011.
216 Understood here as an approach taking into account the logical relationship with the problematic situation.
217 As a substitute to psychologists.
218 Practice shows that the divers in the units may also be subject to other occupational diseases, not only associated with the diving, for example, disorders of the spine associated with running a high-speed boat.

REFERENCES

Allen T.T. 2006. *Introduction engineering statistics and six sigma*. London: Springer Ltd. ISBN 10: 1-85233-955-1; ISBN 13: 978-1-85233-955-5

Anderson B., Saltzman H.A., Barbee J.Y. 1965. Retinal vascular and functional response to hyperbaric oxygenation. In *Proceedings of the third international conference on hyperbaric medicine*, ed. Brown I.W., Cox B.G., 276–280. Durham: Duke University Press.

ASTM. 1997. *Flammability and sensitivity of materials in oxygen-enriched atmospheres*. Philadelphia: American Society for Testing and Materials. ISBN: 0-8031-1844-9.

Automotive Industry Task Force. 2010. *Measurement systems analysis reference manual*. 4th Edition. Chrysler Group LLC, Ford Motor Company, General Motors Corporation. ISBN: 978-1-60-534211-5.

Bachrach A.J., Desiderati B.M., Matzen M.M. 1988. *A pictorial history of diving*. San Pedro: Best Publishing Co. ISBN: 0-941332-09-8.

Bartosz G. 2008. *Druga twarz tlenu—wolne rodniki w przyrodzie*. Warszawa: Wydawnictwo Naukowe PWN. ISBN: 978-83-01-13847-9.

Bennett P.B., Elliott D.H. 1993. *The physiology and medicine of diving*. London: W.B. Saunders Company Ltd. ISBN: 0-7020-0538-X.

Berghage T.E., Vorosmarti Jr. J., Barnard E.E.P. 1978. *Recompression treatment tables used throughout the world by government and industry*. Bethesda, MD: US Naval Medical Research Center. NMRI 78-16.

Betts E.A. 1992. *The application of enriched air mixtures*. Freeport: American Nitrox Divers Inc.

Betts E.A. 1994. *Introduction to enriched air diving*. Freeport: American Nitrox Divers Inc.

BIML. 2000. *International vocabulary of terms in legal metrology*. Paris: Bureau International de Métrologie Légale.

Bitterman N. 2004. CNS oxygen toxicity. *Undersea and Hyperbaric Medicine Journal*. 31, 63–72.

Bøe J., Hartung K.-H. 1983.1983. Employment of the polycom 101 gas mixing unit for divers in major project in Norway. *Dräger Review*. 51, 26–28.

Brubakk A.O., Neuman T.S. 2003. *Bennett and Elliott's physiology and medicine of diving*. Philadelphia: Saunders. ISBN: 0-7020-2571-2.

Bühlmann A.A. 1975. Decompression theory: Swiss practice. In: *The physiology and medicine of diving and compressed air work*, ed. Bennett P.B. Elliott D.H. London: Baillière Tindall Ltd.

Bühlmann A.A. 1984. *Decompression-decompression sickness*. Berlin: Springer-Verlag. ISBN: 3-540-13308-9; ISBN: 0-387-13308-9.

Burdick R.K., Borror C.M., Montgomery D.C. 2005. *Design and analysis of gauge R&R studies: making decisions with confidence intervals in random and mixed ANOVA models*. Philadelphia: Society for Industrial and Applied Mathematics. ISBN: 0-89871-588-1.

Burdick R.K., Graybill F.A. 1992. *Confidence intervals on variance components*. New York: Marcel Dekker. ISBN: 0-8247-8644-0.

Butler F.K., Knafelc M.E. 1986. Screening for oxygen intolerance in U.S. Navy divers. *Undersea Biomedical Research*. 13, 91–98.

Butler F.K., Thalmann E.D. 1984. CNS oxygen toxicity in closed-circuit SCUBA diving. In *Proceedings of the eighth symposium on underwater physiology*, ed. Bachrach A.J., Matzen M.M., 15–30. Bethesda: Undersea Medical Society.

Butler F.K., Thalmann E.D. 1986a. Central nervous system oxygen toxicity in closed-circuit SCUBA divers II. *Undersea Biomedical Research*. 13, 193–223.

Butler F.K., Thalmann E.D. 1986b. *CNS oxygen toxicity in closed-circuit SCUBA divers III*. Panama City: USN Experimental Diving Unit. Report No. 5-86.

CGA. 1985. *Cleaning equipment for oxygen service*. Arlington: Compressed Gas Association Inc. CGA G-4.1.

Clark J.M., Thom S.R. 2003. Oxygen under pressure. In *Bennett and Elliott's physiology and medicine of diving*, ed. Brubakk A.O. Neuman T.S., 358–418. Philadelphia: Saunders. ISBN: 0-70-20-2571-2.

Comex Marseille. 1986. *Medical book*. Marseille: Comex.

DMW. 1982. *Tabele dekompresji i rekompresji nurków*. Gdynia: Dowództwo Marynarki Wojennej. Sygn.Mar.Woj. 860/81.

DMW. 1984. *Przepisy nurkowania*. Gdynia: Dowództwo Marynarki Wojennej. Mar. Woj. 913/84.

DMW. 2007. *Tymczasowa instrukcja: standardowy test ciśnieniowy i tolerancji tlenowej*. Gdynia: Dowództwo Marynarki Wojennej. Załącznik 2 do rozkazu Dowódcy Marynarki Wojennej nr 30/SRM z dn., 2 April.

Donald K.W. 1947. Oxygen poisoning in man part I. *British Medical Journal*. 17, 667–672.

Donald K.W. 1992. *Oxygen and the diver*. Harley Swan: The SPA Ltd. ISBN: 1-85421-176-5.
Donald K.W. 1955. Oxygen bends. *Journal of Applied Physiology*. 7, 639–644.
Doolette D.J., Mitchell S.J. 2003. Biophysical basis for inner ear decompression sickness. *Journal of Applied Physiology*. 94, 2145–2150.
Drägerwerk A.G. 2011. *Dräger-tubes & CMS handbook: soil, water, and air investigations as well as technical gas analysis*. 16th Edition. Lübeck: Dräger Safety AG & Co. KGaA.
Edmonds C., Lowry C., Pennefather J. 1992. *Diving and subaquatic medicine*. Oxford: Butterworth Heinemann Ltd.
Farmer J.C., Thomas W.G. 1976. Ear and sinus problems in diving. In *Diving medicine*, Strauss R.H., 109–134. New York: Grune & Stratton, Inc. ISBN: 0-8089-0699-2.
Fire Research, Test, Development and Education Centre. 1993. *Test report: safety at 8% oxygen—15 minutes*. Copenhagen: Fire Research, Test, Development and Education Centre.
Flook V. 1992. Basic respiratory physiology. In *Lung physiology and divers' breathing apparatus*, ed. Flook V., Brubakk A.O., 1–16. Ballater, Scotland: Ballater: BPCC-AUP Aberdeen Limited. Proceedings from the International Workshop, 1–4 November. ISBN: 82-595-7125-0
Flook V. 2002. Breathing gas contamination by volatile hydrocarbons. *European Journal of Underwater and Hyperbaric Medicine*. 32, 17–27.
Fowler B. 1991. Narcosis. *AquaCorps*. 3, 48–50.
Gilliam B., von Maier R. 1992. *Deep diving—an advanced quide to physiology, procedures and systems: scientific applications of deep diving*. San Diego: Watersport Publishing, Inc. ISBN: 0-922769-31-1.
Glenn J.B. 1992. *Oxygen safety in the production of enriched air nitrox breathing mixtures*, ed. Mastro R.W., Hamilton R.W. Wilmington: University of North Carolina. Proceedings of the American Academy of Underwater Sciences 12th Annual Scientific Diving Symposium.
Gussmann J. 1984. *Człowiek zdobywa głębiny*. Gdańsk: Wydawnictwo Morskie. ISBN: 83-215-4047-3.
Hamilton R.W. 1989. Tolerating exposure to high oxygen levels: repex and other methods. *Marine Technology Society Journal*. 23, 19–25.
Hamilton R.W., Adams G.M., Harvey C.A., Knight D.R. 1982. *SHAD-NISAT: a composite study of simulated shallow saturation diving*. Groton: Naval Submarine Medical Research Laboratory Submarine Base. Report No. 985.
Hamilton R.W., Kenyon D.J., Peterson R.E. 1988. *REPEX habitat diving procedures: repetitive vertical excursions, oxygen limits, and surfacing techniques*. Washington: National Oceanic and Atmospheric Administration. Technical Report No. 88-1B.
Hamilton R.W., Kenyon D.J., Peyerson R.E., Butler G.J., Beers D.M. 1988. *REPEX: development of repetitive excursions, surfacing techniques, and oxygen procedures for habitat diving*. Washington: National Oceanic and Atmospheric Administration. Technical Report No. 88-1A.
Harabin A.I., Survanshi S.S. 1993. *A statistical analysis of recent NEDU single-depth human exposures to 100% oxygen at pressure*. Bethesda: Naval Medical Research Institute. NMRI 93-59; AD-A237-488.
Harabin A.L., Survanshi S.S., Homer L.D. 1994. *A model for predicting central nervous system toxicity from hyperbaric oxygen exposure in man: effects of immersion, exercise, and old and new data*. Bethesda: Naval Medical Research Institute. NMRI 94-0003; AD-A278-348.
Harvey C.A. 1977. *Shallow saturation hyperbaric exposures to nitrogen-oxygen environments and isobaric switches to helium-oxygen*. Toronto: Hyperbaric Medical Society, Inc.
Hills B.A. 1966. *A thermodynamic and kinetic approach to decompression sickness*. Adelaide: Libraries Board of South Australia. Occasional Papers in Physiology No. 1.
Hills B.A. 1997. *Decompression sickness—the biophysical basis of prevention and treatment: vol. 1*. Chichester: John Willey & Sons. ISBN: 0-471-99457-X.
Honeywell. 2022. *Gas detection tubes and sampling handbook*. www.honeywellanalytics.com/.

Hyldegaard O., Jensen T. 2007. Effect of heliox, oxygen and air breathing on helium bubbles after heliox diving. *UHMS Journal*. 34, 107–122.

Imbert J.-P. 1975. *Proposition of a perfusion limited model for isobaric counterdiffusion*. Philadelphia: University of Pennsylvania Medical Center. Report No. 7-1.

Jakus B., Kłos R. 2007. Zdolność procesu pomiarowego. *Polish Hyperbaric Research*. 19, 33–46.

Karreman G., Lambertsen C.J. 1977. Kinetics of isobaric counterdiffusion. *Bulletin of Mathematical Biology*. 39, 587–595.

Keller H., Bühlmann A.A. 1965. Deep diving and short decompression by breathing mixed gases. *Journal of Applied Physiology*. 20, 1267–1270.

Kenny J.E. 1973. *Business of diving*. Houston: Gulf Publishing Co. ISBN: 0-87201-183-6.

Kłos R. 1989. Zastosowanie metod termokonduktometrycznych do pomiaru składu mieszanin gazowych używanych w komorach hiperbarycznych. *Zeszyty Naukowe AMW*. 100, 135.

Kłos R. 1990. *Metoda pomiarów składu mieszanin oddechowych w nurkowych kompleksach hiperbarycznych*. Gdynia: Akademia Marynarki Wojennej.

Kłos R. 2000. *Aparaty Nurkowe z regeneracją czynnika oddechowego*. Poznań: COOPgraf. ISBN: 83-909187-2-2.

Kłos R. 2007a. *Mathematical modelling of the normobaric and hyperbaric facilities ventilation*. Gdynia: Wydawnictwo Polskiego Towarzystwa Medycyny i Techniki Hiperbarycznej. ISBN: 978-83-924989-0-2.

Kłos R. 2007b. *Zastosowanie metod statystycznych w technice nurkowej*. Gdynia: Polskie Towarzystwo Medycyny i Techniki Hiperbarycznej. ISBN: 978-83-924989-26.

Kłos R. 2008. *Systemy podtrzymania życia na okręcie podwodnym*. Gdynia: Polskie towarzystwo medycyny i Techniki Hiperbarycznej. ISBN: 978-83-924989-4-0.

Kłos R. 2012. *Możliwości doboru ekspozycji tlenowo-nitroksowych dla aparatu nurkowego typu AMPHORA—założenia do nurkowań standardowych i eksperymentalnych*. Gdynia: Polskie Towarzystwo Medycyny i Techniki Hiperbarycznej. ISBN: 978-83-924989-8-8.

Kłos R. 2014. *Helioksowe nurkowania saturowane—podstawy teoretyczne do prowadzenia nurkowań i szkolenia*. Gdynia: Polskie Towarzystwo Medycyny i Techniki Hipernbarycznej. ISBN: 978-83-938322-1-7.

Kłos R. 2015. *Katalityczne utlenianie wodoru na okręcie podwodnym*. Gdynia: Polskie Towarzystwo Medycyny i Techniki Hiperbarycznej. ISBN: 978-83-938-322-3-1.

Kłos R. 2023. *Oxygen diving*. Boca Raton, FL: CRC Press (Taylor & Francis Group, LLC). ISBN: 978-1-032-31389-4 (hbk) ISBN: 978-1-032-31390-0 (pbk) ISBN: 978-1-003-30950-5 (ebk). DOI: 10.1201/9781003309505.

Kłos R., Konarski M., Olszański R. 2004. The implementation of factor analysis for the evaluation of selected blood parameter changes induced by hyperbaric exposure. *International Maritime Health*. 55, 87–101.

Lambertsen C.J. 1986. *Studies in isobaric counterdiffusion*. Filadelfia: Institute for Environmental Medicine.

Lambertsen C.J., Idicula J. 1975. A new gas lesion syndrome in man, induced by "isobaric gas counterdiffusion". *Journal of Applied Physiology*. 39, 434–443.

Lamberdsen C.J., Kough R.H., Cooper D.Y., Emmel G.L., Loeschcke H.H., Schmidt C.F. 1953. Oxygen toxicity: effects in man of oxygen inhalation at 1 and 3,5 atmospheres upon blood gas transport, cerebral circulation and cerebral metabolism. *Journal of Applied Physiology*. 9, 471–486.

Miller J.W., Koblick I.G. 1984. *Living and working in the sea*. New York: Van Nostrand Reinhold Company Inc. ISBN: 1-886699-01-1.

Montgomery D.C. 2009. *Statistical quality control*. 6th Edition. New York: John Wiley & Sons Inc. ISBN: 978-0-470-16992-6.

NFPA. 1999. *Recommended practice on materials, equipment, and systems used in oxygen-enriched atmospheres*. Quincy: National Fire Protection Association. NFPA 53.

NOAA. 2001. *NOAA diving manual—diving for science and technology*. Flagstaff: Best Publishing Co. ISBN: 0-941332-70-5.
NOAA. 2017. *NOAA diving manual—diving for science and technology*. Flagstaff: Best Publishing Co. ISBN: 9781930536883.
NSO. 2020. *Allied guide to diving medical disorders—national information*. Edition A Version 1. Brussels: NATO Standardization Office (NSO). Standards Related Document ADivP-02.1 (STANAG 1432).
Palmer R. 1994. *An introduction to technical diving*. Middlesex: Underwater World Publication Ltd. ISBN: 0-946020-23-X.
Przylipiak M., Torbus J. 1981. *Sprzęt i prace nurkowe-poradnik*. Warszawa: Wydawnictwo Ministerstwa Obrony Narodowej. ISBN: 83-11-06590-X.
Pyzdek T. 2003. *The six sigma handbook*. New York: The McGraw-HIll Companies, Inc. ISBN: 0-07-141596-3. DOI: 10.1036/0071415963.
Rutkowski D. 1990. *Nitrox manual*. Key Largo: Hyperbarics International, Inc.
Sharley P., Pyle R. 1992. *The twilight zone: the potential, problems and theory behind using mixed gas, surface-based SCUBA for research diving between 200 and 500 feet*. Diving for Science, Proceedings of the American Academy of Underwater Sciences (12th Annual Scientific Diving Symposium).
Shilling C.W. 1981. *A history of the development of decompression tables*. Bethesda: Undersea Medical Society, Inc.
Strauss R.H. 1976. *Diving medicine*. New York: Grune & Stratton Inc. ISBN: 0-8089-0699-2.
Torbati D., Church D.F., Keller J.M., Pryor W.A. 1992. Free radical generation in the brain precedes hyperbaric oxygen induced convulsions. *Free Radical Biology and Medicine*. 13, 101–106.
UHMS 22 Workshop. 1979. *Isobaric inert gas counterdiffusion*, ed. Lambertsen R.C., Bornmann C. J. UHMS 22 Workshop. Philadelphia: Undersea Medical Society. UMS Publication No. 54WS(IC)1-11-82.
UNIPHOS. 2022. *Gas detector tube handbook*. http://www.uniphos.eu.
USN. 1993. *US Navy diving manual*. Washington: The Direction of Commander, Naval Sea Systems Command.
USN. 2008. *US Navy diving manual*. Washington: The Direction of Commander Naval Sea Systems Command. 0910-LP-106-0957.
USN. 2011. *US Navy diving manual*. Washington: The Direction of Commander Naval Sea Systems Command. SS521-AG-PRO-010 0910-LP-115-1921.
USN. 2016. *US Navy diving manual*. Washington: The Direction of Commander Naval Sea Systems Command. SS521-AG-PRO-010 0910-LP-115-1921.
Vann R.D. 1989. *The physiology of NITROX diving*, ed. Hamilton R.W., Crosson D.J., Hulbert A.W. Workshop on Enriched Air NITROX Diving. Washington: National Oceanic and Atmospheric Administration.
Walters K.C., Gould M.T., Bachrach E.A., Butler F.K. 2000. Screening for oxygen sensitivity in US Navy combat swimmers. *Undersea and Hyperbaric Medicine*. 27, 21–26.
Wendling J., Nussberger P., Schenk B. 1999. Milestones of the deep diving research laboratory Zurich. *SPUMS Journal*. 29, 91–98.
Wienke B.R. 2001. *Technical diving in depth*. Flagstaff: Best Publishing Co. ISBN: 0-941332-97-7.
Wienke B.R. 2003. *Basic decompression theory and application*. Flagstaff: Best Publishing Co., 2003. ISBN: 1-930536-14-3.
Военное Издательство. 1983. *Руководство по борьбе за живуесть подводной лодки РБЖ-ПЛ-82*. Москва: Военное Издательство.
Чикер Н.П. 1968. *Справочник специалиста аварийно-спасательной службы ВМФ*. Москва: Военное издателство МО СССР.

5 Ventilation

The mathematical models of ventilation in a semi-closed rebreather *SCR* and the constant dosage of premix have been presented in the literature dealing with the ventilation of hyperbaric chambers, submarines, and the mining excavation. However, they are not precise enough (Reimers and Hansen, 1972; Haux, 1982). Most of the results presented were obtained during investigations on the construction of the semi–closed circuit rebreather *SCR* described in Chapter 3 (Kłos, 2021).

The mathematical models of ventilation for *SCR* are based on the mass balance of oxygen in the respiratory mixture inhaled by a diver.[1] The mathematical models of ventilation for *SCR* are based on the mass balance of oxygen in the respiratory mixture inhaled by a diver.[2] For time approaching infinity $t \rightarrow \infty$, the mathematical models proposed here are the same as the models proposed by other authors (Reimers and Hansen, 1972; Williams, 1975; Haux, 1982; Frånberg, 2015). The full-time dependent models are useful during the design of a diving apparatus and suitable decompression. In many cases, they are good enough to be used for illustrating the important phenomena occurring in a diving apparatus. Frequently, conclusions concerning the composition stabilization rate of the diving apparatus atmosphere after disturbances caused by a change in the oxygen consumption level[3] or caused by periodical ventilation can be drawn based on that model.

The mathematical models for designing *SCR*s with constant dosage of *premix* that have been presented in the literature ignore the accuracy of the values affecting the composition of a breathing medium. The mathematical model proposed here takes into account the metering accuracy of a breathing medium and the accuracy in calculating the percentage of oxygen in the *premix*.

The classic design of *SCR*s can cause errors leading even to diving accidents. Diving accidents were recorded during tests carried out at medium to high work rate performed by the diver.

5.1 MODELLING

The problem of ventilation is treated here more widely than in most cases, in which the prevailing opinion is that ventilation involves only the removal of contaminants. Despite the comprehensive description of ventilation, the problems presented and analyzed in detail are limited to ensuring proper oxygen partial pressure in the breathing medium inhaled by a diver.

As for the *SCR*, ventilation problems are related more to ensuring the oxygen partial pressure than to removing carbon dioxide. The problem of carbon dioxide removal is more related to chemical regeneration of the atmosphere.[4] The problems of carbon dioxide removal from an *SCR* and the experimental results will be described later in Chapter 6.

DOI: 10.1201/9781032649207-5

5.1.1 One-Bag SCR

Oxygen is supplied to the breathing bag with the fresh 1 and regenerated breathing medium 2. The oxygen from the breathing bag is removed with the inhaled gas 3 through relief valve 4. Part of the oxygen from the inhaled medium is assimilated by the diver's organism 5, and after carbon dioxide removal,[5] the breathing medium flows back to the breathing bag 2. Based on Figure 5.1, the oxygen-and-breathing-medium balance can be calculated. This has been prepared and presented in Table 5.1 and Table 5.2.

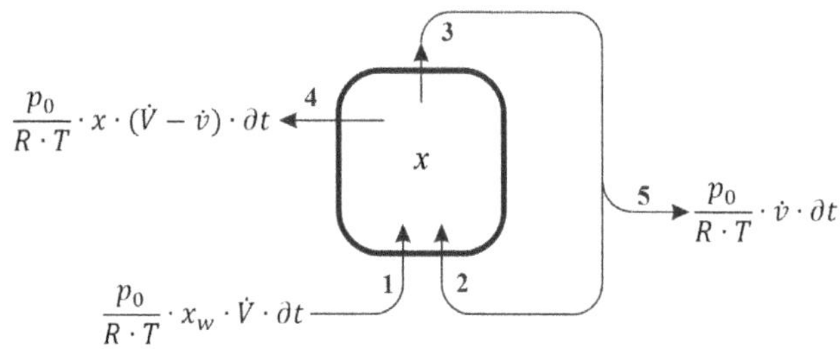

FIGURE 5.1 Oxygen molar balance in the breathing bag, where $\dot{V}$ = is metering of the fresh breathing medium; $\dot{v}$ = is oxygen consumption; x_w = is oxygen molar fraction in the *premix*; ∂t = is elementary time; p_0 = is normal pressure; R = is universal gas constant; T is the absolute temperature; and V = is breathing bag volume.

TABLE 5.1
Oxygen Molar Balance in the Breathing Bag

		With the Fresh Breathing Medium	With Breathing Circulation	Through the Relief Valve
Oxygen	Increase	$\frac{p_0}{R.T} \cdot \dot{V} \cdot x_w \cdot dt$	$\frac{p}{R \cdot T}\left[\dot{V}_E \cdot x(i) - \frac{p_0}{p} \cdot \dot{v}\right] dt$	—
	Decrease	—	$\frac{p}{R \cdot T} \cdot \dot{V}_E \cdot x(i) \cdot dt$	$\frac{p_0}{R \cdot T} \cdot (\dot{V} - \dot{v}) \cdot x(i) \cdot dt$
The whole breathing medium	Increase	$\frac{p_0}{R \cdot T} \cdot \dot{V} \cdot dt$	$\frac{p}{R \cdot T} \cdot \dot{V}_E - \frac{p_0}{R \cdot T} \cdot \dot{v} \cdot dt$	—
	Decrease	—	$\frac{p_0}{R \cdot T} \cdot \dot{V}_E \cdot dt$	$\frac{p_0}{R \cdot T} \cdot (\dot{V} - \dot{v}) \cdot x(i) \cdot dt$

Where $\dot{V}$ is the metering of the fresh breathing medium; $\dot{v}$ is oxygen consumption; $\dot{V}_E$ is pulmonary ventilation; x_w is oxygen molar fraction in the *premix*; ∂t is elementary time; i is the successive time interval ∂t for substance balance; $x(i)$ is oxygen molar fraction in the breathing bag at the moment $t = i \cdot \partial T$; p_0 is normal pressure; R is universal gas constant; and T is absolute temperature.
Source: Based on Figure 5.1.

TABLE 5.2

Relationship between Oxygen Molar Fraction x and Time t for the Single-Bag Semi–Closed Circuit Diving Apparatus *SCR* with the Constant Metering of *premix* by Means of Limits Calculus[6]

A	1°	$\dot{V} \neq f(H);\ \dot{\upsilon} \neq f(H);\ \dot{V}_E \neq f(H)$	
	2°	$\dot{V} = \dot{V}_0 = \dot{V}(p = p_0);\ \dot{\upsilon} = \dot{\upsilon}_0 = \dot{\upsilon}(p = p_0)$	
T		$x(O_2) = f(t)$	
P	1°	For elementary time $d\tau$ at time $(i+1)\cdot d\tau$, it can be written as: $n(i+1) = n(i) + \frac{p_0}{R\cdot T}\cdot\left[x_w\cdot\dot{V} + x(i)\cdot\dot{V} - \dot{\upsilon} + x(i)\cdot\dot{\upsilon}\right]\cdot\partial t$	From oxygen balance
	2°	$x(i+1) = \left[x(i) - \frac{x_w\cdot\dot{V} - \dot{v}}{\dot{V} - \dot{v}}\right]\cdot\left(1 - \frac{p_0}{p}\cdot\frac{\dot{V} - \dot{\upsilon}}{V}\cdot\partial t\right) + \frac{x_w\cdot\dot{V} - \dot{v}}{\dot{V} - \dot{v}}$	From 1° divided by $n = \frac{p\cdot V}{R\cdot T}$ and $x(i) = \frac{n(i)}{n}$
	3°	$x(i+2) = \left[x(i+1) - \frac{x_w\cdot\dot{V} - \dot{v}}{\dot{V} - \dot{v}}\right]\cdot\left(1 - \frac{p_0}{p}\cdot\frac{\dot{V} - \dot{\upsilon}}{V}\cdot\partial t\right) + \frac{x_w\cdot\dot{V} - \dot{v}}{\dot{V} - \dot{v}}$	From 2°
	4°	$x(i+2) = \left[x(i) - \frac{x_w\cdot\dot{V} - \dot{v}}{\dot{V} - \dot{v}}\right]\cdot\left(1 - \frac{p_0}{p}\cdot\frac{\dot{V} - \dot{\upsilon}}{V}\cdot\partial t\right)^2 + \frac{x_w\cdot\dot{V} - \dot{v}}{\dot{V} - \dot{v}}$	From 3° and 2°
	5°	$x(i+j) = \left[x(i) - \frac{x_w\cdot\dot{V} - \dot{v}}{\dot{V} - \dot{v}}\right]\cdot\left(1 - \frac{p_0}{p}\cdot\frac{\dot{V} - \dot{\upsilon}}{V}\cdot\partial t\right)^j + \frac{x_w\cdot\dot{V} - \dot{v}}{\dot{V} - \dot{v}}$	From 2° to 4°
	6°	For $i = 0$, $j = t$, and $x(0) = x_0$, it can be written as: $x(t) = \frac{x_w\cdot\dot{V} - \dot{v}}{\dot{V} - \dot{v}} + \left[x_0 - \frac{x_w\cdot\dot{V} - \dot{v}}{\dot{V} - \dot{v}}\right]\cdot\lim_{j\to\infty}\left(1 - \frac{p_0}{p}\cdot\frac{\dot{V} - \dot{\upsilon}}{V}\cdot\partial t\right)^j$	From 5°
	7°	$x(t) = a + b\cdot\lim_{j\to\infty}\left[\left(1 + \frac{c}{j}\right)^{j/c}\right]^c$ where $a = \frac{x_w\cdot\dot{V} - \dot{v}}{\dot{V} - \dot{v}}$, $b = x_0 - a$, $c = -\frac{p_0}{p}\cdot\frac{\dot{V} - \dot{\upsilon}}{V}\cdot j\cdot\partial t$	From 6° $\lim_{j\to\infty}(j\cdot dt) = t$
	8°	$x(t) = exp(c) + a$, because $\forall_{a(n)\neq 0}\ \lim_{n\to\infty} a(n) = 0 \Rightarrow \lim_{n\to\infty}\left[1 + a(n)\right]^{1/a(n)} \equiv e$	From 7° $a(n) = \frac{c}{j}$
	9°	$x(t) = \frac{x_w\cdot\dot{V} - \dot{\upsilon}}{\dot{V} - \dot{\upsilon}} + \left(x_0 - \frac{x_w\cdot\dot{V} - \dot{\upsilon}}{\dot{V} - \dot{\upsilon}}\right)\cdot\exp\left(-\frac{p_0}{p}\cdot\frac{\dot{V} - \dot{\upsilon}}{V}\cdot t\right)$	From 7° to 8° q.e.d.

Where $\dot{V}$ = is the metering of the fresh breathing medium; $\dot{v}$ = is oxygen consumption; $\dot{V}_E$ = is pulmonary ventilation; x_w = is oxygen molar fraction in the *premix*; x_0 = is initial composition of breathing medium in breathing bag; ∂t = is elementary time; i = is the successive time interval ∂t for substance balance; $x(i)$ = is oxygen molar fraction in the breathing bag at the moment $t = i\cdot\partial t$; $n(i)$ = is the number of oxygen moles in the breathing bag at the moment $t = i\cdot\partial t$; p_0 is normal pressure; p = is pressure at the depth of diving; R = is universal gas constant; T = is the absolute temperature; and V is breathing bag volume.

Note: A = assumption; T = thesis, P = proof (evidence).

Source: Based on oxygen molar balance in the breathing bag presented in Table 5.1.

The function describing the relationship between oxygen content in the breathing bag and time is derived from the substance balance in Table 5.2: $x(O_2) = f(t)$. This relationship can be derived on the assumption that premix metering $\dot{V}$, oxygen consumption $\dot{\upsilon}$, and pulmonary ventilation $\dot{V}_E$ are not diving depth–dependent H: $\dot{V} \neq f(H)$, $\dot{\upsilon} \neq f(H)$, $\dot{V}_E \neq f(H)$. In reality, a slight dependence on the depth of diving is recorded (Brennan and Bdonchuk, 1975; Kłos, 2021).

Before equation $x(O_2) = f(t)$ is derived, it is worthy of note that the values of premix metering $\dot{V}$, oxygen consumption $\dot{\upsilon}$, and pulmonary ventilation $\dot{V}_E$ are related to normal pressure p_0; therefore, the number of oxygen moles should be related to the expanded breathing medium. According to Figure 5.1 and Table 5.1, the formula describing the oxygen molar fraction in the breathing bag $x(t)$ of *SCR* can be derived in Table 5.2. The same result was obtained when the problem was being solved by means of the integral calculus and limits calculus (Kłos, 2021). Two conclusions can be drawn from the analysis of equation (9°) from Table 5.2:

$$\forall_{t=0}\, x(0) = x_0 \tag{5.1}$$

$$\forall_{t\to\infty}\, x(t) = x_s = \frac{x_w \cdot \dot{V} - \dot{\upsilon}}{\dot{V} - \dot{\upsilon}} \tag{5.2}$$

where $x_s =$ is the stable value of oxygen molar fraction in the inhalation bag $x_s = \lim_{t\to\infty} x(t)$.

Equation (5.1) is the simple result of the assumption (6°) from Table 5.2, whereas equation (5.2) is presented in the literature as the formula obtained from the mass balance of oxygen and breathing medium in the breathing bag (Williams, 1975; Haux, 1982).

5.1.2 Two-Bag SCR

Equation (5.2) can be used for the two-bag version of the semi–closed circuit apparatus with constant breathing medium metering at the normal operation circumstances. It can be proved using the formula for the stable value of oxygen molar fraction in the breathing bag of the two-bag version of *SCR* (Kłos, 2021):

$$\forall_{t\to\infty}\, x(t) = x_s = \frac{x_w \cdot \dot{V} - \dot{\upsilon}}{\dot{V} - \dot{\upsilon}} + \frac{p_0 \cdot x_w \cdot \dot{V}}{p \cdot \dot{V}_E} \tag{5.3}$$

where $\dot{V}_E$ is pulmonary ventilation.

For *SCR* with constant *premix* dosage in the hyperbaric environment, $p_0 < p$, $\dot{V} \ll \dot{V}_E$, and $0 \ll x_w < 1$. It follows that the stable value of oxygen molar fraction in the breathing bag x_s of *SCR* can be written as equation (5.2) because $\frac{p_0 \cdot x_w \cdot \dot{V}}{p \cdot \dot{V}_E} \ll \frac{x_w \cdot \dot{V} - \dot{\upsilon}}{\dot{V} - \dot{\upsilon}}$.

Stabilization and homogenization of the breathing medium composition is ventilation-dependent. It is essential for safety, effectiveness, and economy of diving.

First of all, it has a direct bearing on decompression safety. These problems are most important in the design phase of the diving apparatus. An example of the theoretical breathing medium composition in the inhalation/exhalation bag of *SCR* is presented in Figure 5.2.

The composition of gas is the function of the time and oxygen consumed stream during breathing on the surface: $p = p_0$.

In practice, the stabilization time of the breathing medium composition is exceeded by the washout time of the additional breathing space, not the breathing bag, that is, the inhalation hose, scrubber, etc.

Based on equation (5.2), the formulas enabling calculation of time necessary for stabilization of the inhale bag atmosphere can be derived.[7] Two criteria for stabilization evaluation can be applied, as in the case of error evaluation. The criterion of absolute stabilization is the first of them: $\Delta x = \left|x(t) - x(\infty)\right|$, where Δx represents accepted limited deviation from the asymptotic value of the oxygen content in the breathing bag. Using equation (9°) from Table 5.2, it can be written as $\Delta x = \left|x(t) - x(\infty)\right| = \left(x_0 - \frac{x_w \cdot \dot{V} - \dot{\upsilon}}{\dot{V} - \dot{\upsilon}}\right) \cdot \exp\left(-\frac{p_0}{p} \cdot \frac{\dot{V} - \dot{\upsilon}}{V} \cdot t_s\right)$, where t_s represents the time necessary for the stabilization of the breathing medium composition. The

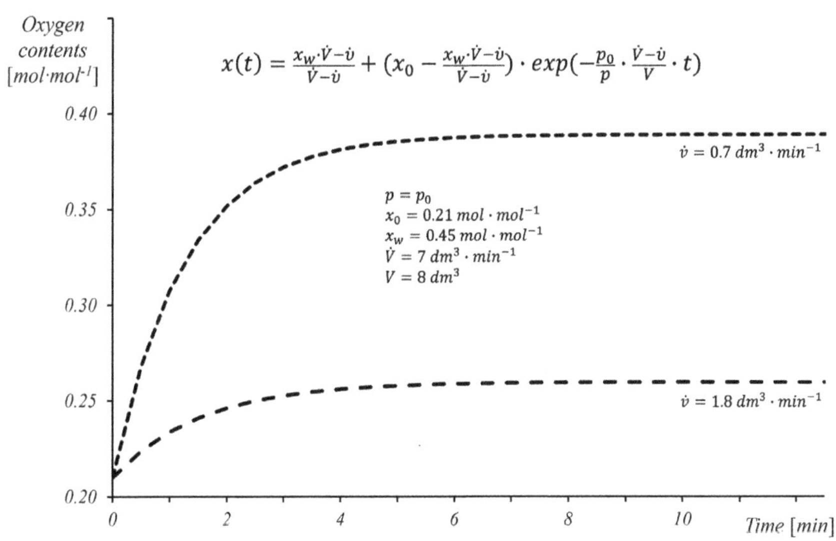

FIGURE 5.2 Theoretical composition of the breathing medium in the breathing bag of a single-bag *SCR* as the function of time and stream of oxygen consumed during breathing on the surface, where $p =$ is pressure at the depth of diving; $p_0 =$ is atmospheric pressure; $\dot{V} =$ is metering of the fresh breathing medium; $\dot{v} =$ is oxygen consumption; $\dot{V}_E =$ is pulmonary ventilation; $x_w =$is oxygen molar fraction in the *premix*; $x_0 =$ is the initial composition of the breathing medium in the breathing bag.

criterion of absolute stabilization is not convenient to compare breathing medium composition stabilization times. For this task, it is more convenient to apply the relative stabilization criterion δ:

$$\frac{\delta}{100\%} = \frac{|x(t) - x(\infty)|}{x(\infty)} = \frac{\dot{V} - \dot{\upsilon}}{x_w \cdot \dot{V} - \dot{\upsilon}} \cdot \left(x_0 - \frac{x_w \cdot \dot{V} - \dot{\upsilon}}{\dot{V} - \dot{\upsilon}}\right) \cdot \exp\left(-\frac{p_0}{p} \cdot \frac{\dot{V} - \dot{\upsilon}}{V} \cdot t_s\right) \quad (5.4)$$

where δ relative stabilization criterion, and t_s –is time necessary for the stabilization of the breathing medium composition.

Based on the analysis of equation (5.4), it can be stated that stabilization time is proportional to the pressure that is exerted on the diver: $t_s(p > p_0) = \frac{p}{p_0} \cdot t_s(p_0)$. As it follows, at the lower pressure, the stabilization time of the breathing medium composition t_s is greater than when under higher pressure (Figure 5.3).

The diving technology should take into account the stabilization time t_s of the breathing medium composition in the breathing space of the diving apparatus. In cases where the diver is transported to the worksite by means of a diving bell, up to a certain depth, he breathes the air contained in the bell. After exceeding the safety

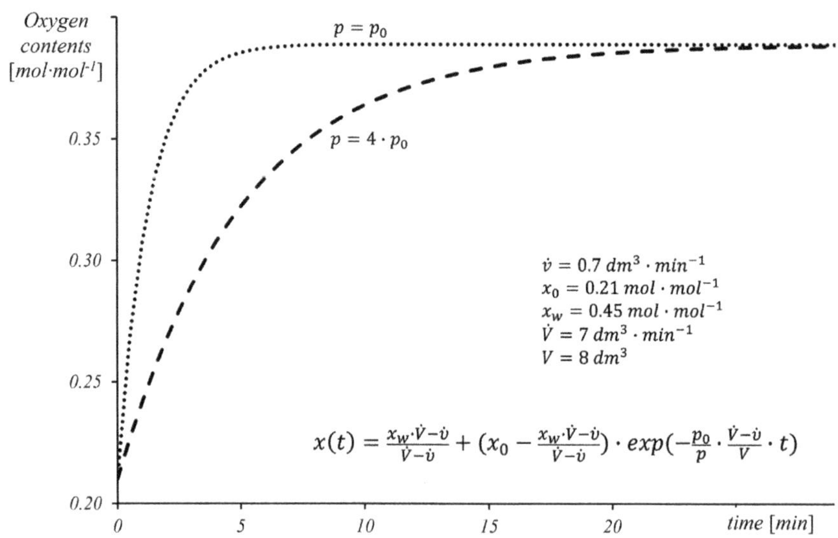

FIGURE 5.3 The theoretical composition of the breathing medium in the breathing bag of *SCR* as a function of time and pressure at the depth of diving at the same stream of oxygen consumed $\dot{\upsilon}$, where $p =$ is pressure at the depth of diving; $p_0 =$ is atmospheric pressure; $\dot{V} =$ is metering of the fresh breathing medium; $\dot{\upsilon} =$is oxygen consumption; $\dot{V}_E =$ is pulmonary ventilation; $x_w =$ is oxygen molar fraction in the *premix*; and $x_0 =$ is the initial composition of the breathing medium in the breathing bag.

depth, at which the breathing medium is not hypoxic, the diver can start breathing with the diving apparatus. In order to accelerate the breathing medium composition stabilization, frequent washing of the breathing space is recommended.

Worthy of note is the fact that stabilization of the breathing medium composition in the breathing space is associated with dynamic equilibrium between the fresh and exhaled breathing medium. Washing the whole breathing space with a fresh breathing medium does not guarantee immediate equilibrium; however, it will cause the oxygen content in the inhalation bag to approximate to the stable value when the diver does not exercise.

As before, the oxygen-and-breathing-medium molar balance is necessary. The balance is presented in Table 5.3.

According to this, the balance can be written as follows:

$$x(i+1)=\frac{\frac{p}{p_0}\cdot\left[V_z\cdot x(i)(V_c-V_z)\cdot x_w\right]}{\frac{p}{p_0}\cdot V_c}=\frac{V_z}{V_c}\cdot\left[x(i)-x_w\right]+x_w \tag{5.5}$$

where V_c = is the total breathing space volume, and V_z = the total residual volume in the breathing space.[8]

Using equation (5.5), the formula, which describes the oxygen content in the breathing space after $i+2$ washing, can be written as follows:

TABLE 5.3
Molar Balance of Oxygen and the Breathing Medium in the Course of Washing of the Closed Breathing Space

		Emptying	Filling
Oxygen	Increase	-	$\frac{p}{p_0}\cdot(V_c-V_z)\cdot x_w$
	Decrease	$\frac{p}{p_0}\cdot(V_c-V_z)\cdot x(i)$	-
	Remainder	$\frac{p}{p_0}\cdot V_z\cdot x(i)$	$\frac{p}{p_0}\cdot\left[V_z\cdot x(i)(V_c-V_z)\cdot x_w\right]$
Breathing medium	Increase	-	$\frac{p}{p_0}\cdot(V_c-V_z)$
	Decrease	$\frac{p}{p_0}\cdot(V_c-V_z)$	-
	Remainder	$\frac{p}{p_0}\cdot V_z$	$\frac{p}{p_0}\cdot V_c$

Where V_c = is the total breathing space volume, and V_z = the total residual volume in the breathing space.

$x(i+2) = V_z^2 \cdot V_c^{-2} \cdot \left[x(i) - x_w\right] + x_w$. The comparison of this equation with equation (5.5) can be generally written as:

$$x(i,j) = \left(\frac{V_z}{V_c}\right)^j \cdot \left[x(i) - x_w\right] + x_w \tag{5.6}$$

where $i =$ is the quantity of the moment of time, expressed by the basic episode equal to Δt, and $j =$ the number of breathing space rinses.

Equation (5.6) enables us to draw an important conclusion that can be written as follows: $x(j) \neq f(H)$; it means that washing of the breathing space is of the same effectiveness for each depth H. The theoretical composition of the breathing medium in the inhalation bag of *SCR* as a function of time after and without the preliminary washing of the breathing space is shown in Figure 5.4.

On assumption that for $i = 0 \Rightarrow x(0) \equiv x_0$, equation (5.6) can be written as:

$$x(j) = \left(\frac{V_z}{V_c}\right)^j \cdot \left[x_0 - x_w\right] + x_w \tag{5.7}$$

where $V_c =$ is the total breathing space volume, and $V_z =$ the total residual volume in the breathing space.

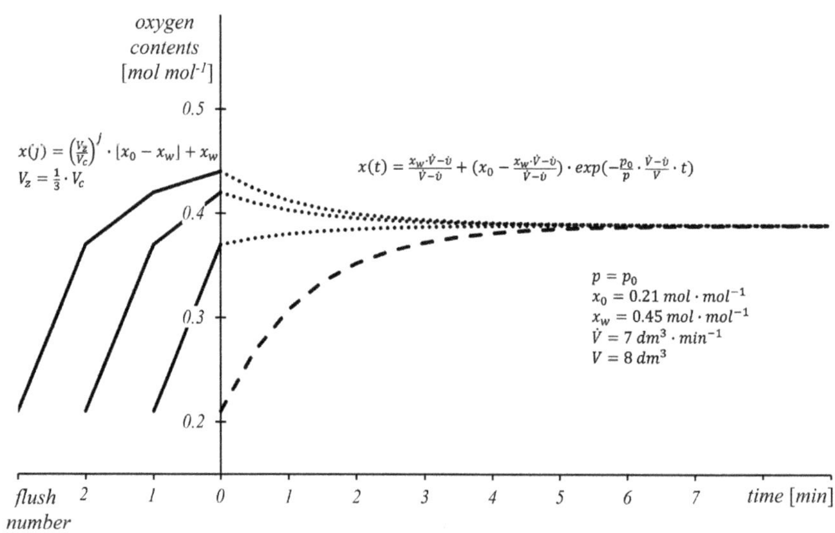

FIGURE 5.4 The theoretical composition of the breathing medium in the breathing bag of *SCR* as a function of time in the course of breathing on the surface after the washing and without the washing of the breathing space, where $p =$ is pressure at the depth of diving; $p_0 =$ the atmospheric pressure; $\dot{V} =$ the metering of the fresh breathing medium; $\dot{v} =$ the oxygen consumption; $\dot{V}_E =$ the pulmonary ventilation; $x_w =$ the oxygen molar fraction in the *premix*; and $x_0 =$ the initial composition of the breathing medium in the breathing bag.

5.1.3 Metering Bellows Dispenser

The *CRABE* is an *SCR* with two breathing bags placed one inside the other. It is powered by *nitrox*, *heliox*, and *trymix* as *premix* or oxygen during a part of decompression (Kłos, 2021). Its construction was described in Chapter 3 (Figure 3.14).

The breathing gas that remains in circulation is always a breathing mixture, because even during oxygen decompression, the diver does not breathe pure oxygen,[9] as in the circuit there is always some circulating nitrogen N_2 or helium *He* left from the breathing gas, which comes from places that are not ventilated sufficiently in the breathing space of the apparatus and from the diver's tissue cleansing with oxygen. The assessment of the effectiveness of cleansing of the breathing space and the diver's body with oxygen is crucial for the correct design of safe decompression.

During inspiration, part of the breathing mixture is taken: $-V_I$. The elementary volume of inspiration $-dV_I = -dh \cdot (A - a)$ will be the product of the elementary change in the height of bags $-dh$, multiplied by the area of the base of the large bag A minus the area of the small bag's base area $-a$ (Figure 5.5).

At this time, the elementary drop in the height of bags $-dh$ will cause a release of the content of the small bag dV_u to water, equal to the product of the elementary drop in height $-dh$ and surface area of the small bag $a: -dV_u = -dh \cdot a$. The ratio of elementary volume of the gas released from the small sack to water $-dV_u$ to the elementary volume of inspiration $-dV_I$ will be: $\frac{-dV_u}{-dV_I} = \frac{a}{A-a} = \frac{u}{U-u} \equiv r$, where u represents the volume of the small bag, and U the volume of the large bag. It follows from the preceding considerations that the released number of moles of the breathing mixture, and as a consequence of oxygen $-n_1 = f(\dot{V}_E, r)$, is associated with lung ventilation[10] $\dot{V}_E$ by means of the volume ratio of the small bag u and the large bag U: $r = \frac{u}{U-u}$. Thus, the molar stream of oxygen escaping through the small bag relief valve $-\dot{n}_1$ can be estimated using the Clapeyron equation as $-\dot{n}_1 = -\frac{p}{R \cdot T} \cdot \dot{V}_E \cdot r \cdot x$, where p represents the pressure at the diving depth $H: p = f(H)$. The same number of moles $-\dot{n}_1 = -\frac{p}{R \cdot T} \cdot \dot{V}_E \cdot r \cdot x$ decreases during inspiration from the large bag to the small bag (Figure 5.5 and Table 5.4).

During inspiration, the oxygen stream having value $-\dot{n}_2 = -\frac{p_0}{R \cdot T} \cdot \dot{\upsilon}_0$ is consumed from the breathing mixture drawn in. In the same phase, the dosing system resupplies the volume of the large bag with fresh premix, delivering oxygen having the value $\dot{n}_3 = \left(\frac{p}{R \cdot T} \cdot \dot{V}_E \cdot r + \frac{p_0}{p} \cdot \dot{\upsilon}_0 \right) \cdot x_w$. Thus, the elementary change in oxygen content ∂x in volume of the large bag $(U - u)$ at time ∂t can be written as $\frac{p}{R \cdot T} \cdot (U - u) \cdot \frac{\partial x}{\partial t} = \dot{n}_3 - \dot{n}_2 - \dot{n}_1$ (Figure 5.5). After setting the values of streams, the ordinary first-order differential equation can be written[12] as:

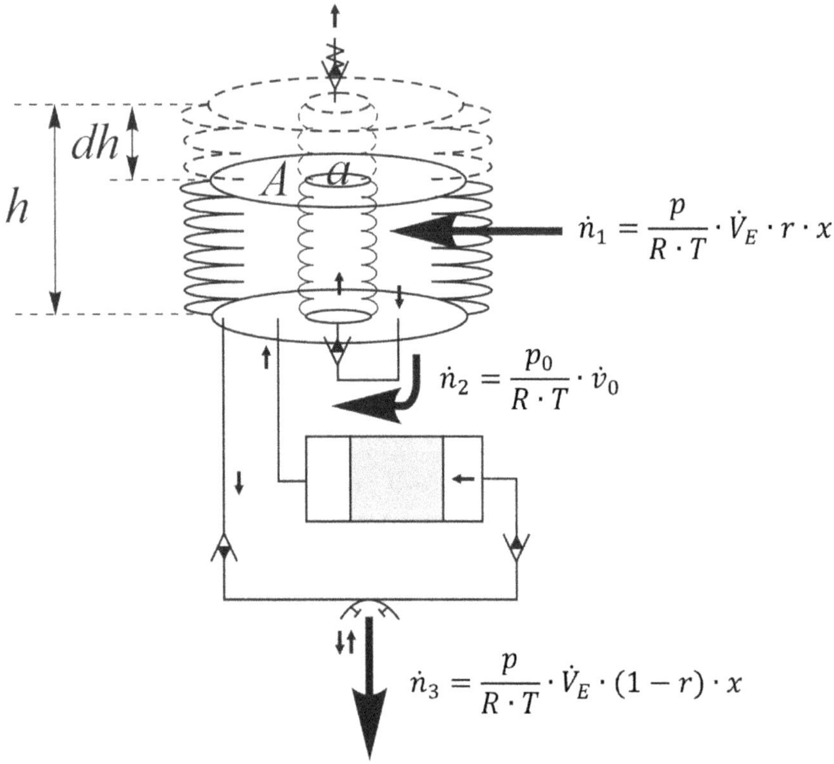

FIGURE 5.5 Diagram of operation of the bag system in the bag during inspiration, where A = is the area of the base of the large bag; a = the area of the base of the small bag; dh = the elementary change in height of bags; h = the standard height of bags; p = the pressure at the diving depth; p_0 = the normal pressure; R = the universal constant of gas; T = the absolute temperature; U = the volume of the large bag; u =the volume of the small bag; r =the volume ratio of the small bag to the large bag $r = \frac{u}{U-u}$; $\dot{V}_E$ = the ventilation of lungs; $\dot{v}_0$ = the stream of consumed oxygen; x = the molar fraction of oxygen in the fresh breathing gas; x_w = the molar fraction of oxygen in the fresh breathing gas; and x_0 = the molar fraction of oxygen in the breathing space before the diving apparatus is started.

$$\frac{p}{R \cdot T} \cdot (U - u) \cdot \frac{\partial x}{\partial t} = \frac{p}{R \cdot T} \cdot \left(\dot{V}_E \cdot r + \frac{p_0}{p} \cdot \dot{v}_0 \right) \cdot x_w - \frac{p_0}{R \cdot T} \cdot \dot{v}_0 - \frac{p}{R \cdot T} \cdot \dot{V}_E \cdot r \cdot x \cdot \quad (5.8)$$

where $\frac{\partial x}{\partial t}$ = is elementary change in oxygen content ∂x in the large bag for elementary time change ∂t, and other designations as in Figure 5.5.

TABLE 5.4
Balance of Streams of Mixture and Oxygen Mass[11] for a Semi-Closed Rebreather with a System of Bags Placed One above the Other

Particulars	Mixture		Oxygen	
	Escapes	Adds	Escapes	Adds
Inspiration	$\frac{p}{R\cdot T}\cdot\dot{V}_E\cdot(1-r)$	—	$\frac{p}{R\cdot T}\cdot\dot{V}_E\cdot(1-r)\cdot x$	—
To small bag	$\frac{p}{R\cdot T}\cdot\dot{V}_E\cdot r$	—	$\frac{p}{R\cdot T}\cdot\dot{V}_E\cdot r\cdot x$	—
Addition valve	—	$\frac{p}{R\cdot T}\cdot\dot{V}_E\cdot r+\frac{p_0}{R\cdot T}\cdot\dot{v}_0$	—	$\frac{p}{R\cdot T}\cdot\left(\dot{V}_E\cdot r+\frac{p_0}{p}\cdot\dot{v}_0\right)\cdot x_w$
Expiration	—	$\frac{p}{R\cdot T}\cdot\dot{V}_E\cdot(1-r)-\frac{p_0}{R\cdot T}\cdot\dot{v}_0$	—	$\frac{p}{R\cdot T}\cdot\dot{V}_E\cdot(1-r)\cdot x-\frac{p_0}{R\cdot T}\cdot\dot{v}_0$
Sum	$-\frac{p}{R\cdot T}\cdot\dot{V}_E$	$\frac{p}{R\cdot T}\cdot\dot{V}_E$	$-\frac{p}{R\cdot T}\cdot\dot{V}_E\cdot x$	$\frac{p}{R\cdot T}\cdot[\left(\dot{V}_E\cdot r+\frac{p_0}{p}\cdot\dot{v}_0\right)\cdot x_w+$ $+\dot{V}_E\cdot(1-r)\cdot x-\frac{p_0}{p}\cdot\dot{v}_0]$
Total	0		$\frac{p}{R\cdot T}\cdot\left(\dot{V}_E\cdot r+\frac{p_0}{p}\cdot\dot{v}_0\right)\cdot x_w-\frac{p_0}{R\cdot T}\cdot\dot{v}_0-\frac{p}{R\cdot T}\cdot\dot{V}_E\cdot r\cdot x$	

Where $x_w =$ is the molar fraction of oxygen in fresh respiratory gas; $x =$ the molar fraction of oxygen in inhaled breathing gas; $x_0 =$ the molar fraction of oxygen in breathing space before the diving apparatus is started; $p =$ the pressure at diving depth; $p_0 =$ the normal pressure; $r =$ the ventilated volume ratio of the small bag to the large bag $r=\frac{u}{U-u}$ $R =$ universal gas constant; $T =$ the absolute temperature; $U =$ the volume of the large bag; $u =$ the volume of the small bag; $\dot{V}_E =$ the ventilation of the lungs; and $\dot{v}_0 =$ the stream of consumed oxygen.

The solution of equation (5.8) in the form of mathematical proof is presented in Table 5.5:

$$x(t) = x_w - \varepsilon_p \cdot \varepsilon \cdot \varepsilon_k + \left(x_0 - x_w + \varepsilon_p \cdot \varepsilon \cdot \varepsilon_k\right) \cdot exp\left(-\frac{\dot{\upsilon}_0 \cdot r^2}{\varepsilon_0 \cdot u} \cdot t\right) \tag{5.9}$$

where x_0 = is molar fraction of oxygen in the breathing space before the diving apparatus is started; $\varepsilon_k = \frac{1 - x_w}{r}$ = is *module*[13] representing the structural solution of the breathing space in the diving apparatus; $\varepsilon = \frac{\dot{\upsilon}}{\dot{V}_E}$ = is *breathing module*;[14] and $\varepsilon_p = \frac{p_0}{p}$ = is *pressure module.*

It follows from equation (5.9) that oxygen content x in the large bag for time $t = 0$ from the definition is $x(t = 0) \cdot x_0$, and for the time going to infinity $t \to \infty$, it is $x(t \to \infty) = x_s = x_w - \varepsilon_p \cdot \varepsilon \cdot \varepsilon_k$, where x_s means stable value $x_s(t_s)$ obtained after stabilization value t_s, which is no longer a function of time t with assumed accuracy ϵ: $x_s(\epsilon) \neq f(t)$ for dive parameters $\varepsilon_p(t)$; $\varepsilon(t) = idem$ invariable in time t.

In equation (5.9), the coefficient $\varepsilon_k = \frac{1 - x_w}{r}$ is a formally dimensionless ratio[15] constituting a *criterion number*[16] describing the structural solution of the breathing space in the diving apparatus, whose value for a given solution is constant during the diving process $\varepsilon_k = idem \neq f(t)$, where time t refers to time of dive. It will be referred to here as structural module ε_k.

The ratio $\varepsilon_p = \frac{p_0}{p}$ is called here the pressure module, which is the *criterion number* of external conditions. It decreases with diving depth H: $H \nearrow \Rightarrow \varepsilon_p \searrow$. With the increase of depth H, the *pressure module* ε_p has less and less effect on the decrease in the stable oxygen content in the breathing gas, which is breathed by the diver x_s, due to its inverse proportionality to pressure p: $x_s \sim \varepsilon_p = \frac{p_0}{p}$. For the constant depth $H = idem$, the *pressure module* ε_p is a constant value: $\varepsilon_p = idem \big| H = idem$.

The module $\varepsilon = \frac{\dot{\upsilon}}{\dot{V}_E}$ is called here the *breathing module*; for standard pressure p_0, it will be marked as ε_0, connecting oxygen consumption $\dot{\upsilon}_0$ and ventilation of lungs $\dot{V}_E$ with work W_0 done by a man on surface[17] $\dot{\upsilon}_0 = f(W_0)$. The use of the *breathing module* related to normal conditions gives an opportunity to develop a draft design of a diving apparatus, because these values are determined with good accuracy, but in general, it should be assumed that the *breathing module* ε can be dependent on the pressure $\varepsilon = f(p)$. This assumption must, however, be verified experimentally, as the workload under water W consists of additional types of physical work which do not occur on the surface.

TABLE 5.5
Derivation of the Ventilation Model Related to Oxygen Content x in Function of Time $x = f(t)$ for a Semi–Closed Circuit Rebreather with a System of Bags Placed One above the Other

A		$\dot{\upsilon}_0 \neq f(H)$	
T		$x \rightarrow x = f(t)$	
P	1°	$\frac{p}{R\cdot T}\cdot(U-u)\cdot\frac{\partial x}{\partial t} = \frac{p}{R\cdot T}\cdot\left(\dot{V}_E\cdot r + \frac{p_0}{p}\cdot\dot{\upsilon}_0\right)\cdot x_w - \frac{p_0}{R\cdot T}\cdot\dot{\upsilon}_0 - \frac{p}{R\cdot T}\cdot\dot{V}_E\cdot r\cdot x$	From mole balance of oxygen
	2°	$\frac{\partial x}{\partial t} = \frac{\dot{V}_E\cdot r}{U-u}\cdot x_w - \frac{p_0}{p}\cdot\frac{\dot{\upsilon}_0}{U-u}\cdot(1-x_w) - \frac{\dot{V}_E\cdot r}{U-u}\cdot x$	Ordinary first-order differential equation
	3°	$\forall_{a=\frac{\dot{V}_E\cdot r}{U-u}x_w - \frac{p_0}{p}\frac{\dot{\upsilon}_0}{U-u}(1-x_w)\ \wedge\ b=\frac{\dot{V}_E\cdot r}{U-u}} \quad dx = (a - b\cdot x)\cdot dt$	From 2°
	4°	$\frac{dx}{b\cdot x - a} + dt = 0$	From 3°
	5°	$\int\frac{dx}{b\cdot x - a} + \int dt = const = c$	From 4° indefinite integral definition
	6°	$\frac{1}{b}\cdot\ln\lvert b\cdot x - a\rvert + t = c$	From 5°
	7°	$\ln\lvert b\cdot x - a\rvert \equiv b\cdot(c-t) = c' - b\cdot t$	c' – new constant of integration
	8°	$e^{(c'-b\cdot t)} = c''\cdot e^{-b\cdot t} \equiv b\cdot x - a$ c'' – new constant of integration	From 7° and natural logarithm definition
	9°	$\forall_{t\rightarrow 0}\ x \rightarrow x_0 \Rightarrow c'' = b\cdot x_0 - a \Rightarrow (b\cdot x_0 - a)\cdot e^{-b\cdot t} = b\cdot x - a$ $\exists_{\frac{a}{b}=x_w-\varepsilon_p\cdot\varepsilon\cdot\varepsilon_k} \quad x(t) = \left(x_0 - \frac{a}{b}\right)\cdot e^{-b\cdot t} + \frac{a}{b}$	From 8° and 3° $\varepsilon_k = \frac{1-x_w}{r}$ $\varepsilon_p = \frac{p_0}{p}$ $\varepsilon = \frac{\dot{\upsilon}}{\dot{V}_E}$
	10°	$x(t) = x_w - \varepsilon_p\cdot\varepsilon\cdot\varepsilon_k + (x_0 - x_w + \varepsilon_p\cdot\varepsilon\cdot\varepsilon_k)\cdot exp\left(-\frac{\dot{\upsilon}_0\cdot r^2}{\varepsilon_0\cdot u}\cdot t\right)$	From 2°, 8°, and 9° q.e.d.

Where $p =$ is the pressure at diving depth; $p_0 =$ the normal pressure; $R =$ the universal gas constant; $T =$ the absolute temperature; $U =$ the volume of the large bag; $u =$ the volume of the small bag; $r =$ the volume ratio of the small bag to the large bag $r = \frac{u}{U-u}$; $\dot{V}_E$ the ventilation of lungs; $\dot{\upsilon}_0 =$ the stream of consumed oxygen; $x =$ the molar fraction of oxygen inhaled breathing gas; $x_w =$ the molar fraction of oxygen in fresh breathing gas; $x_0 =$ the molar fraction of oxygen breathing space before diving apparatus is started; $\varepsilon_k = \frac{1-x_w}{r}$ – the module representing structural solution of breathing space in diving apparatus; $\varepsilon = \frac{\dot{\upsilon}}{\dot{V}_E}$ – the breathing module; $\varepsilon_p = \frac{p_0}{p}$ – the pressure module; and $\frac{\partial x}{\partial t}$ – the elementary change in oxygen content ∂x in the large bag with elementary time change ∂t.

Note: A– = assumption; T = = thesis; P = = proof (evidence).

For example, the work done to overcome greater than in the air resistance associated with an increased pressure difference between the pulmonary centroid[18] and the center of buoyancy of the breathing space in the apparatus in immersion. Expending extra effort to overcome the resistance associated with the higher density of breathing gas ρ, causing an increase in flow resistance Δp posed by elements of the breathing space in the apparatus,[19] is also significant. In the breathing space of the diving apparatus, there are also spatial obstructions that decrease the ventilation in some spaces, so-called *dead spaces*, whose impact on the safety of decompression should also be taken into account. Another condition that is difficult to determine unequivocally is an individual person–dependent phenomenon known as *bradycardia*,[20] associated with this slowed-down breathing f occurring in immersion, reducing ventilation of lungs $\dot{V}_E$ and, consequently, the breathing space in the diving apparatus (Birch, MacLaren and George, 2009). Hence the assumption that the *breathing module* ε should depend on pressure $\varepsilon = f\left(\varepsilon_p\right)$. It follows from many years of experience that the *breathing module* ε_i is also an individualized function of *morphology*[21] of a diver i.

An important factor is the speed of the system's response to the stepwise forced change in the partial pressure of oxygen $p(t)$ in the circulating breathing gas.[22] For this purpose, the ventilation model (5.9) can be multiplied by pressure p present at diving depth H: $p(t) = p \cdot x(t)$. After ordering, the following can be written:

$$p(t) = p \cdot x_w - p_0 \cdot \varepsilon_0 \cdot \varepsilon_k + \left(p \cdot x_0 - p \cdot x_w + p_0 \cdot \varepsilon \cdot \varepsilon_k\right) \cdot exp\left(-\frac{\dot{v}_0 \cdot r^2}{\varepsilon \cdot u} \cdot t\right) \quad (5.10)$$

where $p(t)$ – is partial pressure of oxygen in function of time t, and p – the pressure present at diving depth H.

Using formula (5.10), it is possible to estimate stabilization times $t_s\left(\Delta p_{kr}\right)$ needed to achieve a stable value of oxygen partial pressure p_s in the breathing space of the diving apparatus with accuracy up to the specified critical pressure difference $\Delta p_{kr} = p(t) - p(t \to \infty)$. The stabilization times calculated in this way are close to stabilization times $t_s\left(\Delta p_{kr}\right)$ in *SCR*, with a nozzle constant dosing system – see previous chapters (Kłos, 2000, 2021).

The theoretical process of oxygen partial pressure stabilization $p(t)$ in the breathing space of the *SCR CRABE SCUBA* is shown in Figure 5.6, and an example of real stabilization is shown in Figure 5.7.

The equation $(t \to \infty) = x_s = x_w - \varepsilon_p \cdot \varepsilon \cdot \varepsilon_k$, derived from an analysis of the model (5.10), can be written as:

$$x_s = x_w - \frac{1 - x_w}{r} \cdot \frac{\dot{v}}{\dot{V}_E} \cdot \frac{p_0}{p} = x_w - \varepsilon_k \cdot \varepsilon \cdot \varepsilon_p \Big| t \to \infty \quad (5.11)$$

For the assumption $\varepsilon = idem$, dependence $x_s = f_{x_w, \varepsilon, r}\left(\varepsilon_p\right)$ is equation of straight-line $y = \alpha \cdot x + \beta$ with negative directional coefficient $\alpha = -\varepsilon_k \cdot \dot{v}$. Thus, for the domain of positive numbers, *it* is a monotonically decreasing function. The transformed equation (5.11) leads to dependence $x_w - x_s = f_{x_w, \varepsilon, r}(p)$, which is the

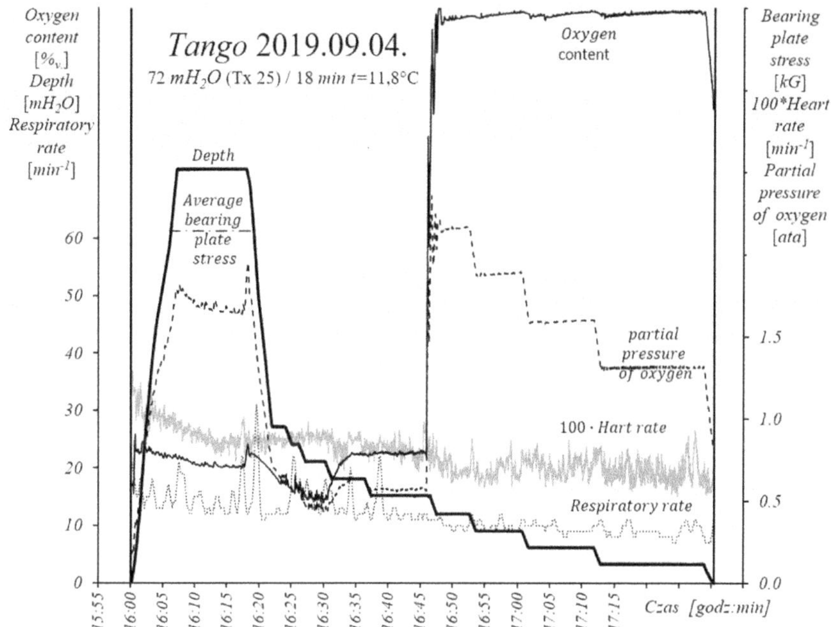

FIGURE 5.6 Example process of stabilizing the oxygen content $x(t)$ and its partial pressure $p(t)$ during a dive with *SCR CRABE SCUBA* apparatus.

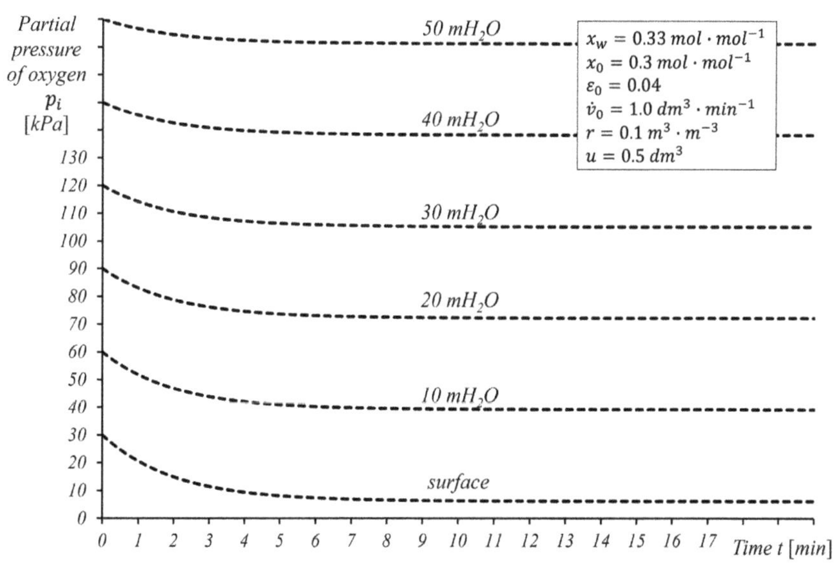

FIGURE 5.7 Theoretical times of partial pressure stabilization $p(t)$ of oxygen in the large bag.

hyperbolic power function[23] decreasing together with depth H, always greater than zero, $x_w - x_s = \varepsilon_k \cdot \varepsilon \cdot \frac{p_0}{p} > 0$, because $\varepsilon_k \geq 0 \wedge \varepsilon \geq 0$.

Transforming boundary mathematical model $t \to \infty$ (5.11), the following equation can be obtained: $\frac{p \cdot x_w - p_s}{p_0} = \frac{\Delta p}{p_0} = \frac{1 - x_w}{r} \cdot \frac{\dot{\upsilon}}{\dot{V}_E}$. This equation can be further transformed into the form $\frac{p \cdot x_w - p_s}{p_0} = \frac{\Delta p}{p_0} = \frac{1 - x_w}{r} \cdot \frac{\dot{\upsilon}}{\dot{V}_E}$, where p represents the pressure at diving depth H; p_s is the actual partial pressure of oxygen in the breathing gas inhaled by the diver for determined structural conditions $\varepsilon_k = idem$ and diving conditions ε_p, $\varepsilon = idem$, and $\Delta p = p \cdot x_w - p_s$ is the difference in partial pressure of oxygen in the *premix* and the breathing gas inhaled by the diver. For different depths H, the *breathing module* $\varepsilon = \frac{\dot{\upsilon}}{\dot{V}}$ of oxygen consumption $\dot{\upsilon}$ for lung ventilation $\dot{V}_E$ is a function of diving depth $\varepsilon = f(H)$:

$$\forall_{\varepsilon_p, \varepsilon, \varepsilon_k = idem} \, p \cdot x_w - p_s = \Delta p = \varepsilon \cdot \frac{1 - x_w}{r} \cdot p_0 \tag{5.12}$$

where $\varepsilon =$ is the breathing module, being the ratio of oxygen consumption to lung ventilation $\frac{\dot{\upsilon}}{\dot{V}_E}$ at diving depth H.

Interesting from the point of view of applications used for predicting safe decompression is a model combining total pressure p at diving depth H with oxygen partial pressure p_s, which can be obtained by multiplying stable content x_s from dependence (5.12) by hydrostatic pressure p at diving depth H:

$$p_s = x_w \cdot p - \varepsilon_k \cdot \varepsilon \cdot p_0 \tag{5.13}$$

For $x_w, \varepsilon, r = idem$, dependence (5.13) is an equation of straight line $y = \alpha \cdot x + \beta$ having positive directional coefficient $\alpha = x_w$. Therefore, it should be a monotonically increasing function, with positive values for positive $\mathbb{R}_+$ arguments, which is confirmed by the results of our own research, as shown in Figure 5.5 (Kłos, 2021).

5.1.4 Conclusions

The presented analyses were used to develop a deterministic model of algebraic ventilation of the breathing apparatus space.

5.2 SIMULATION

The implementation of the new diving technologies and constructions developed in the Diving Gear and Underwater Work Technology Department of the Naval Academy carried some risk related to manned experiments. Therefore, a laboratory stand which could be used to carry out unmanned research on the underwater breathing apparatus *UBA* was built. Its use should reduce risk in such research. Models of similar laboratory stand prototype have been developed in some countries. The most similar one is used in Sweden (Loncar, 1992; Kłos, 2002; Frånberg, 2015).

The typical laboratory stand is *a hyperbaric breathing machine*. The additional device should enable oxygen uptake from the breathing atmosphere, then humidifying and replacing oxygen taken from the atmosphere by a suitable amount of carbon dioxide emitted. The solution to this problem will enable modelling the processes occurring in *SCRUBA*. This will enable saving time, costs of diving equipment design, and costs of developing decompression schedules. As there is no verified mathematical model available, it was necessary to reconstruct the prototype after each series of the preliminary investigations. Moreover, the model to be developed and the verified model will make it possible to evaluate the diving systems in use.

5.2.1 Assumptions

To simulate the respiratory process, a mechanical representation of pulmonary ventilation and gas exchange in the course of breathing is needed. The processes should be precisely and repeatedly represented; however, there is no need for each detail of the real respiratory cycle to be represented.

Usually, it is assumed that an approximated shape of respiration should be sinusoidal (Kłos, 2007, 2021). When the body gas exchange is simulated, attention should be focused on oxygen consumption and its replacement by the carbon dioxide emitted. A respiratory quotient defines the volumetric ratio for the emitted carbon dioxide that replaces the oxygen consumed. Two values of *respiratory quotient*[24] $_{RQ} = \{1.0, 0.8, 0.75\}_v$ are most often used.

5.2.2 Research on Chemical Reactions

The process is simulated by combustion of catalyst chemical substances in a reactor. The respiratory simulator makes volumetric uptake of chemical substances from the breathing space and catalyst combustion possible, then the volume containing the combustion products is recycled to the *SCR* breathing loop. The following chemical reactions describe the process:

$$
\begin{aligned}
&2 \cdot CH_3CHO + 5 \cdot O_2 \xrightarrow{catalyst} 4 \cdot CO_2 + 4 \cdot H_2O \quad \varepsilon_{RQ} = 0.80 \\
&(CH_3)_2CO + 4 \cdot O_2 \xrightarrow{catalyst} 3 \cdot CO_2 + 3 \cdot H_2O \quad \varepsilon_{RQ} = 0.75
\end{aligned}
\tag{5.14}
$$

Acetone $(CH_3)_2CO$ was used in the research on the catalytic combustion process. To maintain the required rate of acetone, a metering pump was used. The diver's oxygen consumption was assumed to be $\dot{v} = (0.3, 2.4)\, dm^3 \cdot min^{-1}$ (Table 4.15). Therefore, acetone metering was maintained within the range of $\dot{u} = (0.2, 1.8)\, cm^3 \cdot min^{-1}$.

To evaluate the efficiency of the catalyst, a measuring stand was built. It is presented in Figure 5.8.

The metering pump shown in Figure 5.8 is to maintain satisfactory accuracy of counterpressure liquid metering. To check the pump characteristics, indirect measurements of the counterpressure approximately $20\, MPa$ were carried out (Kłos, 2021).

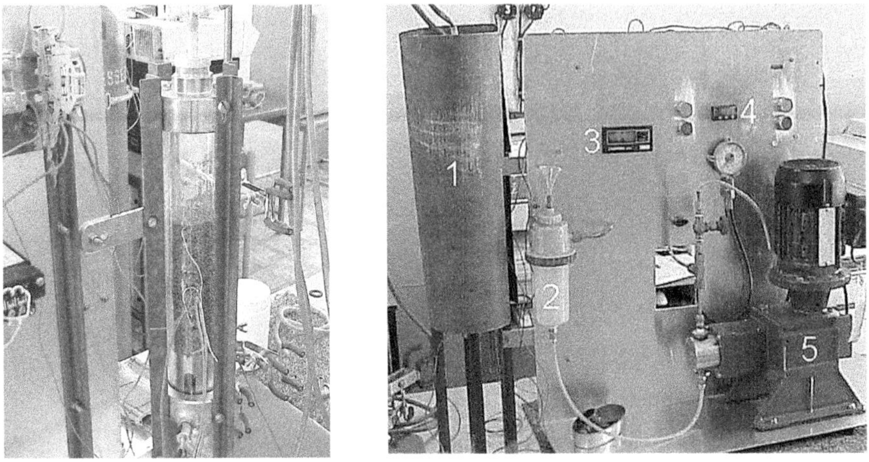

FIGURE 5.8 Laboratory stand for research on catalyst efficiency. *Left*, reactor; *right*, measuring stand: (1) reactor in casing (version 1); (2) reacting substance container; (3) temperature multipoint measuring set display; (4) temperature regulator; (5) metering pump.

The preliminary research on acetone $(CH_3)_2CO$ oxidation was carried out with the use of the catalyst,[25] 0.5% platinum deposited on ca. $3\,mm$ diameter rollers of alumina type 0.3R73. A glass pipe having $70\,mm$ in diameter was filled with $600\,g$ of the catalyst. The catalyst bed in the glass pipe had approximately $18.5\,cm$ in height. The airflow rate was maintained at the level of $(2.050.05)\,m^3\cdot h^{-1}$. From the glass reactor, the air then passed through a gas chromatograph metering valve coupled with a mass spectrometer GC/MS. The details concerning the type of operating parameters of gas chromatograph and mass spectrometer GC/MS were presented earlier (Kłos, 2007, 2021). Based on these measurements, it can be concluded that the catalyst used in the experiment can also be used for constructing an oxygen uptake attachment, which also enables carbon dioxide and water vapor emission. The theoretical relationship between the oxygen and carbon dioxide content during a complete combustion can be calculated taking into account the variation of volume. The theoretical relationships concerning liquid water are as follows:

$$(CH_3)_2CO + 4\cdot O_2 \xrightarrow{catalyst} 3\cdot CO_2 + 3\cdot H_2O \qquad \varepsilon_{RQ} = \frac{V_{CO_2}}{V_{O_2}} = \frac{3}{4}$$

$$\left.\begin{array}{r} V_{O_2} \cong 0.209\cdot V \\ V_{CO_2} \cong 0.000\cdot V \\ V_{N_2} \cong 0.780\cdot V \\ V_{residual} \cong 0.011\cdot V \end{array}\right\} \xrightarrow[reaction]{cathalitic} \left\{\begin{array}{l} V_{O_2} \cong 0.209\cdot V - \dfrac{4}{3}\cdot V_{CO_2} \\ V_{CO_2} \cong V_{CO_2} \\ V_{N_2} \cong 0.780\cdot V \\ V_{rest} \cong 0.011\cdot V \end{array}\right.$$

$$Sum: \qquad V \qquad\qquad V - \frac{1}{3}\cdot y$$

where $V =$ is the volume of reagents, V_{CO_2} = the volume of carbon dioxide, and V_{O_2} = the volume of oxide.

$$\begin{cases} x_{CO_2} \cong \dfrac{V_{CO_2}}{V - \frac{1}{3}\cdot V_{CO_2}} \quad \Rightarrow \quad V \cong \dfrac{V_{CO_2}}{x_{CO_2}} + \dfrac{1}{3}\cdot V_{CO_2} \\ x_{CO_2} \cong \dfrac{0.209\cdot V - \frac{4}{3}\cdot V_{CO_2}}{V - \frac{1}{3}\cdot V_{CO_2}} \end{cases}$$

$$x_{O_2} \cong 0.209 - 1.264\cdot x_{CO_2} \tag{5.15}$$

where x_{CO_2} = are the contents of carbon dioxide, and x_{O_2} = the contents of oxide.

In experimental conditions, water usually occurs in vapor form;[26] however, water vapor can condense into liquid water into a sampling line, because it is long enough. Therefore, some water can be removed. In the case of water vapor in reaction products, derivation of the relationship between the oxygen contents and carbon dioxide is as follows:

$$(CH_3)_2CO + 4\cdot O_2 \xrightarrow{catalyst} 3\cdot CO_2 + 3\cdot H_2O \qquad \frac{V_{O_2}}{V_{CO_2}} = \frac{V_{O_2}}{V_{H_2O}} = \frac{4}{3}$$

$$\left.\begin{matrix} V_{O_2} \cong 0.209\cdot V \\ V_{CO_2} \cong 0.000\cdot V \\ V_{N_2} \cong 0.780\cdot V \\ V_{H_2O} \cong 0.000\cdot V \\ V_{residual} \cong 0.011\cdot V \end{matrix}\right\} \xrightarrow{\text{cathalitic reaction}} \begin{cases} V_{O_2} \cong 0.209\cdot V - V_{CO_2+H_2O} \\ V_{CO_2} \cong \frac{3}{4}\cdot V_{CO_2} \\ V_{N_2} \cong 0.780\cdot V \\ V_{H_2O} \cong \frac{3}{4}\cdot V_{CO_2} \\ V_{rest} \cong 0.011\cdot V \end{cases}$$

$$Sum: \qquad V \qquad\qquad V - \frac{1}{3}\cdot y$$

$$\begin{cases} x_{CO_2} \cong \dfrac{\frac{3}{4}\cdot V_{CO_2}}{V - \frac{1}{2}\cdot V_{CO_2}} \quad \Rightarrow \quad V \cong \left(\dfrac{0.75}{x_{CO_2}} - 0.5\right)\cdot V_{CO_2} \\ x_{CO_2} \cong \dfrac{0.209\cdot V - V_{CO_2}}{V - \frac{1}{2}\cdot V_{CO_2}} \end{cases}$$

$$x_{O_2} \cong 0.209 - 1.473\cdot x_{CO_2} \tag{5.16}$$

Figure 5.9 shows the relationship between oxygen O_2 and carbon dioxide CO_2 contents obtained from the experiments.

It is evident from Figure 5.9 that there is a good correlation[27] between oxygen O_2 and carbon dioxide CO_2 contents in the post-reaction mixture. As it follows from the preceding text, the post-reaction mixture contained water vapor. However, it is evident from Figure 2.3 that for 0.95 confidence level,[28] there are some points lying beyond confidence level. This can suggest that, in some cases, water can be at least partly condensed.

Sometimes, significant carbon monoxide CO contents were recorded during the experiments. The increase in the carbon monoxide CO contents was a result of the increase in the catalyst temperature in the reaction zone. This may have been caused by incomplete combustion or carbon dioxide CO_2 dissociation to carbon monoxide CO and oxygen O_2 that can occur at a high temperature on a catalyzer. A rapid increase in temperature in the reaction zone during the work of the reactor was observed. This made cooling of the reactor necessary. Otherwise, the oxygen uptake simulation would have continued for only $1 min$ before thermal destruction. This amount of time is insufficient to carry out research on ventilation of the SCR breathing space. This problem was solved by means of *exploratory factor analysis*[29] EFA.

5.2.3 Dimensionality Reduction

The *exploratory factor analysis EFA* technique was used to analyze the data obtained from the experiment. This method can be used to reduce the problem dimensions,

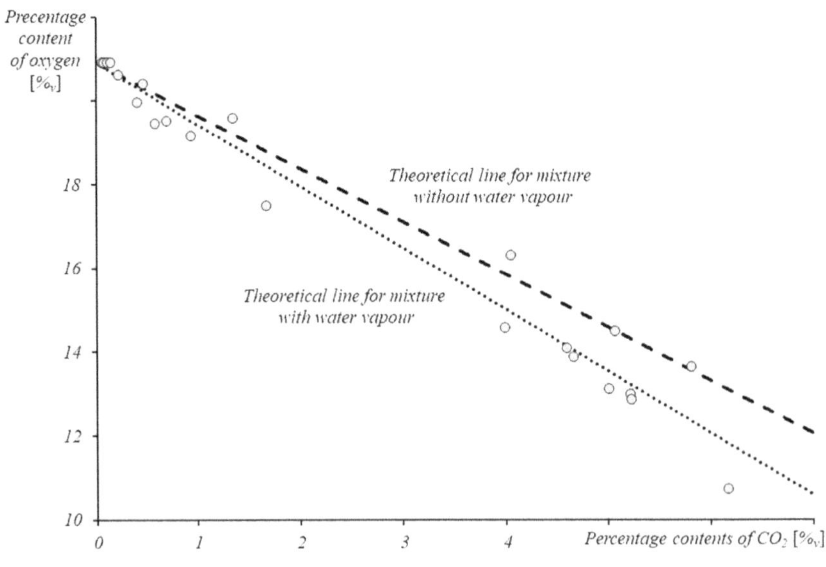

FIGURE 5.9 Comparison of measurement results with mathematical models of oxidation reaction (5.15) and reaction (5.16).

which makes it easier to determine the mechanisms[30] of chemical reaction occurring in the catalyzer.[31] In the first phase of *EFA*, *principal component analysis*[32] *PCA* was used as a method for factor extraction.

The *EFA* is used for two purposes: to reduce the number of the problem dimensions being analyzed and to determine the problem structure. Determination of the problem structure involves transforming the existing variables into the orthogonal coordinate system and hierarchic classification. Reduction of the number of variables is applied to many technical problems when a complex measurement is needed to describe many features of the object being investigated. A reduction of the number of variables describing the problem is performed due to analytical limitations. In many cases, reduction is possible without significant loss of information. It takes place, for example, in the case of a two-dimensional problem, when data scatter diagram is similar to that presented in Figure 5.10. Of course, the example of a two-dimensional problem can be extended to multidimensional space.

From Figure 5.10, the presented data is strongly correlated, and replacement of the $X \perp Y$ coordinate system with concentration ellipse axis $X' \perp Y'$ makes it possible to characterize measuring points with the use of variable X' and partial loss of information only. In this case, the reduction of dimensions will be connected with the rotation of the coordinate system by $45°$. At the beginning, covariance matrix that determines the mutual relationship between variables should be built. Prior to that, data standardization should be done, as data is usually expressed in the different units. It involves subtracting from each measured value its mean value and dividing the difference by

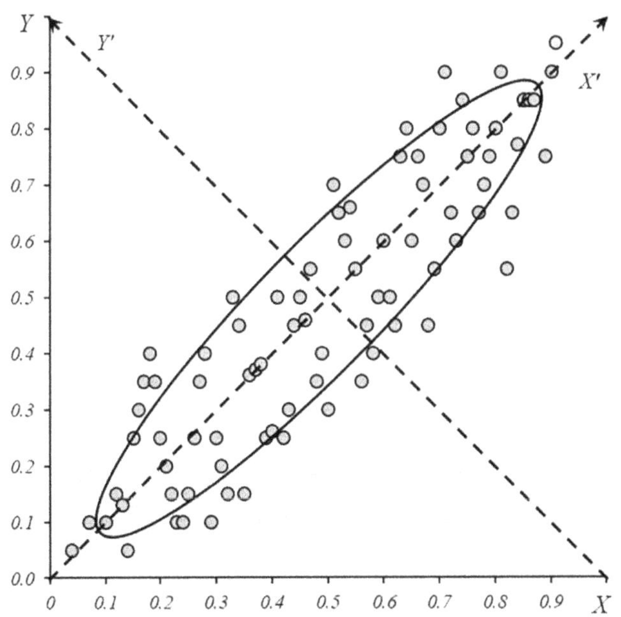

FIGURE 5.10 An example of a two-dimensional scatter plot with a selected ellipse of concentration and its main axes for translated data on standard variability in the range [0, 1].

the standard deviation. For the data prepared this way, the covariance matrix becomes the correlation matrix. For the two-dimensional case, it will be as follows:[33]

$$\forall_{\bar{x}\equiv 0;\sigma\equiv 1}\underline{C}=\begin{pmatrix} V_x & C_{xy} \\ C_{xy} & V_y \end{pmatrix}=\begin{pmatrix} 1 & r \\ r & 1 \end{pmatrix}=\underline{r} \tag{5.17}$$

where $\sigma =$ is standard deviation; $V =$ is standard deviation $V\equiv\sigma^2$; $C =$ is covariance; $\bar{x} =$ is mean value; $\underline{C} =$ is covariance matrix; and $\underline{r} =$ is correlation matrix.

The values lying beyond the diagonal of the covariance matrix (5.17) determine the mutual correlation between the values. The picture of the diagonal matrix of the covariance[34] of the coordinate system rotation by 45° was applied in this case. The rotation establishes the axes of the coordinate system along the principal axis of the scatter ellipse:

$$\underline{r}=\begin{pmatrix} 1 & r \\ r & 1 \end{pmatrix}\xrightarrow{45^\circ\ rotation}\begin{pmatrix} a^2 & 0 \\ 0 & b^2 \end{pmatrix}$$

It follows from the assumption that after completing the operation, the total scatter of data[35] cannot change, in this case, $a^2+b^2=1^2+1^2=2$. The new variables formed from the initial variables after the rotation of the coordinate system are called the *principal components*. Due to their attributes, these variables are not correlated,[36] their mean values are equal to zero $\bar{x}\equiv 0$, and they differ from each other only by the scatter.[37] The relationship between the principal components and the original attributes can be written for the particular attribute as follows:

$$X_i=\sum_k a_{ik}\cdot X_k^{'} \tag{5.18}$$

Where X_i previous variable; $X_i^{'}$ is the new variable; $i =$ is the number of original variables; $k =$ is the number of main components; and $a_{i1} =$ the factor loads.

An analysis that takes into account all the principal components, however, does not lead to reduction of the problem dimensions. If some principal components are neglected, for example, if equation (5.21) is valid for $k>2$, it can be simplified for only two new variables, as follows: $X_i=a_{i1}\cdot X_1^{'}+a_{i2}\cdot X_2^{'}+E_i$, where E_i represents peculiar contribution for i th attribute.

It is convenient to subject the rest of variables $X_1^{'}$ and $X_2^{'}$ to standardization. After the standardization is executed, the variables are known as factors. Variance of the variable X_i is given as:

$$V(X_i)=a^2+b^2=1^2+V(E_i) \tag{5.19}$$

where a^2+b^2 is the common variation reserves, and $V(E_i)$ the peculiar variance.

However, after the operations have been completed, the interpretation of the relationships between the attributes measured can still be difficult. To facilitate the interpretation, some strategies of simplification of the factor structure were developed. The factor loads can be treated as correlation coefficients between the factors and variables; therefore, they can be repeatedly subjected to rotation performed in such a way that, finally, some of the values are high and the others are close to zero. Such a structure is known to be the simple structure, and it is convenient to employ this method[38] using computer software. *Varimax* rotation is the most often used strategy. It involves maximizing the variance for the particular attribute. After this rotation is completed, the original variables are displaced near the axis of the next factor in the space.

5.2.4 Checking

The standardized experimental verification was analyzed earlier, and it is thus not presented here (Kłos, 2021). The experimental verification was confirmed by the fact that carbon monoxide occurrence was observed after a certain time in the course reaction. In this case, cooling of the reactor will lead to eliminating CO from the post-reaction mixture (Figure 5.11).

To obtain the equations describing the relationship between O_2 and CO_2 contents, three experimental runs were carried out for extended content ranges of O_2 and CO_2. A good enough agreement of the experimental results obtained with the theoretical models can be noticed comparing the results with equations (5.15) and (5.16). Negligible differences in the slope are probably due to different moisture contents in the gas analyzed. The different moisture contents could have resulted from the different temperatures of the gas analyzed.

To validate the method, air was used as a breathing medium. The $SCR\,APW-6M\,SCUBA$ constant gas metering system maintained the gas flow at a level of $\dot{V} = (8.61 \pm 0.06)dm^3 \cdot min^{-1}$.

The gas constant metering measurements were performed on the measuring stand. The tested attachment that enables oxygen uptake and CO_2 emission was connected to the breathing machine presented in Figure 5.12 and Figure 1.12, *top*.

When the oxygen contents decrease below[39] approximately $12\%_v$, there occur problems with reaction stabilization, as the combustion process in the catalyst gets weaker and the reaction stalls. This is a typical phenomenon known from the fire thermodynamics. To elaborate experimental results, the same method as the method used for the preliminary model verification was applied. Such an approach was necessary as the metering nozzle placed in the reaction zone could change its characteristics; therefore, acetone metering could be different due to the conditions established. The verification of the model was performed using equation (5.14). Assuming that $x_s = 0.117\,mol \cdot mol^{-1}$, the oxygen consumption is as follows:

$$\dot{v} = \frac{0.209 - 0.117}{1 - 0.117} \cdot 8.61 = (0.90 \pm 0.07)\,dm^3 \cdot min^{-1}.$$

Equation 9° from Table 5.2, $SCR\,APW-6M\,SCUBA$ data on the volume of the breathing loop, and the experimental relationship between the mean oxygen contents and time make it possible to compare the mathematical model with the experimental results (Figure 5.13).

It is worthy of note that despite the sufficient fitting of the theoretical curve and the experimental results and the good representation of the theoretical line formed by measurement points,[40] the curve is turned relatively to the pivot point. According to the propagation of error theory, the existence of random errors is manifested by

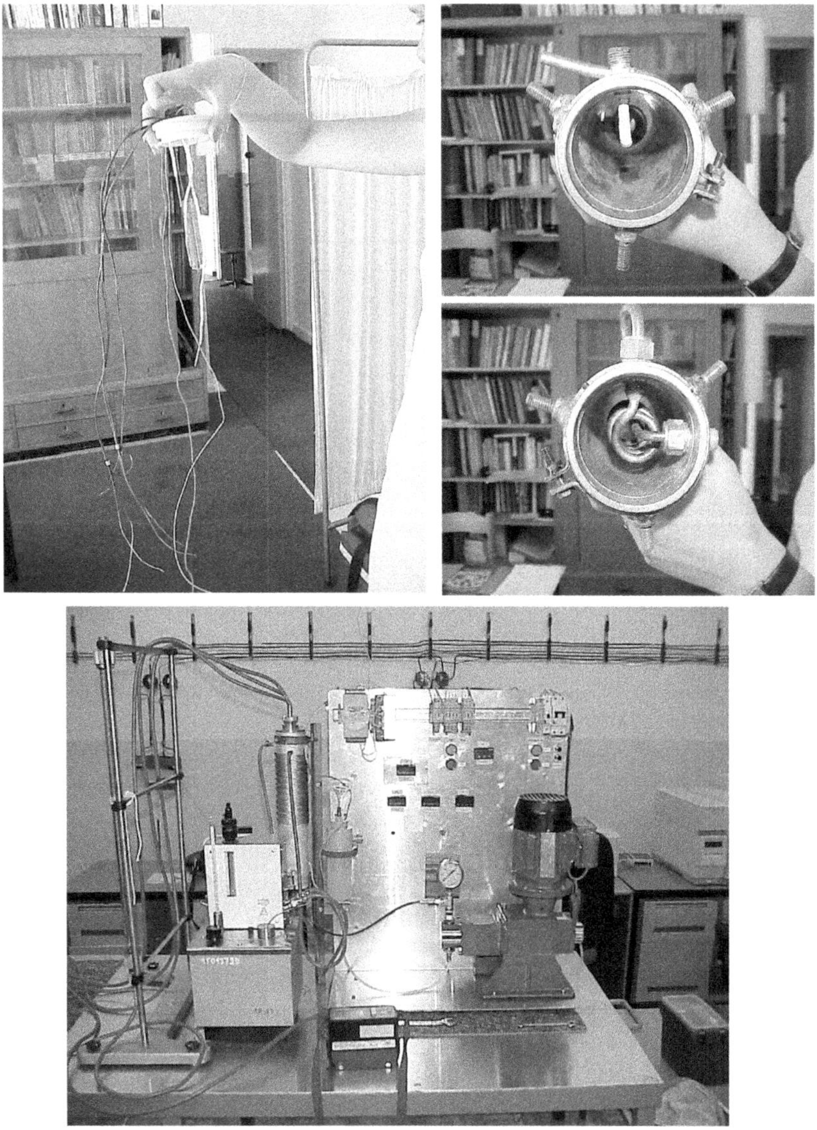

FIGURE 5.11 Metal reactor used for testing *CO* emission as the temperature function: *top left*, measuring sensors and heater; *top right*, side view from the side internal of metering nozzle and side view from the side internal of the water cooler; *bottom*, main view of reactor at the measuring stand.

statistical scatter that was not observed here. The existence of systematic statistical error connected with the linear effect is manifested by displacement of the experimental curve and the theoretical curve. However, here a different kind of displacement of the experimental curve and the theoretical curve is observed. This suggests

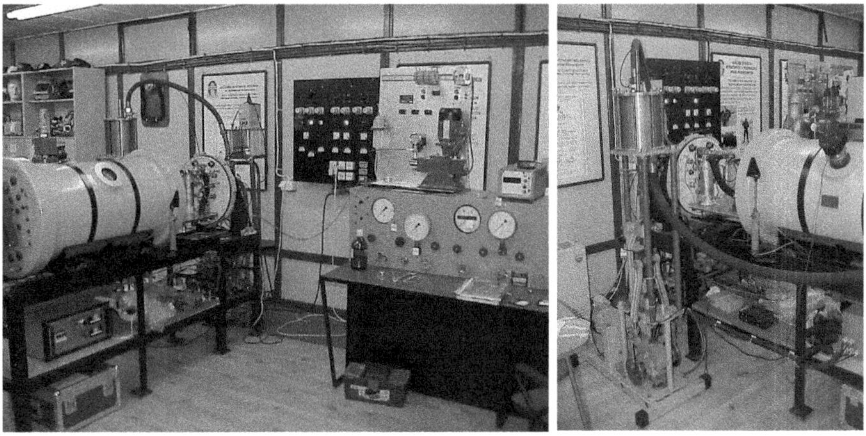

FIGURE 5.12 Experimental respiratory and metabolic simulator stand: *left*, the reactor supplying and control set; *right*, the respiratory breathing simulator.

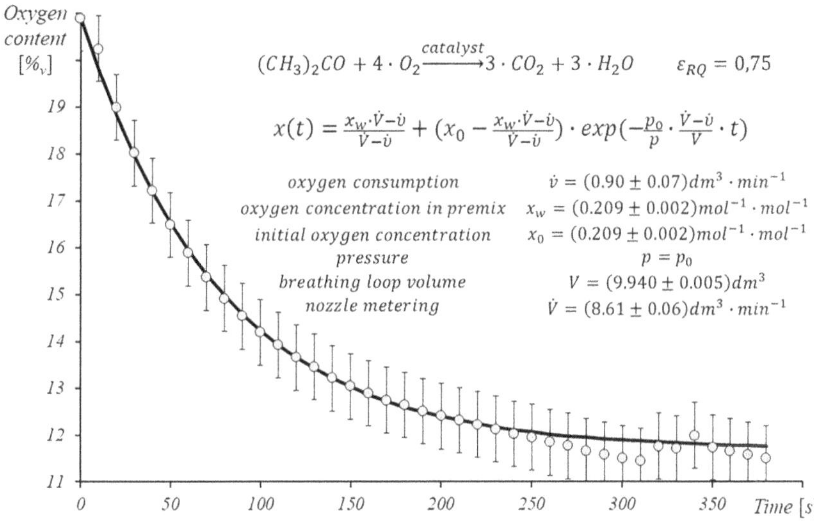

FIGURE 5.13 Comparison of oxygen contents measured in the inhalation bag of *SCR APW – 6M SCUBA* with the mathematical model represented by a solid line.

inaccuracy of the theoretical model applied. The model that was earlier developed (Table 5.2) for the two-bag underwater breathing apparatus[41] can be alternatively used (5.3).

It seems that the conclusions drawn earlier, concerning negligible, from the practical viewpoint, differences between models (5.2) and (5.3), in fact, may not be correct, as these differences can be measured during experiments. Transformation of equation (5.3) with respect to oxygen consumption gives

$$\dot{v} = \frac{\frac{p}{p_0} \cdot \frac{x_w - x_s}{x_w} \cdot \dot{V}_E + \dot{V}}{1 + \frac{p}{p_0} \cdot \frac{1 - x_s}{x_w} \cdot \frac{\dot{V}_E}{\dot{V}}} \tag{5.20}$$

where p_0 = is the normal pressure; p = is pressure at the depth of diving; $\dot{V}$ = is metering of the fresh breathing medium; $\dot{v}$ = is oxygen consumption; $\dot{V}_E$ = is pulmonary ventilation; x_w = is oxygen molar fraction in the *premix*; x_s = is the stable value of oxygen molar fraction in the inhaled breathing bag $x_s = \lim_{t \to \infty} x(t)$.

The following experimental schedule was adopted:

- After gas mixture composition has been stabilized at the depth of $0\,mH_2O$, experiment is switched to the depth of $10\,mH_2O$.
- After gas mixture composition has been stabilized at the depth of $10\,mH_2O$, experiment is switched to the depth of $20\,mH_2O$.
- After gas mixture composition has been stabilized at the depth of $20\,mH_2O$, experiment is switched back to the depth of $10\,mH_2O$.
- After gas mixture composition has been stabilized at the depth of $10\,mH_2O$, experiment is switched back to the depth of $0\,mH_2O$.

Theoretically, calculated from equation (5.20), the maximum difference between oxygen contents in the breathing mixture contained in the inhalation bag cannot be higher than $4\%_v$ within the depth H range of $H \in [0,20]\,mH_2O$. But for the same depth, the oxygen contents in the breathing mixture contained in the inhalation bag (5.20) will be the same. As follows from the preceding text, differences in the oxygen contents should be well visible during an experiment planned in this manner.

The calculated oxygen consumption shown in Table 5.6 should be treated as measurement of the same parameter, and the population investigated has the normal distribution.[42]

The mean oxygen consumption is $\dot{v} = (1.22 \pm 0.02)\,dm^3 \cdot min^{-1}$. The high accuracy of measurements obtained in the real experiments, including five runs for three depths, suggests a good match of the theoretical model developed for the two-bag SCR to the real object, known in this case as $SCR\,APW - 6M\,SCUBA$.

5.2.5 Validation

As has been mentioned, the mapping of the breathing process in hyperbaric conditions is connected not only with the possibility of carrying out investigations under increased pressure and mechanical simulation of lung ventilation but also with mapping of gas exchange. These processes should be reproduced with the reproducibility, accuracy, and precision required by the scientific research method. However, they cannot simulate all the details of the real human–machine ergonomic

TABLE 5.6
Variation of Oxygen Contents in the Inhalation Bag and Oxygen Consumption Resulting from It

		Pressure Increasing		Pressure Decreasing	
	Depth	Oxygen Contents	Oxygen Consumption	Oxygen Contents	Oxygen Consumption
	(mH_2O)	$(\%_v)$	$(dm^3 \cdot min^{-1})$	$(\%_v)$	$(dm^3 \cdot min^{-1})$
Experiment No.	$\pm 1\, mH_2O$	$\pm 0.2\%_v$	–	$\pm 0.2\%_v$	–
1	0	14.8	1.142	13.1	1,279
	10	12.0	1.126	–	–
	20	10.7	1.150		
2	0	13.5	1.247	12.8	1,302
	10	10.4	1.255	10.4	1,255
	20	9.4	1.255		
3	0	14.1	1.199	13.6	1,239
	10	10.5	1.247	10.7	1,231
	20	9.9	1.215		
4	0	13.4	1.255	12.9	1,295
	10	10.2	1.271	10.3	1,263
	20	9.6	1.239		
5	0	13.8	1.223	13.8	1,223
	10	11.4	1.175	11.3	1,183
	20	10.8	1.142		

interaction, offering only a simplified model, but adequate to the research project undertaken. In the case of simulation of gas exchange taking place in the body, the most important is oxygen consumption of $\dot{v}_{O_2}$ and the emission of carbon dioxide $\dot{v}_{CO_2}$ in its place. Built was a new simulator with extremely short lines between the *UBA* and the breathing/metabolic simulator and a much bigger hyperbaric chamber (Figure 5.14).

The validation process was based on volumetric measurements, which for liquid such as *propanone* Me_2CO are dependent on temperature t and for oxygen O_2, also from pressure p. Most often, the measurements were made at temperature of approximately $t \cong 25$. During the measurements, constant pressure value $p = idem$ was maintained. Oxygen consumption calculations were referenced to the temperature and pressure present at the level of inhalation from the mouthpiece of the dive apparatus $\{t, p\}$, and referenced to standard conditions $T_0 \cong 273.15K; p_0 \cong 101.325kPa$. For each measurement, the molar volume of oxygen was determined depending on the temperature present in the respiratory circuit of the diving apparatus.

FIGURE 5.14 Stand of combined simulators: respiratory, metabolic, and hyperbaric.

Propanone is characterized by molar mass $M_{Me_2CO} \cong 58.08\,kg \cdot kmol^{-1}$, density $d_{20} \cong 0.791\,kg \cdot dm^{-3}$, melting point $t_t \cong -95°C$, and boiling point $t_w \cong 56°C$. It follows from reaction (5.14) that $1\,mol$ of Me_2CO reacts with 4 moles of oxygen O_2; hence, mass stream of consumed oxygen $\dot{m}_i$ is proportional to molar mass of oxygen M_i, molar mass of *propanone* M_{Me_2CO}, volume flow of *propanone* $\dot{V}_{Me_2CO}$, and Me_2CO density d_{25} : $\dot{m}_i = 4 \cdot \frac{M_i}{M_{Me_2CO}} \cdot \dot{V}_{Me_2CO} \cdot d_{25°C}$. Hence, mole stream of oxygen $\dot{n}_i = \frac{\dot{m}_i}{M_i} = 4 \cdot \frac{1}{M_{Me_2CO}} \cdot \dot{V}_{Me_2CO} \cdot d_{25°C}$. Thus, the stream of consumed oxygen referred to the normal conditions $\dot{v}_0$ will be $\dot{v}_0 = v_0 \cdot \dot{n}_i = 4 \cdot \frac{v_0}{M_{Me_2CO}} \cdot \dot{V}_{Me_2CO} \cdot d_{25°C}$, where molar volume of oxygen under normal conditions $v_0 \cong 22.39\,dm^3 \cdot mol^{-1}$. Ventilation flow referred to standard conditions $\dot{V}_E$ will be proportional to cylinder capacity V_E of the piston breathing device, its frequency $\dot{n}$, the pressure conditions p, and measurement temperature p: $\dot{V}_E = V_E \cdot \dot{n} \cdot \frac{p}{p_0} \cdot \frac{T_0}{T}$, where $T_0 \cong 273.15K$ and $p_0 \cong 101.325kPa$. It follows from the preceding considerations that the breathing module $\varepsilon = \frac{\dot{v}_0}{\dot{V}_E}$ will be:

$$\varepsilon = 4 \cdot \frac{v_0 \cdot \dot{V}_{Me_2CO} \cdot d_{25°C}}{M_{Me_2CO} \cdot V_E \cdot \dot{n} \cdot \frac{p}{p_0} \cdot \frac{T_0}{T}} \tag{5.21}$$

where $v_0 =$ is molar volume of oxygen under normal conditions; $\dot{V}_{Me_2CO} =$ is volume flow of *propanone*; $d_{25^\circ C} =$ is density of *propanone* at temperature $t = 25^\circ C$; $M_{Me_2CO} =$ is molar mass of *propanone*; $V_E =$ is cylinder capacity of the piston breathing device; $\dot{n} =$ is piston frequency of the respiratory device; $p =$ is pressure under measurement conditions; and $t =$ is temperature under measurement conditions.

As already mentioned in reaction (5.1), 1 *mole* of propanone Me_2CO reacts with $4\,mol$ of oxygen O_2. Thus, 1 molar volume of acetone $V_{Me_2CO} \cong 73.89\,cm^3$ reacts with 4 molar volumes of oxygen $V = 4 \cdot v_0 \cong 97.76\,dm^3$. Thus, volume flow of consumed oxygen $\dot{v}$ can be calculated from proportion: $\dot{v} \cong \frac{97.76}{73.89} \cdot 10^3 \cdot \dot{V}_{Me_2CO} \cong 1323 \cdot \dot{V}_{Me_2CO}$. Inverse dependence can be written as: $\dot{V}_{Me_2CO} \cong 0.756 \cdot 10^{-3} \cdot \dot{v}$. For *propanone* stream $\dot{V}_{Me_2CO}$ expressed in $[\dot{V}_{Me_2CO}] = cm^{-3} \cdot min^{-1}$ and flow of consumed oxygen $\dot{v}$ expressed in $[\dot{v}] = dm^3 \cdot min^{-1}$, the proportion takes the form $\dot{v} \cong 1.323 \cdot \dot{V}_{Me_2CO}$. Assuming in compliance with Table 4.15 that the diver's oxygen consumption may be between $\dot{v} \in [0.3, 2.4]\,dm^3 \cdot min^{-1}$, the corresponding dosing of propanone Me_2CO should be within $\dot{V}_{Me_2CO} \in [0.23, 1.81]\,cm^3 \cdot min^{-1}$.

The current volumetric efficiency of the pump was measured during each measurement indirectly, by measuring the time with absolute error in time measurement $\Delta\tau = 0.2s$ used to empty the volume of the pipette with accuracy $\Delta V = \pm 0.2\,cm^3$.

Air and breathing mixtures were used to supply the *SCR CRABE SCUBA*. The detailed results of the simulation tests were summarized earlier and thus not presented here (Kłos, 2021). Figure 5.15 shows examples of the results from a single study.

5.2.6 Conclusions

The diving technology is based on the knowledge of the composition of the breathing mixture, so to commence initial design of any diving equipment, the breathing loop ventilation model can be used.

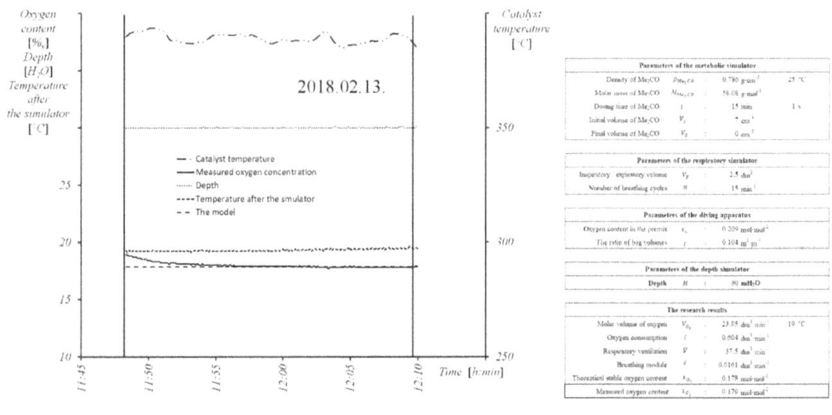

FIGURE 5.15 Example results of investigations on ventilation in *SCR CRABE SCUBA*.

The algebraic models of the *SCR* breathing loop ventilation presented in Section 5.1 have been validated with the consolidated depth, respiratory action, and metabolic process simulator with satisfactory repeatability and reproducibility, as described in this section.

5.3 CONSTRUCTION

The diving technology is built on the breathing mixture composition knowledge; hence, the breathing loop ventilation model forms the basis for the initial design of any diving equipment, but it is crucial for the *SC* R design process.

5.3.1 Separate Bags

According to the ventilation model of the single-bag and constant *premix* metering *SCR* , the function describing the relation between oxygen molar fraction in the breathing gas mixture and time can be presented as equation (9°) from Table 5.2. As time t elapses, oxygen molar fraction in the breathing bag $x(t)$ at the constant stream of the oxygen consumed $\dot{v}$ asymptotically approaches to the constant value limit: $\lim_{t\to\infty} x(t) = \frac{x_w \cdot \dot{V} - \dot{v}}{\dot{V} - \dot{v}} = x_s = const$. The constant value of oxygen molar fraction in the breathing gas mixture x_s continues for several minutes when gas metering is relatively sufficient. Based on this assumption, oxygen partial pressure is as follows:

$$\forall_{\dot{v}=const}\ p_s = x_s \cdot p = \frac{x_w \cdot \dot{V} - \dot{v}}{\dot{V} - \dot{v}} \cdot p \tag{5.22}$$

where $p_s =$ is oxygen partial pressure in the breathing gas mixture; $p =$ the absolute pressure at the diving depth H ; $\dot{V} =$ the metering of the *premix*; $\dot{v} =$ the oxygen consumption; $x_w =$ the oxygen molar fraction in the *premix*; and $p =$ the absolute pressure at the diving depth H .

Therefore, it is possible to calculate optimum metering values $\dot{V}$ and oxygen fractions x_w in the *premix* at the assumed value limits of oxygen consumption $\dot{v}$, partial pressure of oxygen p_i , and absolute pressure at the diving depth p. Maximum oxygen partial pressure p_s^{max} occurs when the stream of oxygen consumed is minimum $\dot{v}^{min}$ and the absolute pressure at the diving depth is maximum[43] p^{max}. Minimum oxygen partial pressure p_s^{max} is related to $\dot{v}^{max}$ and p^{min}. Based on that, the following system of equations can be written:

$$\begin{cases} p_s^{max} = \dfrac{\dot{V} \cdot x_w - \dot{v}^{min}}{\dot{V}_0 - \dot{v}^{min}} \cdot p^{max} \\ \\ p_s^{min} = \dfrac{\dot{V} \cdot x_w - \dot{v}^{max}}{\dot{V} - \dot{v}^{max}} \cdot p^{min} \end{cases} \tag{5.23}$$

where $p_s^{max} =$ is maximum oxygen partial pressure in the breathing gas mixture; $p_s^{min} =$ the minimum oxygen partial pressure in the breathing gas mixture; $p^{max} =$

the maximum absolute pressure at the diving depth H; p^{min} = the minimum absolute pressure at the diving depth H; and $\dot{v}^{min}$ = the minimum, and $\dot{v}^{max}$ = the maximum, oxygen consumption.

The ranges of values of oxygen partial pressure in the breathing gas mixture p_s, oxygen consumption $\dot{v}$, and absolute pressure at the diving depth p are assumed for the system of equations (5.23). The values of absolute pressure at the diving depth p follow from the assumed permissible range of diving depths H. Values of the oxygen stream consumed $\dot{v}$ are experimentally determined as a function of the work rate. Table 4.15 presents some estimated values of the oxygen stream consumed $\dot{v}$ which are found in the literature (Kłos, 2021). The permissible oxygen partial pressures are experimentally determined and published in the literature. A broader discussion of this problem was presented earlier and will not be presented here (Kłos, 2023). Often, in order to shorten the decompression process, the maximum permissible values of oxygen molar fraction p_s^{max} are applied. The values of the *premix* $\dot{V}$ and oxygen molar fraction in the *premix* x_w are determined from the equation system (5.23). The scope of the issue relating to a proper choice of optimum gas composition in the *premix* x_w and its metering $\dot{V}$ should be broadened. Permissible deviations Δ from assumed values of *premix* $\dot{V}$ and oxygen molar fraction in *premix* x_w should be taken into account. The reason for introducing the additional parameters is that an accurate preparation of the breathing medium is very troublesome and expensive. Determination of permissible limits of breathing gas metering rates makes it possible to make a decision whether or not the diving apparatus is suitable for use. Taking into consideration the permissible deviations from metering value $\dot{V}$ and oxygen content x_w, the following system of equations can be derived:

$$\begin{cases} p_s^{max} = \dfrac{\left(\dot{V}+\Delta\dot{V}\right)\cdot\left(x_w+\Delta x_w\right)-\dot{v}^{min}}{\dot{V}+\Delta\dot{V}-\dot{v}^{min}}\cdot p^{max} \\ p_s^{min} = \dfrac{\left(\dot{V}-\Delta\dot{V}\right)\cdot\left(x_w-\Delta x_w\right)-\dot{v}^{max}}{\dot{V}-\Delta\dot{V}-\dot{v}^{max}}\cdot p^{min} \end{cases} \tag{5.24}$$

where Δx_w = is the maximum permissible upward or downward deviation from oxygen partial pressure that is fixed as an optimal value in the *premix* $\Delta\dot{V}$ = maximum permissible upward or downward deviation from metering stream of breathing medium that is fixed as an optimal value permissible deviation of *premix* metering.

After solving the system of equations (5.24) with reference to optimal value metering of *premix* $\dot{V}$, a quadratic equation is obtained. The choice of solution to the equation must take into account the physical nature of the phenomenon. Using software, the system of equations (5.24) should be solved with the computer software (Figure 5.16).

The relationship between oxygen partial pressure in the inhalation breathing mixture p_s and the stream of oxygen consumed $\dot{v}$ and the depth of diving H calculated on the assumptions is presented in Figure 5.17.

It is not possible to design an effective and safe kind of diving apparatus that could be adapted for the wide range of diving depths H and supplied with only one

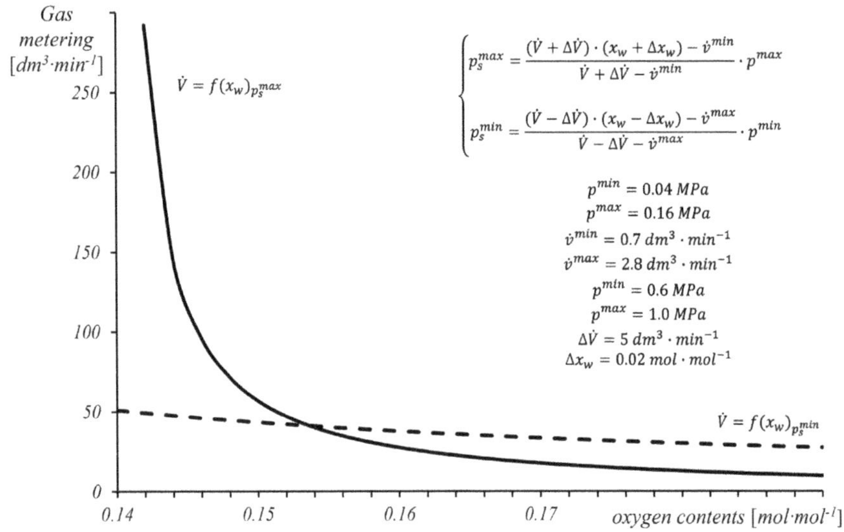

FIGURE 5.16 An example of calculating an optimal value metering of premix $\dot{V}$ and oxygen molar fraction in premix x_w for an SCR with t premix constant metering, where p_s^{max} = is maximum oxygen partial pressure in the breathing gas mixture; p_s^{min} = the minimum oxygen partial pressure in the breathing gas mixture; p^{max} = the maximum absolute pressure at the diving depth H; p^{min} = the minimum absolute pressure at the diving depth H; $\dot{v}^{min}$ = the minimum; $\dot{v}^{max}$ = the maximum oxygen consumption; Δx_w = the maximum permissible deviation of oxygen contents in the premix; and $\Delta\dot{V}$ = the maximum permissible deviation of the gas metering.

breathing medium having constant oxygen content x_w. A diving apparatus is usually designed for use of several different breathing mediums and metering (Table 5.7).

Because there is no one optimal solution to the sufficiently wide range of safe diving depth[44] H, an *SCR* is usually designed for use of several different *premixes*. Optimum solutions to the design of *SCR* with constant *premix* dosage are presented in Table 5.7. In order to obtain the results from the equation system (5.24), the same method was used as in the case of the example calculations of optimal values of *premix* $\dot{V}$ and oxygen molar fraction in *premix* x_w, presented in Figure 5.16 to Figure 5.17. The intersection point of the function (5.24) is the optimal solution that must be found.

5.3.2 Metering Bellows Dispenser

The diving apparatus *design module* $\varepsilon_k = \frac{1 - x_w}{r}\varepsilon$ can be determined from direct measurement[45] of the volume ratio of the small bag and the large bag U, $r = \frac{u}{U - u}$, and oxygen content in premix x_w. *Breathing modules* ε can be found in the literature concerning atmospheric pressure p_0: $\varepsilon_0 \in [0.035; 0.055]$ (PN-EN-14143, 2005). Experiments showed that the breathing module ε slightly depends on pressure p: $\varepsilon = f(p)$ (Kłos, 2021).

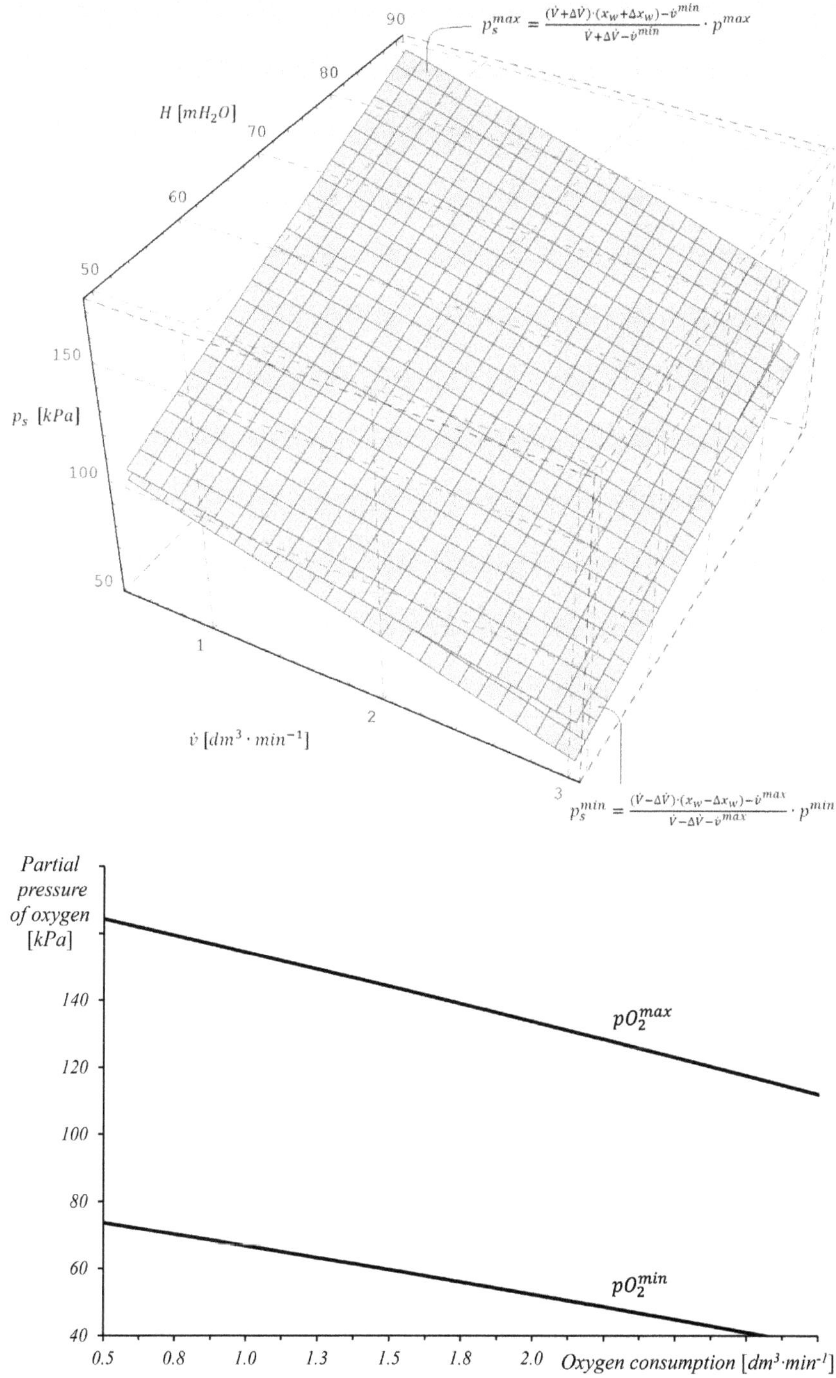

FIGURE 5.17 The relationship between oxygen partial pressure p_s in the breathing mixture, the stream of oxygen consumed $\dot{v}$, and the depth of diving H.

TABLE 5.7
Optimum Gas Metering and Oxygen Molar Fractions in the Prepared Breathing Medium

Minimum oxygen partial pressure	$= 20 \ kPa$
Maximum oxygen partial pressure	$= 160 \ kPa$
Minimum oxygen consumption	$= 0.7 \ dm^3 \cdot min^{-1}$
Maximum oxygen consumption	$= 2.8 \ dm^3 \cdot min^{-1}$
Metering tolerance	$= \pm 2.0 \ dm^3 \cdot min^{-1}$
Oxygen contents tolerance	$= \pm 0.02 \ mol \cdot mol^{-1}$

H^{max} (mH_2O)		H^{min} (mH_2O) 0	10	20	30	40	50	60	70	80	90	100	110
150	x_w				0.099	0.101	0.103	0.104	0.105	0.105	0.106	0.106	0.106
	$\dot{V}$				70.82	54.56	47.59	43.71	41.24	39.53	38.28	37.32	36.56
140	x_w				0.107	0.109	0.111	0.112	0.113	0.113	0.114	0.114	0.114
	$\dot{V}$				58.62	47.29	42.09	39.10	37.17	35.81	34.80	34.02	33.41
130	x_w			0.112	0.116	0.119	0.120	0.121	0.122	0.122	0.123	0.123	0.123
	$\dot{V}$			75.33	49.16	41.15	37.27	34.98	33.46	32.39	31.59	30.96	30.47
120	x_w			0.123	0.127	0.129	0.131	0.132	0.132	0.133	0.133	0.133	0.134
	$\dot{V}$			58.04	41.61	35.91	33.01	31.26	30.08	29.24	28.61	28.11	27.72
110	x_w			0.136	0.140	0.142	0.143	0.144	0.144	0.145	0.145	0.145	
	$\dot{V}$			46.09	35.44	31.37	29.22	27.89	26.99	26.33	25.83	25.44	
100	x_w		0.143	0.151	0.154	0.156	0.157	0.158	0.158	0.159	0.159		
	$\dot{V}$		78.67	37.31	30.31	27.41	25.82	24.82	24.13	23.63	23.25		
90	x_w		0.161	0.168	0.171	0.173	0.174	0.174	0.175	0.175			
	$\dot{V}$		51.40	30.60	25.96	23.91	22.75	22.01	21.50	21.12			
80	x_w		0.183	0.189	0.192	0.193	0.194	0.195	0.195				
	$\dot{V}$		36.68	25.29	22.22	20.79	19.97	19.43	19.05				
70	x_w		0.209	0.214	0.217	0.218	0.219	0.219					
	$\dot{V}$		27.45	20.98	18.98	18.00	17.42	17.04					
60	x_w		0.242	0.246	0.248	0.249	0.250						
	$\dot{V}$		21.10	17.40	16.12	15.48	15.09						
50	x_w	0.269	0.285	0.288	0.290	0.290							
	$\dot{V}$	40.05	16.44	14.36	13.58	13.18							
40	x_w	0.331	0.342	0.345	0.346								
	$\dot{V}$	20.49	12.86	11.75	11.31								
30	x_w	0.418	0.425	0.427									
	$\dot{V}$	12.74	10.00	9.47									
20	x_w	0.554	0.557										
	$\dot{V}$	8.50	7.63										
10	x_w	0.840											
	$\dot{V}$	3.50											

x_w = oxygen contents ($mol \cdot mol^{-1}$)
$\dot{V}$ = *premix* metering ($dm^3 \cdot min^{-1}$)

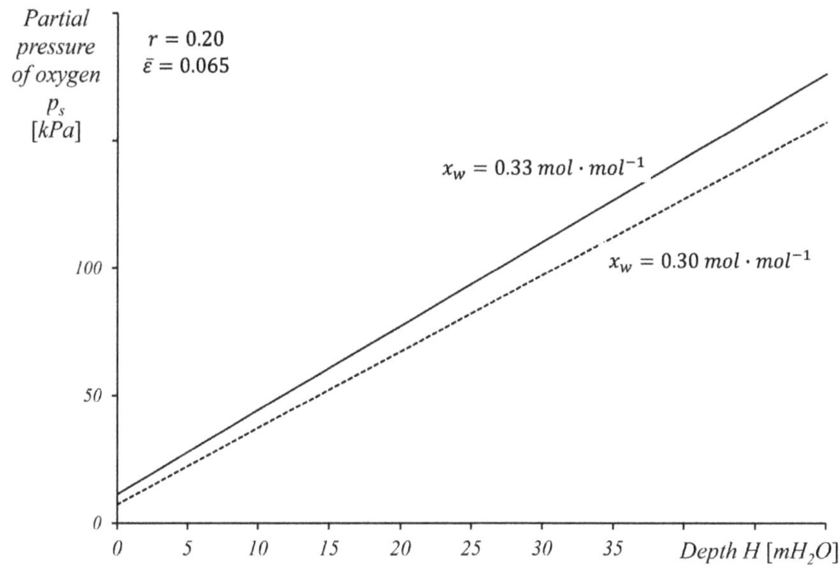

FIGURE 5.18 Example dependence of stabilized partial pressure of oxygen $p_s = f_{x_w,\bar{\varepsilon},r}(H)$ in function of depth H for the *design module* $\varepsilon_k = \frac{1-x_w}{r}$ and the mean value $\bar{\varepsilon}$ of the *breathing module* $\varepsilon = \frac{\dot{v}}{\dot{V}_E}$.

The dependence of stable partial pressure of oxygen p_s in function of depth H for the mean value of the *breathing module*[46] $\bar{\varepsilon} = 0.065$ in depth range $H \in [0,50]\, mH_2O$ and the volume ratio of bags $r = 0.2$ are shown in Figure 5.18.

In order to reduce the value of stable partial pressure of oxygen p_s at maximum depth $H = 50\, mH_2O$, the manufacturer of *SCR CRABE SCUBA* applied a different technical solution for the apparatus, adopting the volume ratio of bags as $r = 0.1\frac{m^3}{m^3}$. For such a structure and *premix* with oxygen content $x_w = 0.33\frac{mol}{mol}$, there will be a decrease in stable partial pressure of oxygen on the surface $p_s(p = p_0)$ to the hypoxic level $p_s < 20\, kPa$. Only for *premix* with oxygen content $x_w \cong 0.52\frac{mol}{mol}$, partial pressure of oxygen p_s will increase to the value $p_s \cong 18\, kPa$ acceptable under emergency conditions. The manufacturer has additionally introduced a constant *premix* dosing system to support apparatus ventilation, which switches on within depth range $H \in [22,24]\, mH_2O$. The ejector support is switched on/off during the dive within the depth range $H \in [22,24]\, mH_2O$ marked in Figure 5.19.

5.3.3 Conclusions

Diving technology is based on the knowledge of the composition of the gas mixture inhaled by the diver as a function of diving parameters.

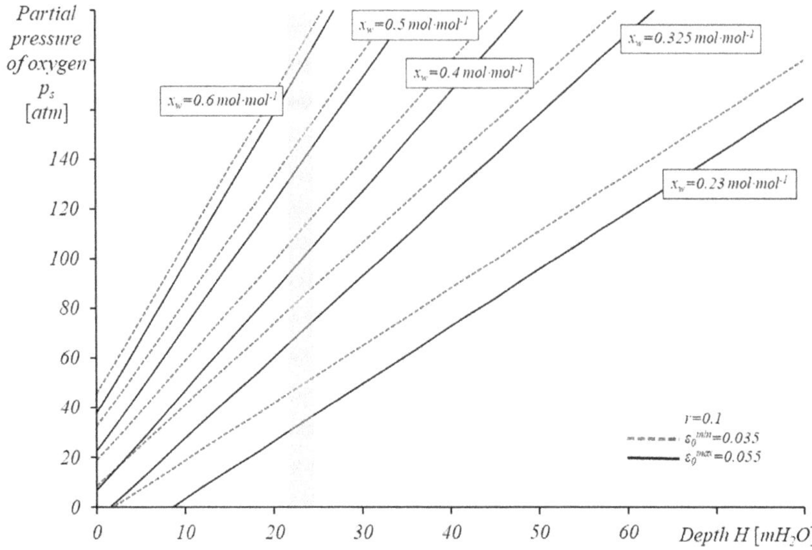

FIGURE 5.19 Examples[47] of theoretical ranges of stable oxygen partial pressures p_s in inhaled breathing gas in function of depth H and various premixes with oxygen content x_w as used by the manufacturer of the apparatus;[48] for *premixes* with oxygen content $x_w \leq 0.5 mol \cdot mol^{-1}$ for depths within the range $H \in \{21, 24\} mH_2O$, the ejector support for dosing of fresh *premix* is used.

Algebraic models of the SCR breathing loop ventilation should be validated by means of a consolidated simulation of depth, respiratory action, and metabolic process in order to confirm the values of the assumed parameters of the breathing space ventilation model in a diving apparatus. This can be made by comparing the results of measurements of stable values of oxygen partial pressure with the theoretical values derived from the model having satisfactory repeatability and reproducibility.

NOTES

1 For the time approaching infinity $t \to \infty$
2 In most cases, estimated for the time approaching infinity $t \to \infty$
3 For example, caused by change in work rate.
4 Where the gas motion is an important but not essential problem, for example, by the contact time.
5 Regeneration.
6 Derivation of the relationship between oxygen molar fraction x and time t for the single-bag semi–closed circuit diving apparatus *SCR* with the constant metering of *premix* by means of integral calculus was published earlier (Kłos, 2021).
7 Depending upon the assumed criterion.
8 "Dead space."
9 Formally, decompression *Nx* takes place due to the nitrogen content coming from the body cleansing and residual content after the cleansing of the breathing, the content of helium with *Tx* in the knitting conditions of the CRABE apparatus can be neglected.

10 Inspired/expired steam of volume V_E.
11 Moles.
12 In differential equations, the function occurring under the derivative designation is unknown. If the unknown is a function of one variable, then the equation is called *ordinary*. The order of the highest derivative in the equation is the *order of the equation*.
13 Here the *module* is referred to as a formally dimensionless quantity; the modules will also be referred to as *criterion numbers*.
14 Ventilatory equivalent for oxygen $V_E(O_2)$.
15 Sometimes the formally dimensionless ratio is dependent on the values on the basis of which it was obtained; that is why their designation is sometimes given – for example, for real gases, molar fraction $x_n x$ having dimension $[x_n] = \frac{mol}{mol}$ is not equal to volume fraction x_v having the dimension $[x_v] = \frac{m^3}{m^3} : \chi_n \neq \chi_v$ With regard to the numerical value, therefore, often the designations of the criterion numbers are also marked.
16 As previously noted, dimensionless relations will be called *modules* or *criterion numbers*, except the breathing module ε, which is commonly referred to as the *breathing module*.
17 Related to the measured values of load W_0 on the surface.
18 *Centroid* is understood here as a geometric center from space lying in the geometric center of its shape equivalent to the center of buoyancy.
19 Usually, in investigations on oxygen consumption dependent on effort, resistance of the breathing space in a measuring apparatus was negligible in relation to breathing resistance generated by a diving apparatus already on the surface.
20 Slowed heart rate, tachycardia, because in immersion, the pressure on limbs increases and there occurs a contraction of skin vessels due to cold, causing an increase in venous return, which causes an increase in stroke volume, which leads to a reduced heart rate.
21 Understood here as a science of the form, shaping, and constitution of organisms.
22 Speed of stabilization.
23 As $\chi_s \sim p^{-1}$
24 *Respiratory quotient* $\varepsilon_{RQ} = \frac{\dot{\nu}'}{\dot{\nu}}$ describes the ratio of carbon dioxide output $\dot{\nu}'$ to oxygen intake $\dot{\nu}$; when carbohydrates are metabolized in the human body, it is determined at the level $\varepsilon_{RQ} = 1.0\ dm^3 \cdot dm^{-3}$, for proteins $\varepsilon_{RQ} \cong 0.8\ dm^3 \cdot dm^{-3}$ and for fats $\varepsilon_{RQ} \cong 0.7\ dm^3 \cdot dm^{-3}$.
25 0.5% platinum deposited on ca. $3mm$ diameter rollers of alumina.
26 Because of high reaction temperature.
27 Correlation coefficient $r > |-0.98|$
28 Confidence level is not marked in the figure.
29 *EFA* is a statistical method used to describe variability among observed, correlated variables in terms of a potentially lower number of unobserved variables called factors.
30 Therefore, unfavorable *CO* production mechanism.
31 Catalyzed acetone oxidation process.
32 *PCA* is a statistical procedure that uses an orthogonal transformation to convert a set of observations of possibly correlated variables into a set of values of linearly uncorrelated variables, called principal components.
33 For standard values $\forall_{\bar{x}\equi0;\ \sigma\equiv1}\ r \equiv \frac{c_{xy}}{\sigma_x \sigma_y} \Rightarrow r = C_{xy}$
34 Correlation.
35 Equals the sum of all variances in the data.
36 Are orthogonal.
37 Variance.
38 Strategies.
39 It means "at sufficiently large acetone dosage."
40 The curve is contained within the limits of calculated maximum systematic errors.

41 As a matter of fact, the diving apparatus type APW-6M is two-bag SCR.
42 Liliefor's, Shapiro–Wilk's, and K-S tests.
43 Maximum depth of diving H^{max}
44 That is, optimal composition of *premix* x_w and its metering $\dot{V}$
45 For example, by flooding with water.
46 $\varepsilon(p) \cong 0{,}082 - 2.7 \cdot 10^{-5} \cdot \rho \rightarrow \varepsilon(\rho = 100kP\alpha) = 0{,}08 \wedge \varepsilon(\rho = 600kP\alpha) = 0.06.$
47 *Premix* having oxygen content $\chi_w = 0.23\ mol\cdot mol^{-1}$ is *trimix*, whereas the other ones are *nitrox* mixtures.
48 When the values of the breathing module determined for normal pressure are applied, $\varepsilon_0 \in [0.035, 0.05]$.

REFERENCES

Birch K., MacLaren D., George K. 2009. *Fizjologia sportu-krótkie wykłady*. Warszawa: Wydawnictwo Naukowe PWN. ISBN: 978-83-01-15460-8.

Brennan D.M.A., Bdonchuk W.W. 1975. Oxygen consumption of SCUBA divers—a technique for measurement and analysis. In *Proceedings of the 6th international conference on underwater education, October 1974*. San Diego, CA: National Association of Underwater Instructors.

Frånberg O. 2015. *Oxygen content in semi-closed rebreathing apparatuses for underwater use: measurements and modeling*. Stockholm: School of Technology and Health. ISSN: 1653-3836.

Haux G. 1982. *Subsea manned engineering*. London: Bailliére Tindall. ISBN: 0-7020-0749-8.

Kłos R. 2000. *Aparaty Nurkowe z regeneracją czynnika oddechowego*. Poznań: COOPgraf. ISBN: 83-909187-2-2.

Kłos R. 2002. Metabolic simulator supports diving apparatus researches. *Sea Technology*. 12, 53–56.

Kłos R. 2007. *Mathematical modelling of the normobaric and hyperbaric facilities ventilation*. Gdynia: Wydawnictwo Polskiego Towarzystwa Medycyny i Techniki Hiperbarycznej. ISBN: 978-83-924989-0-2.

Kłos R. 2021. *Ventilation of normobaric and hyperbaric objects*. Boca Raton, FL: CRC Press (Taylor & Francis Group, LLC). ISBN: 978-0-367-67523-3 (hbk) ISBN: 978-0-367-67524-0 (pbk) ISBN: 978-1-003-13164-9 (ebk).

Kłos R. 2023. *Oxygen diving*. Boca Raton, FL: CRC Press (Taylor & Francis Group, LLC). ISBN: 978-1-032-31389-4 (hbk); ISBN: 978-1-032-31390-0 (pbk); ISBN: 978-1-003-30950-5 (ebk). DOI: 10.1201/9781003309505.

Loncar M. 1992. Breathing simulator simulates human oxygen intake. *Offshore*. 10, 82.

Reimers S.D., Hansen O.R. 1972. *Environmental control for hiperbaric applications*. Panama City: US Navy Experimental Diving Unit. NEDU Report No. 25-72.

Williams S. 1975. Engineering principles of underwater breathing apparatus. In *The physiology and medicine of diving*, ed. Bennett P.B., Elliott D.H., 34–46. London: Baillière Tindall.

6 Chemisorption

Chemisorption is a process which is used for removing carbon dioxide CO_2 from breathing gases in *SCR*s. Another field of application is as a bed in scrubbers built into ventilation systems for cleaning ecologically closed breathing atmosphere. These systems are used in medicine and military applications.

6.1 METHODS

Theoretically, there are many methods of removing carbon dioxide CO_2 from a closed or semi-closed breathing atmosphere. The methods used for CO_2 removal are as follows: chemisorption or physical methods as absorption and adsorption. Mixed physical/chemical methods can also be used (Nuckols, Sarich and Tucker, 1996; Marshall, Nuckols, Sarich, Marr and Robillard, 2015). In chemical methods, CO_2 is removed with the use of liquid-[1] or solid-state[2] substances. Chemisorption with oxygen emission[3] is one of the most interesting methods for breathing atmosphere regeneration:

$$\begin{array}{l} 2 \cdot KO_2 + H_2O \rightarrow 2 \cdot KOH + 1.5 \cdot O_2 + 14.2\ kcal \\ 2 \cdot KO_2 + CO_2 \rightarrow 2 \cdot K_2CO_3 + H_2O + 35\ kcal \\ \hline 2 \cdot KO_2 + CO_2 \rightarrow K_2CO_3 + 1.5 \cdot O_2 + 49.2\ kcal \end{array}$$

From the stoichiometry equation, it is evident that the reaction of 2 moles of CO_2 with alkali superoxide KO_2 theoretically[4] causes emission of 3 moles of oxygen O_2. Additionally, moisture can discompose alkali superoxide. The reaction of potassium superoxide together with carbon dioxide CO_2 and water vapor H_2O is highly exothermic.

Physical methods include absorption in liquid absorbents or adsorption on a solid package, for example, adsorption on active carbon. The mixed methods can be based on an electrochemical mechanism. At first, the method relies on CO_2 dissolving and absorption in water or alkaline solution – chemisorption. The carbonate and hydrogen carbonate anions are separated from the other anions with the use of an electric field and membrane partition. On anode, separated anions are reduced to gas.

In most solutions, chemical methods are used for removing CO_2 from closed and semi–closed circuit diving apparatuses and submarine atmosphere. First of all, chemical methods use solid sorbents based on hydroxides of alkali metals. Calcium hydroxide $Ca(OH)_2$ activated with sodium hydroxide $NaOH$ or potassium hydroxide KOH is the basic component of the sorbent:

$$2 \cdot NaOH + CO_2 \rightarrow Na_2CO_3 + H_2O + 28\ kcal \tag{6.1}$$

$$Ca(OH)_2 + CO_2 \rightarrow CaCO_3 + H_2O + 19\ kcal \tag{6.2}$$

DOI: 10.1201/9781032649207-6

The reactions described with equations (6.1) and (6.2) are summary reactions, but the mechanism of CO_2 absorption is more complicated and will be described further. Chemisorption of carbon dioxide absorption in hydroxides of alkali metals is highly exothermic.

Baralime/Baralyme, soda lime,[5] lithium hydroxide $LiOH$, etc. are the typical CO_2 used in CCRs/SCRs and submarine atmosphere regeneration systems. The absorbents are manufactured as flakes, regular granules produced in a pelleting process, and irregular granules obtained as a result of crushing and sieving. These types of absorbents dissolved in water give highly caustic solutions that may cause dangerous burns if in contact with mucosal or skin.

6.1.1 Baralime

Baralime consists of $Ba(OH)_2 \cdot 8 \cdot H_2O$ and $Ca(OH)_2$. Hydroxides of alkali metals as $NaOH$ or KOH are often used as catalytic additions. There are two kinds of Baralime (Furevik, 1972):

- Noncaustic: $80\%Ca(OH)_2$, $20\%Ba(OH)_2 \cdot 8 \cdot H_2O$
- Caustic: $75\%Ca(OH)_2$, $20\%Ba(OH)_2 \cdot 8 \cdot H_2O$, $5\%KOH$

The following equations summarize the chemical reactions of Baralime with CO_2:

$$Ba(OH)_2 \cdot 8 \cdot H_2O + CO_2 \rightarrow BaCO_3 + 9 \cdot H_2O \tag{6.3}$$

$$H_2O + CO_2 \leftrightarrow H_2\,CO_3 \tag{6.4}$$

$$H_2CO_3 \leftrightarrow H^+ + HCO_3^- \; K = 4.3 \cdot 10^{-7} \tag{6.5}$$

$$HCO_3^- \leftrightarrow H^+ + CO_3^{2-} \; K = 5.6 \cdot 10^{-11} \tag{6.6}$$

$$2 \cdot H^+ + CO_3^{2-} + Ca(OH)_2 \rightarrow CaCO_3 + 2 \cdot H_2O \tag{6.7}$$

$$2 \cdot H^+ + CO_3^{2-} + KOH \rightarrow K_2CO_3 + 2 \cdot H_2O \tag{6.8}$$

$$K_2CO_3 + Ca(OH)_2 \rightarrow CaCO_3 + 2 \cdot KOH \tag{6.9}$$

Carbonic acid H_2CO_3, as the product of carbon dioxide CO_2 dissolved in water H_2O, is neutralized by hydroxides of alkaline metal. As the barium hydroxide $Ba(OH)_2 \cdot 8 \cdot H_2O$ contains crystallization water needed to create the water environment, therefore the first reaction (6.3) begins the sequence, which constitutes chemisorption of CO_2.

H_2CO_3 is the product of a reaction between H_2O and CO_2 from reaction (6.4). H_2CO_3 is chemically bonded on the surface of the absorbent pellets, forming a thin superficial layer. Reaction of dissociation is the kinetic barrier of CO_2 chemisorption by hydroxides of alkaline metals, because the necessary carbonic ions will be generated only after the reaction of the ions generated earlier in the dissociation reaction of H_2CO_3. The concentration of ions in the solution is relatively low with respect to H_2CO_3, whose content blocks further dissolution of CO_2.

As follows from the preceding text, the process is limited by the process of diffusion. H_2CO_3 can be neutralized according to reaction (6.7) or (6.8). But reaction (6.8) is more probable because of low $Ca(OH)_2$ solubility in water. Exchange reaction (6.9) will proceed because of the low solubility of $Ca(OH)_2$ with respect to K_2CO_3 (see Table 6.1 and Table 6.2), causing transportation of bonded CO_2 into the granules.

As mentioned earlier, the water contained in $Ba(OH)_2 \cdot 8 \cdot H_2O$ is essential in CO_2 chemisorption; however, water is lost in a temperature of $650°C$. For this reason, Baralime should be stored far from heaters and protected against solar radiation. It should be noted that the caustic form of Baralyme is twice more efficient than noncaustic Baralime. It follows from the preceding that the catalytic role of *KOH* is shown in the reaction.

TABLE 6.1
Solubility of Solid Sorbent Components Used in Carbon Dioxide Chemisorption of Chosen Absorbers

Kind of Sorbent	Metal Hydroxide	Approximate Solubility in Cold Water ca. 30°C	Approximate Solubility in Warm Water ca. 100°C	Solubility in Water (0°C)	Solubility in Water (20°C)
			$(g/100g\,H_2O)$		
Baralyme	$Ca(OH)_2$	5.6	94.7	1.67	3.89
	$Ba(OH)_2 \cdot 8 \cdot H_2O$	126	178	97	112
	KOH	0.153	0.077	0.185	0.165
Sodasorb	$Ca(OH)_2$	0.153	0.077	0.185	0.165
	KOH	126	178	97—	112
	$NaOH$	119	347		109
Lithium hydroxide	$LiOH$	12.8	17.5	12.7	12.8

TABLE 6.2
Solubility of Some Carbonates

Kind of Carbonate	$CaCO_3$	K_2CO_3	Na_2CO_3
Solubility in water $0°C$ $(g\,dm^{-3})$	0.0015	112	7.1

6.1.2 Soda Lime

Soda lime consists of calcium hydroxide $Ca(OH)_2$, sodium hydroxide $NaOH$, and/or potassium hydroxide KOH. The approximate contents C of $NaOH$ and/or KOH is $C \in [2,5]\%$ of the solid phase. Soda lime is the most often used CO_2 absorbent. In order to obtain efficient CO_2 removal, water is added to the mixture of hydroxides during the manufacture of absorbent. Water can be lost in cases of high temperature, solar radiation, or storage in leaky containers.

Soda lime is manufactured most often in the form of irregular granulate, cylindrical granulate, or semispherical granulate. The following equations summarize the chemical reactions of soda lime with CO_2:

$$H_2O + CO_2 \leftrightarrow H_2CO_3 \tag{6.10}$$

$$H_2CO_3 \leftrightarrow H^+ + HCO_3^- \; K = 4.3 \cdot 10^{-7} \tag{6.11}$$

$$HCO_3^- \leftrightarrow H^+ + CO_3^{2-} \; K = 5.6 \cdot 10^{-11} \tag{6.12}$$

$$2 \cdot H^+ + CO_3^{2-} + NaOH \rightarrow Na_2CO_3 + 2 \cdot H_2O \tag{6.13}$$

$$Na_2CO_3 + Ca(OH)_2 \rightarrow CaCO_3 + 2 \cdot NaOH \tag{6.14}$$

The preceding reactions are similar to those presented in the case of the Baralime chemisorption process. The only difference depends on the beginning of the reaction. The reaction is initiated by carbon dioxide diffusion from the gaseous phase to the water layer on the package surface. The water layer is formed by water adsorption from the gaseous phase or comes from the sorbent. All phases of chemisorption mechanism run in a water environment.

Better solubility of $KOH/NaOH$ compared to $Ca(OH)_2$ is a very important parameter of CO_2 absorption (Table 6.1). Water and $KOH/NaOH$ are regenerated and released in the next stages of the reaction of catalyzed CO_2 bonding process. In catalytic reactions, the highest yield of reaction is reached at the proper concentration of catalyst; therefore, it is very important to maintain the concentration of $KOH/NaOH$ at the proper level, ensuring the highest efficiency.

6.1.3 Lithium Hydroxide

A relatively high yield of CO_2 chemisorption per mass unit of absorbent makes lithium hydroxide $LiOH$ useful in special applications, for example, from the atmosphere of spacecraft.

The following equations summarize the chemical reactions of carbon dioxide in $LiOH$:

$$LiOH + H_2O \leftrightarrow LiOH \cdot H_2O \tag{6.15}$$

$$2 \cdot LiOH \cdot H_2O + CO_2 \rightarrow Li_2CO_3 + 3 \cdot H_2O \tag{6.16}$$

The first equation shows the equilibrium reaction of the solid $LiOH$ and the water vapor contained in the absorbate.[6] When $LiOH$ is exposed to a high-humidity atmosphere, the surface of granulate is covered with the saturated solution of $LiOH$. This causes plugging of the absorbent pores, decreasing its active surface and forming the diffusion barrier that influences the chemisorption rate of CO_2. The form of $LiOH$ appearance is the most important fact for the reaction productivity. $LiOH$ should react as monohydrate, not as its saturated solution formed at the absorbent surface (Furevik, 1972).

6.1.4 Structure

Chemisorption efficiency is dependent not only on the chemical properties of the absorbent but also on absorbent porosity. This means that its internal and external surfaces are accessible for the absorbate. The greater active surface causes improvement of CO_2 chemisorption. The active adsorption surface characterizes the adsorption capacity of the adsorbent, which is very difficult to determine properly. The compatibility of the results obtained from the same measuring method can be small. In most cases, the active surface is determined by experiments carried out to investigate pure gas adsorption on adsorbent surface at low pressure. The results obtained may differ depending on the chosen gas.[7]

The less-accurate measurement of the active surface is determined by measuring free space or porosity. The free space is defined as the accessible space in the vessel totally filled up with the sorbent. In order to approximate the determination of this value, the vessel is filled up with the sorbent and flooded with the accurately measured amount of liquid until the vessel is full. The amount of the liquid is controlled all the time.

Sorbent particles have their internal structure, called pores (Figure 6.1).

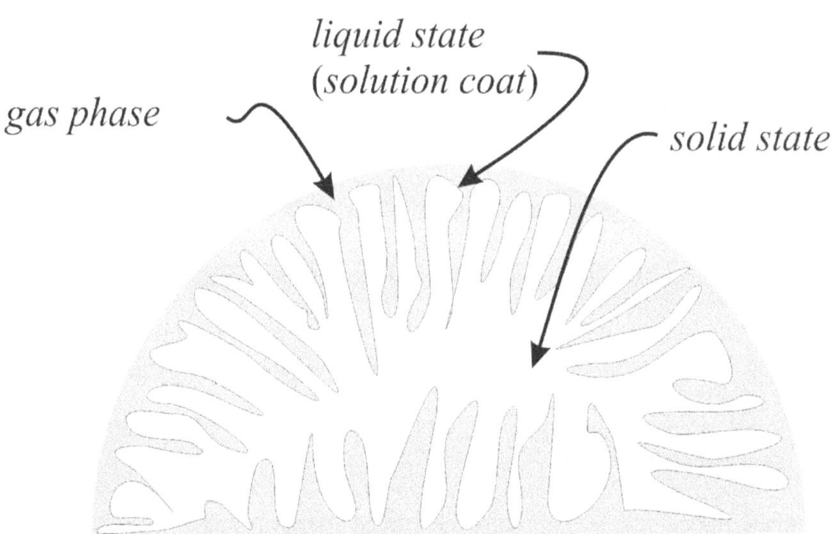

FIGURE 6.1 The scheme of the carbon dioxide sorbent granule.

Pores are extremely small spaces having high resistance to penetration by fluids. The water released during the reaction and the moisture picked up from absorbate can penetrate some accessible pores and condense.[8] This may cause a decrease in the free volume and sorbent active surface. This is due to the shape of the liquid surface inside pores. It is very concave within the thin capillaries. Saturated vapor pressure on such a surface is considerably lower than on the flat surface. In the case of very concave surfaces (as in thin capillaries), nonsaturated vapor over the flat surface becomes supersaturated. As a result, the condensation will be more rapid in thin capillaries.

The absorbent particle size will also exert an influence on sorbent efficiency. Knowledge of the process occurring on the sorbent particle surface enables us to understand the whole chemisorption process. The mechanism of reactions was earlier presented in the description of different kinds of sorbents. The chemisorption reactions have been presented in two stages: the forming of carbonates on the particle surface (it is the first stage of chemisorption) and their penetration of the granule by means of the exchange reaction. The surface layer formed and the water layer are the environment for carbon dioxide chemisorption. The reaction products and dissolved granule elements make the diffusion barrier which disables carbon dioxide reaction with the chemical material contained in the granule. Capillary condensation contributes additionally to a decrease in the active surface and formation of the diffusive barrier.

For this reason, the measured efficiency of the absorbent may be significantly different from the theoretical values calculated on the basis of absorptivity of the sorbent of different form and shape. The reaction between granule and carbon dioxide will be chemically[9] and physically[10] dependent.

Before use, CO_2 absorbents are very often transported. During each transportation stage,[11] the package is exposed to mechanical damage, for example, abrasion, breaking, crushing, shaking, etc. In order to avoid dusting and duct forming,[12] bed particles should be mechanically resistant. Mechanical resistance of the sorbent depends on the structure of granulate and intermolecular bonding power. In order to obtain the repeatable and uniform shape of sorbent particles, tableting of the material is applied. Ensuring accessible, highly active surface of the bed is an important problem in the production of the uniform-shaped granulates. The most important advantages of the use of such granulates are as follows:

- Because of high mechanical resistance to abrasion, crushing, and dusting, they are safe in use.
- The flow resistance through the bed is constant and repeatable.
- The properties of the bed are repeatable and can be covered by quality standards.
- There is an unlimited possibility of using additives – modification of the composition is possible.

The shape and particle size of soda lime formed after crushing and mesh sieve classification are not determined. The bed particles have a relatively large external active surface, but sharp edges are not mechanically resistant. That form of bed is dangerous because of the high tendency to crush and dust, which leads to forming ducts in the bed. Owing to the caustic properties of the bed components, dusting of the bed may cause microscalds of the respiratory system.

6.1.5 Additives

The solid absorbents used for CO_2 chemisorption contain many enriched additives, which are the manufacturer's secret. The additives can be divided as follows:

- Hygroscopic compounds and/or water
- Binder used in forming of granules
- Indicators
- Catalysts

Enriching soda lime and Baralyme by means of water was describer earlier. The addition of binder/tackifier to the aforementioned absorbents is not necessary for this kind of bed; however, they can be used for bonding and stabilizing granules. Many years ago, asbestos dust was used as a binder. Now, because of its carcinogenic properties, it has been withdrawn from use.

Enriching the sorbent by adding strong bases was described earlier. The strong bases are the catalysts for chemisorption of CO_2. The other catalyst additives are used for peroxide mass, for example, manganic dioxide MnO_2. Usually, manufacturers do not give any information concerning the use of catalyst additives; however, it should be taken into account that additives are used.

In order to visualize the bed, a consumption indicator is added. The color changes in the indicator are pH-dependent. During an increase in the concentration of carbon dioxide absorbed, $Ca(OH)_2$ is neutralized by forming H_2CO_3. The formed $CaCO_3$ causes pH to decrease. Among the well-known indicators used in acid–base titration, *p-xylenphthaleine* and *thymolphthaleine* have similar properties[13] as *Nile blue*.[14] Indicator additives are essential, especially for beds used in medicine. It is of no significance in the diving technology because the bed condition cannot be controlled during diving. However, an additive of the indicator should be useful in regeneration installations used in the submersible facilities or shelters.

6.1.6 Conditions

The temperature and humidity effect on the CO_2 chemisorption process on alkali metal hydroxides can be analyzed on the basis of the chemical reactions. The following equations summarize the chemical reactions on different sorbents:

- For chemisorption on Baralyme:

$$\begin{aligned} &a \cdot Ba(OH)_2 \cdot 8 \cdot H_2O + b \cdot Ca(OH)_2 + (a+b) \cdot CO_2 \rightarrow \\ &\rightarrow a \cdot BaCO_3 + b \cdot CaCO_3 + (9 \cdot a + b) \cdot H_2O \end{aligned} \tag{6.17}$$

- For chemisorption on soda lime:

$$Ca(OH)_2 + CO_2 \rightarrow CaCO_3 + H_2O \tag{6.18}$$

- For chemisorption on lithium hydroxide:

$$2 \cdot LiOH + CO_2 \rightarrow Li_2CO_3 + H_2O \tag{6.19}$$

All reactions are exothermic, so the temperature effect is essential. Water vapor is essential for CO_2 absorption, because it is released in all reactions and it is reaction-medium, so the adsorbate humidity will be essential.

Humidity is of utmost importance for CO_2 absorption on $LiOH$. The other cases of humidity effect on absorption are of less importance. The least temperature effect occurs in the case of CO_2 absorption on $LiOH$. In the reactions described earlier, water is released in the same molar proportions as CO_2 absorbed.

In Figure 6.2, the $LiOH - H_2O$ phase diagram is shown. The area[15] between solid lines shows the optimum temperature and the initial water vapor partial pressure for CO_2 chemisorption.

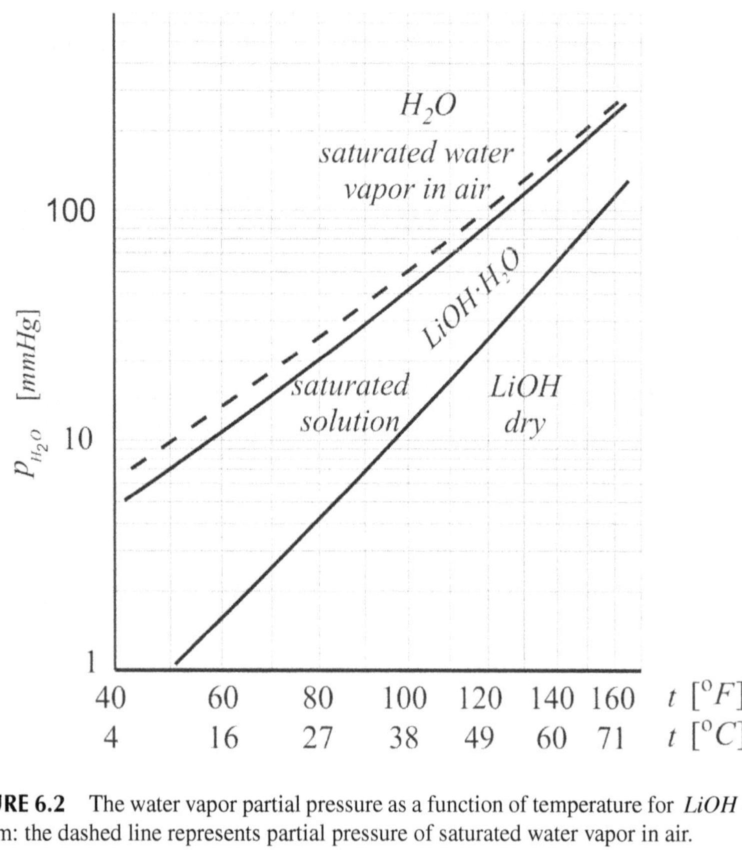

FIGURE 6.2 The water vapor partial pressure as a function of temperature for $LiOH - H_2O$ system: the dashed line represents partial pressure of saturated water vapor in air.

Source: Furevik (1972).

The equimolar amount of H_2O is released in the CO_2 chemisorption reaction. For this reason, the CO_2 partial pressure p_{CO_2} should be correlated with the partial pressure p_{H_2O} of water vapor H_2O contained in the gaseous phase. The maximum effective partial pressure of water vapor p_{H_2O} is approximately equal to $p_{H_2O} \cong 15.3\ mmHg$ at temperature $t \cong 70°F$. At this temperature, relative humidity is equal to $50\%_R$ and corresponds to the partial pressure of water vapor $p_{H_2O} \cong 9.3\ mmHg$. CO_2 partial pressure p_{CO_2} lower than $p_{CO_2} \lesssim 6\ mmHg$ will result in the formation of a saturated solution on the sorbent surface. It is worthy of notice that $LiOH$ is a very strong base, and its solution may cause dangerous burns. This information is very often misinterpreted as a fact of the high $LiOH$ solubility. It is true that lithium is a more alkaline metal than potassium or sodium. Comparing their solubility in Table 6.1 shows that the solubility of $LiOH$ is low as compared to the solubility of $NaOH$ and KOH. For this reason, $LiOH$ has the tendency to dust.

As it follows from research of soda lime package, the maximum absorption capacity for CO_2 occurs at the moisture contents C_{H_2O} in the absorbent approximately equal to $C_{H_2O} \cong 22.5\%$ (Figure 6.3).

No information is given about the humidity of gas mixture used for absorption. During the experiments, probably the dry mixture $5\%CO_2/O_2$ was used. The equimolar amount of water is released in a CO_2 chemisorption reaction. The high CO_2 contents in the absorbate cause an increase in medium humidity.

The increase of water vapor in the absorbate causes an increase in soda lime absorption capacity. At the same mixture flow velocities, the increase in CO_2 contents causes a decrease in duration of the absorber of the same height as soda lime.

The soda lime canister breakthrough times at the different heights of the package for different flow velocities, humidity, and different CO_2 contents in air are presented in Table 6.3.

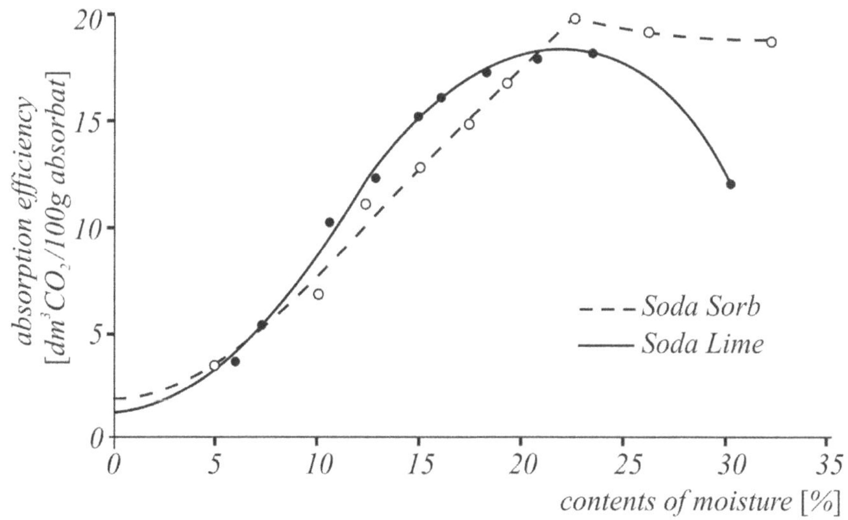

FIGURE 6.3 The effect of the initial water contents in the dry absorbent mass on the chemisorption yield at a constant rate of $5\%_n CO_2/O_2$ for two kinds of soda lime absorbent.

Source: Furevik (1972).

TABLE 6.3
Canister Breakthrough Time after Which the Increase of Carbon Dioxide Contents Exceeded 0.1$\%_v$

CO_2 Contents in Absorbate ($\%_v$)	Absorbate Flow Velocity ($dm^2 \cdot cm^{-2} \cdot min^{-1}$)	Relative Humidity (%)	Canister Breakthrough Time after Which the Increase of Carbon Dioxide Contents Exceeded 0.1$\%_v$ H (cm) 10	20	30	40	50
2	0.4	100	34	210	306	—	—
	0.8	100	12	100	210	338	—
	1.2	100	5	31	70	—	158
	1.4	100	1	16	50	112	147
	1.6	100	2	15	42	68	113
	1.4	80	—	23	45	105	148
	1.2	58	8	25	52	—	—
1	0.4	100	110	344	—	—	—
	0.8	100	24	134	282	—	—
	1.0	100	20	106	188	—	—
	1.6	100	2	31	77	—	—
	0.8	35	17	87	160	—	—
0.5	1.6	100	18	96	230	—	—

CO_2 / air mixtures were used in experiments and soda lime:

Specification	**The Value**
Appearance	White or white-gray pellets of the diameter $\phi = [2.8, 3.5] mm$
Absorbing capacity with respect to carbon dioxide	Not less than $23\%_w$ (by mass)
Absorbing capacity with respect to moisture	Not exceeding $7.5\%_w$ (by mass)
Negative mesh, square mesh dimension $2.0mm$	Not exceeding $5\%_w$ (by mass)
Moisture content	10–18% (by mass)
Drum strength	Not less than $88\%_w$ (by mass)

Source: Majkot and Ambrożek (1980).

From Table 6.3, the increase in water vapor contents in absorbate caused the increase in the absorption capacities of the soda lime, thus contributing to the extension of the absorbent breakthrough time. A decrease of breakthrough time was recorded only for the layers that absorbed water in the first sequence. It was probably due to capillary condensation – for this reason, the absorbent reactive area decreases.

The general opinion is that in hyperbaric/diving condition, CO_2 contents should be kept at a level less than $C_{CO_2} \leq [0.025, 0.5]\%_{SEV}$ of normobaric equivalent. In some stages of diving, and according to some submarine standards, 1% of normobaric equivalent CO_2 content is accepted (Kłos, 2021). In hyperbaric facilities, relative humidity should be kept within the range $\phi \in [50,90]\%_R$, depending upon the diver's comfort. *Equivalent percentage* is defined as CO_2 percentage in the mixture under normal pressure, in which the CO_2 partial pressure is the same in the hyperbaric mixture of the same composition. Approximately $1\%_R$ equivalent corresponds to partial pressure equal to $1\,kPa$.

The optimum moisture contents for the chemisorption process depends upon temperature. The experimental results obtained for three different absorbents presented in Figure 6.4 show that the absorption efficiency for $LiOH$ is less dependent on temperature than for soda lime and Baralyme.

Below the temperature of $60^\circ F$ (approximately $16^\circ C$), the chemisorption efficiency for Sodasorb and Baralyme decreases very rapidly.

$LiOH$ is more expensive than the other two absorbents. Due to $LiOH$ properties, it is the most suitable absorbent used for filling up emergency systems. The solubility of alkali metal hydroxides, excluding $Ca(OH)_2$, increases with temperature, which means that it increases with the progress of the chemical reaction caused by the exothermic effect of the reaction. Table 6.1 shows that $LiOH$ solubility is almost not

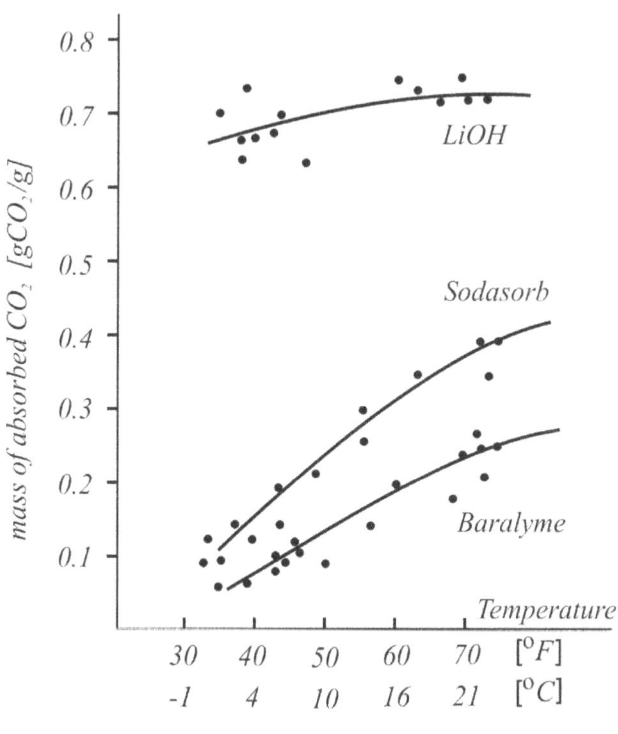

FIGURE 6.4 Comparison of the relationship between carbon dioxide chemisorption efficiency and temperature.

Source: Furevik (1972).

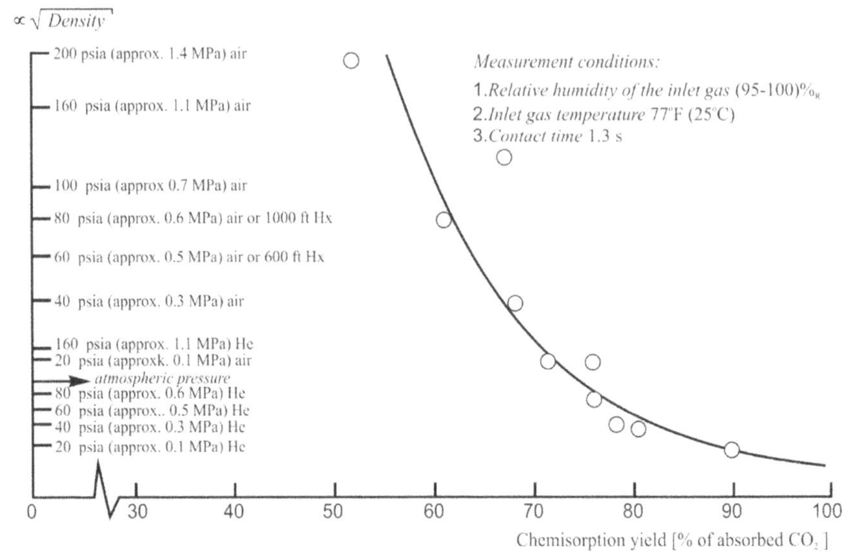

FIGURE 6.5 Effect of absorbate total pressure exerts influence on the CO_2 chemisorption efficiency of the MARK IX absorber.

Source: Furevik (1972).

temperature-dependent. Due to this, small changes in CO_2 chemisorption efficiency with temperature have been found.

As was earlier mentioned, the CO_2 contents in the breathing atmosphere should be kept below a certain level dependent upon the total pressure for soda lime, Baralime, and $LiOH$. The absorbate total pressure exerts influence on the CO_2 chemisorption efficiency. As the pressure increases, the coefficient of carbon dioxide mass transfer to the absorbent grains decreases. Also, due to the increase in absorbate density, the rate of diffusion through the pores into granule decreases (Furevik, 1972). The relationship between carbon dioxide chemisorption efficiency on Sodasorb and the density of absorbate is shown in Figure 6.5.

6.1.7 Conclusions

Soda lime is the most popular carbon dioxide CO_2 absorbent used for military applications. Especially in diving, as packing CCR and SCR canisters, life support systems use with diving complexes/chambers, regeneration of breathing gas mixture systems, and submarine technology for regeneration of submarine atmosphere in normal and emergency situation (Kłos and Kłos, 2004; Kłos, 2009; Marshall, Nuckols, Sarich, Marr and Robillard, 2015).

6.2 RESISTANCE

Carbon dioxide scrubber resistance consists of the resistance imposed by the scrubber construction and the sorbent bed resistance. Normally, the scrubber body resistance is relatively small as compared to the sorbent bed resistance. *Darcy's law* is analogous to Ohm's law in electrostatics and describes the flow of a fluid through a porous medium. *Darcy's equation* describes linearly relating the volume flow rate of the fluid to the hydraulic head difference[16] via the hydraulic conductivity; it describes the sorbent bed resistance (Bear, 2018; Li and Chen, 2021; Allen, 2021):

$$\Delta p = \lambda_f' \frac{w^2}{2\varepsilon^2} \cdot \frac{1}{d_h} \cdot \rho \tag{6.20}$$

where Δp – is fluid pressure drop through the granular bed of height l and hydraulic diameter d_h; λ_f' – is dimensionless resistance factor $\lambda_f' = f(Re)$; Re – is a Reynolds number w –average velocity of the fluid in relation to the total cross section of the scrubber; ε – is the bed porosity; ρ – is the density of the fluid; l – is the height of the bed; and d_h – is the hydraulic diameter.

The *porosity* of the granular bed ε is defined as the ratio of all free spaces V_w to the volume of the granular material V: $= \frac{V_w}{V} = \frac{V_w}{V_s + V_w}$, where V represents the total volume occupied by the granular material; V_w the total free intermolecular volume; and V_s the sum of each granular material particle volume. The hydraulic diameter d_h is defined as $d_h = 4 \cdot r_h = 4 \cdot \frac{V_w}{S}$, where r_h represents the hydraulic radius of a channel, and S is the total area of the bed. As $V_s = V - V_w = (1-\varepsilon) \cdot V$, therefore it can be written as:

$$d_h = 4 \cdot \frac{\varepsilon}{1-\varepsilon} \cdot \frac{V_s}{V} \tag{6.21}$$

where V_s – is the sum of each granular material particle volume, and V – the total volume occupied by the granular material.

Most often, the average particle diameter of the granular bed is determined according to *Sauter's definition*. According to *Sauter's definition*, the set of the granular particles of the volume V_s and area S is hypothetically replaced with the equal volume spheres. The sum of particle volume and area is the same as the particle set being discussed:

$$d_s = 6 \cdot \frac{V_s}{S} \leftarrow \begin{cases} V_s = n_p \cdot \frac{\pi}{6} \cdot d_s^3 \\ S = n_p \cdot \pi \cdot d_s^2 \end{cases} \tag{6.22}$$

where n_p – is the number of spheres, and d_s – the diameter of the set particle according to *Sauter's definition*.

As it follows from equations (6.21) and (6.22):

$$d_h = \frac{2}{3} \cdot \frac{\varepsilon}{1-\varepsilon} \cdot d_s \tag{6.23}$$

For one granulate fraction, the resistance coefficient λ_f' for granulates can be defined according to *Ergun definition*:

$$\begin{cases} \lambda_f' = \frac{4}{3} \cdot \lambda_f \\ \lambda_f = \frac{a}{Re} + 1.75 \, for \, 1 \le Re \le 3000 \\ Re = \frac{1}{1-\varepsilon} \cdot \frac{w \cdot \rho \cdot d_s}{\eta} \end{cases} \tag{6.24}$$

where a – is the dimensionless empirical coefficient (for spheres and cubes, $a = 150$) of absolute viscosity of the fluid ($Pa \cdot s$).

According to the *Blake–Kozena definition*, the resistance coefficient λ_f' for laminar flow can be written as follows:

$$\begin{cases} \lambda_f' = \frac{4}{3} \cdot \lambda_f \\ \lambda_f = \frac{a}{Re} \\ Re = \frac{1}{1-\varepsilon} \cdot \frac{w \cdot \rho \cdot d_s}{\eta} \end{cases} \tag{6.25}$$

From equations (6.20), (6.24), and (6.25), it follows that:

$$\Delta p = \frac{4}{9} \cdot \frac{a}{\varepsilon} \cdot \eta \cdot \left(\frac{l}{d_h}\right)^2 \cdot \frac{w}{l} \tag{6.26}$$

As it follows, from the stream continuity equation for the circular cross-sectional area A of the channel, $\dot{V} = A \cdot w = \frac{V}{l} \cdot w \rightarrow \frac{w}{l} = \frac{\dot{V}}{V}$, where $\dot{V}$ represents the volumetric rate, V the volume of the granular bed, and l represents the height of the granular bed. Thus, equation (6.26) can be transformed to the following equation:

$$\left(\frac{l}{d_h}\right)^2 = \frac{9}{4} \cdot \frac{\varepsilon}{a} \cdot \frac{\Delta p}{\eta} \cdot \frac{V}{\dot{V}} \tag{6.27}$$

Transformation of equation (6.27) gives

$$\frac{l}{d_h} = \frac{3}{4} \cdot \sqrt{\frac{\varepsilon}{a \cdot \eta}} \cdot \sqrt{\Delta p \cdot \frac{V}{\dot{V}}} \tag{6.28}$$

According to the kinetic theory of gases, it can be written as $\eta = f(T)$, $\eta \neq f(p)$ (Pasternack, 1985), and the simplification of equation (6.28) gives

$$\frac{l}{d_h} = const \cdot \sqrt{\Delta p \cdot \frac{V}{\dot{V}}} \tag{6.29}$$

Assuming that $1\,dm^3$ of soda lime can bind approximately $90\,dm^3$ of CO_2, and using equation (6.28), it is transformed to the following form:

$$\frac{l}{d_h} = const \cdot \sqrt{\Delta p \cdot \frac{V_{CO_2}}{90}} \tag{6.30}$$

where V_{CO_2} – is the total volume of carbon dioxide that has to be absorbed dm^3.

6.2.1 Conclusions

Equation (6.30) is interesting because of the joining of the two choice criterion of the optimum scrubber design: scrubber resistance Δp and absorption capacity of V_{CO_2}.

The theoretical diameter of the circular scrubber cross section, the scrubber diameter, and the height of the sorbent bed can be calculated from the following equations:

$$\begin{cases} l = \dfrac{\varepsilon}{1-\varepsilon} \cdot d_s \cdot \sqrt{\dfrac{\varepsilon}{a \cdot \eta}} \cdot \sqrt{\dfrac{\Delta p}{\dot{V}} \cdot \dfrac{V_{CO_2}}{90}} \\ V = \dfrac{V_{CO_2}}{90} = \dfrac{\pi \cdot d_h^2}{4} \cdot l \rightarrow d_h = \dfrac{2}{\sqrt{\pi}} \cdot \sqrt{\dfrac{V_{CO_2}}{90 \cdot l}} \end{cases} \tag{6.30}$$

Equation set (6.30) can be the basis for an introductory design of the carbon dioxide absorber. It should be remembered that the approximate data and mathematical model enable only approximate calculation of absorber parameters. Such calculations are very helpful in designing *SCRs*; however, the results obtained should be experimentally verified by manned and unmanned tests.

6.3 PROPERTIES

As has been earlier proven, porosity ε, bulk density ρ, and hydraulic diameter d_h are the most important in designing a carbon dioxide scrubber.

6.3.1 Porosity

The free volume V_w was determined by flooding the bed of porous material with water. The height of the bed and the diameter should be similar to the ratio as in the carbon dioxide scrubber. As follows from the experiments, the water volume necessary for flooding the bed toward the equation $\varepsilon = \frac{V_w}{V} = \frac{V_w}{V_s + V_w}$, where V represents the total volume occupied by the granular material, V_w the total free intermolecular volume, and V_s the sum of each granular material particle volume, can give porosity ε.

6.3.2 Density

Bulk density was measured by weighing in the known air volume of the bed. In many cases, buoyancy correction cannot be taken into account because its value is less than the mass measuring error. The calculation of the buoyancy correction is presented in Table 6.4.

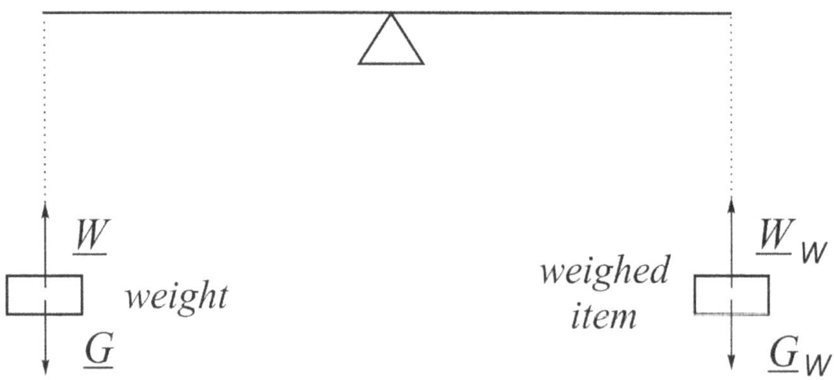

TABLE 6.4
Calculation of the Buoyancy Correction for Weighing in Air

vector
$$\underline{G} - \underline{W} = \underline{G}_w - \underline{W}_w$$

scalar
$$m \cdot g - \rho_{air} \cdot \frac{m}{\rho} \cdot g = m_w \cdot g - \rho_{air} \cdot V \cdot g \rightarrow m_w = m \cdot \left(1 - \frac{\rho_{air}}{\rho_{brass}}\right) + \rho_{air} \cdot V$$

where m – is weight mass; m_w – is mass of the object being weighed; g – is gravitational acceleration; $\underline{G}$ – is weight terrestrial gravity force; $\underline{G}_w$ – is terrestrial gravity force of the object being weighed; V – is the volume of the object being weighted; $\underline{W}$ – is the Archimedes force acting on the weight; $\underline{W}_w$ – is the Archimedes force acting on the subject being weighed; ρ_{brass} – is the density of the weight (for brass: ca. $8400\,kg \cdot m^{-3}$); and ρ_{air} – is air density (ca. $1.293 kg \cdot m^{-3}$).

The error Δm concerning the buoyancy correction neglected at air weighing $V = 1\,dm^3$ weighed material, for typical density of the weighed material $\rho \cong 750\,kg \cdot m^{-3}$, is:

$$\Delta m = m_w - m = \rho_{air} \cdot V \cdot \left(1 - \frac{\rho}{\rho_{brass}}\right) \cong 1.293 g \cdot dm^{-3} \cdot 1 dm^3 \cdot \left(1 - \frac{750 kg\, m^{-3}}{8400\, kg\, m^{-3}}\right) \cong 0.1 g$$

6.3.3 Diameter

Determination of the approximate mean porous material hydraulic diameter can be attained by direct measuring. According to *Sauter's definition*, the mean hydraulic diameter d_h of soda lime can be determined from equation (6.22).

6.3.4 Conclusions

Porosity ε, bulk density ρ, and hydraulic diameter d_h are of the most important properties of a porous material used to design carbon dioxide CO_2 scrubbers for SCRs.

NOTES

1 For example, monoethanolamine $HO-CH_2-CH_2-NH_2$.
2 For example, lithium hydroxide $Li(OH)_2$.
3 For example, with the use of potassium superoxide KO_2.
4 Generation of oxygen excess occurs.
5 Soda Sorb, Sodasorb, Drägersorb, soda lime.
6 Lithium hydroxide is a highly hygroscopic compound.
7 Gas particle size is the measure used for adsorbent surface determination.
8 Capillary condensation.
9 The kind of absorbent, the use of catalyst, moisture contents in the absorbent and absorbate, etc.
10 Porosity, granule structure, etc.
11 From the manufacturer to the storage, from the storage to the user, as the emergency absorbent into regeneration systems, etc.
12 Dropping an absorbent in the scrubber causes the formation of free ducts and relatively easy absorbate flow through them, compared to the rest of the bed, and can cause insufficient removal of carbon dioxide.
13 Discoloration from blue to colorless at pH $pH \in [9.3, 10.5]$
14 Changes color from red to blue at pH ranging $pH \in [10.1, 11.1]$
15 Between solid lines.
16 Which is often just proportional to the pressure difference.

REFERENCES

Allen M.B. 2021. *The mathematics of fluid flow through porous media.* Hoboken: John Wiley & Sons, Inc. ISBN: 9781119663843.

Bear J. 2018. *Theory and applications of transport in porous media.* Cham: Springer International Publishing. ISBN: 978-3-319-72825-4.

Furevik M. 1972. *The reaction principle of the CO2 absorption by alkali—metal hydroxides and superoxides and important factors relative to the design of a hyperbaric CO2 scrubber.* Laksevåg: Norwegian Underwater Institute. NUI Report No. 2-78.

Kłos I., Kłos R. 2004. *Polish soda lime in military applications.* Oświęcim: Chemical Company DWORY S.A. ISBN: 83-920272-0-5.

Kłos R. 2009. *Wapno sodowane w zastosowaniach wojskowych.* Gdynia: Polskie Towarzystwo Medycyny i Techniki Hiperbarycznej. ISBN: 978-83-924989-5-7.

Kłos R. 2021. *Ventilation of normobaric and hyperbaric objects.* Boca Raton, FL: CRC Press (Taylor & Francis Group, LLC), 2021. ISBN: 978-0-367-67523-3 (hbk) ISBN: 978-0-367-67524-0 (pbk) ISBN: 978-1-003-13164-9 (ebk).

Li D., Chen J. 2021. *Mechanics of oil and gas flow in porous media.* Singapore: Petroleum Industry Press and Springer Nature Singapore Pte Ltd. ISBN: 978-981-15-7312-5.

Majkot A., Ambrożek B. 1980. *Badania adsorpcji dwutlenku węgla*. Szczecin: Stocznia im. A. Warskiego. Praca wykonana na zlecenie Biura Projektowo-Konstrukcyjnego.

Marshall L., Nuckols M.L., Sarich A.J., Marr W.J., Robillard D.J. 2015. *Life support systems design: diving and hyperbaric applications*. Annapolis: Ace Sarich Publishing. ISBN: 978-0692604021.

Nuckols M.L., Sarich A.J., Tucker W.C. 1996. *Life support systems design: diving and hyperbaric applications*. Needham Heights: Simon & Schuster Custom Publishing. ISBN-10: 0-536-59616-6; ISBN-13: 978–0536596161.

Pasternack A. 1985. Utilisation of oxygen self-rescues in positive pressure areas. *Dräger Review*. 56, 3–6.

7 Resistance

Breathing resistance and breathing work are the basic parameters that enable evaluating *UBA*s. Using chest muscles to breathe, a human being can produce only a slight negative pressure. At the beginning of respiration, negative pressure Δp_{in} in lungs is maintained within the limit $\Delta p_{in} \in [0.1, 0.3]\,kPa$. Producing higher negative pressure causes muscle fatigue and can be maintained only for a short time. The approximate maximum positive pressure that can be produced by a human being is equal to $20\,kPa$ (Solomon, Martin, Martin and Berg, 2015).

Normally, the difference in pressure between the middle of the lungs and the nasopharynx cavity at standing position in air is, on average, $6\,Pa$ ($\sim 0.6mm\ H_2O$) (Figure 7.1).

In water, this difference is almost 1,000 times higher than in air.

7.1 LOOP

The breathing resistance magnitudes imposed by the construction of the *SCR* are given in Table 7.1.

There are two groups of resistance: the total resistance of the straight ducts and the total local resistance.

Bernoulli's equation, derived from the energy conservation equation for cross sections shown in Figure 7.2, can be written as follows:

$$\frac{w_1^2}{2 \cdot g} + \frac{p_1}{\rho \cdot g} + h_1 = \frac{w_2^2}{2 \cdot g} + \frac{p_2}{\rho \cdot g} + h_2 + \Delta h_r \tag{7.1}$$

where w – is the medium linear gas velocity; p – the static pressure; g – the gravitational acceleration; ρ – the gas density; Δh_r the total flow resistance; and h_i – the height of intersection i, as marked in Figure 7.2.

The total flow resistance can be written as:

$$\Delta h_r = \Delta h_d + \Delta h_l \tag{7.2}$$

where Δh_d – is the straight ducts resistance, and Δh_l – the local resistance.

For laminar flow and circular cross section:

$$\Delta h_d = \lambda \cdot \frac{l}{D} \cdot \frac{w^2}{2 \cdot g} \tag{7.3}$$

where λ – is the resistance coefficient; Re – the Reynolds number; D – the diameter of the duct; and l – the length of duct.

DOI: 10.1201/9781032649207-7

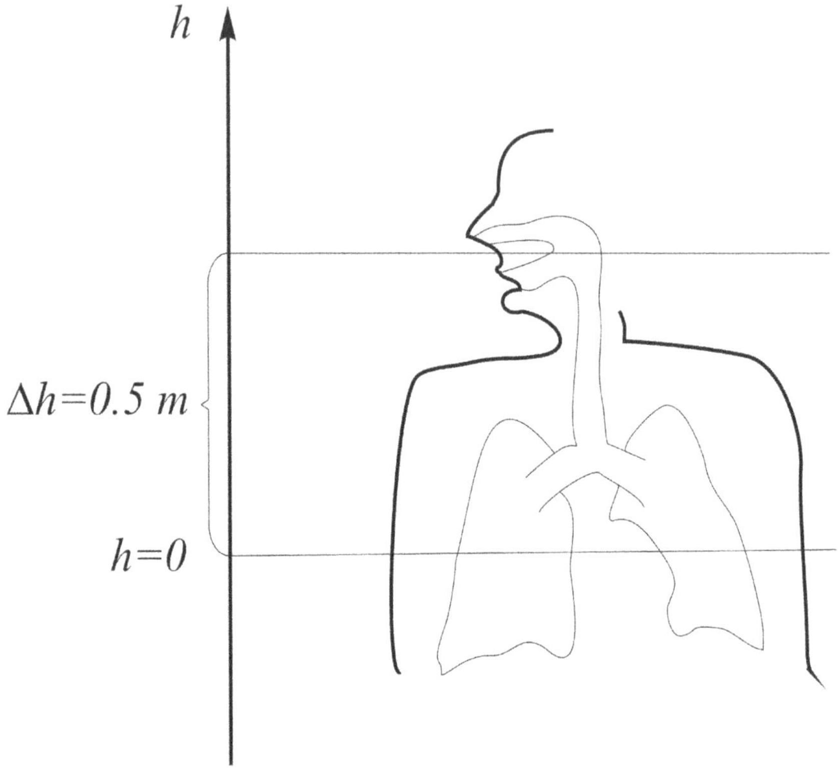

FIGURE 7.1 Breathing resistance with regard to the difference in level of lung middle and nasopharynx cavity in air at sea level, where $\Delta p_{air} = p_0 \cdot \exp\left(-\frac{M \cdot g \cdot \Delta H}{R \cdot T}\right)$ $M \cong 28\, g \cdot mol^{-1}$ $T \cong 273 K$ $R \cong 8.31 J \cdot mol^{-1} \cdot K^{-1}$ $g \cong 9.81\, m \cdot s^{-2}$ $p_0 \cong 101325 Pa$ $\Delta p_{air} \cong 0.6\, mmH_2O$ $\Delta p_{H_2O} \cong 500\, mmH_2O$.

TABLE 7.1
Kinds of Respiratory Resistance in the *SCR*

Inspiration Resistance	Expiration Resistance
• Respiratory resistance connected with geometrical shape of the mouthpiece assembly	• Respiratory resistance connected with geometrical shape of the mouthpiece assembly
• Inspiration valve resistance	• Expiration valve resistance
• Inspiration hose resistance	• Expiration hose resistance
• Resistance connected with flexibility of the inspiration bags	• Resistance connected with flexibility of the inspiration bags
	• Scrubber resistance
	• Relief valve resistance

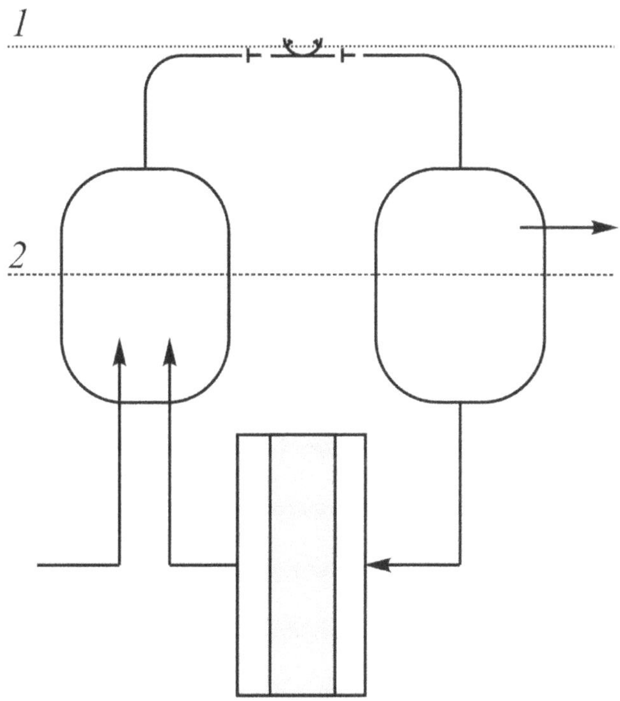

FIGURE 7.2 The cross sections of the diving apparatus for which *Bernoulli's equation* (7.1) was written.

The total local resistance can be written as follows:

$$\Delta h_l = \xi \cdot \frac{w^2}{2 \cdot g} \tag{7.4}$$

where ξ – is the dimensionless local resistance coefficient.

Assuming that, according to Figure 7.2, for the swimming position, $w_1 \gg w_2$ and $h_1 \cong h_2$, equation (7.1) can be written with the use of equations (7.2) ÷ (7.4):

$$\frac{2 \cdot (p_1 - p_2)}{\rho} = \frac{2 \cdot \Delta p}{\rho} = w^2 + \Delta h_r = w^2 \cdot \left(1 + \lambda \cdot \frac{l}{D} \cdot \frac{1}{2 \cdot g} + \xi \cdot \frac{1}{2 \cdot g}\right) \tag{7.5}$$

where w – is the mean linear gas velocity in cross section 1 – (Figure 7.2).

From (7.5) the following can be written: $\Delta p \triangleq w^2 \cdot \rho$. As it follows from the stream continuity law, $\dot{V} = w \cdot A$, where A represents, vertical to the stream, the cross section of gas flow, and $\dot{V}$ is gas stream rate. Finally, equation (7.5) can be written as follows:

$$\Delta p \triangleq \dot{V}^2 \cdot \rho \tag{7.6}$$

where $\dot{V}$ – is lung ventilation, for example, respiratory stream rate.

Two conclusions follow from equation (7.6):

- The respiratory resistance increases $\Delta p \nearrow$ proportionally to the square of the inspired breathing gas stream rate $\dot{V} \nearrow$.
- The respiratory resistance increases $\Delta p \nearrow$ with the increase in gas density $\rho \nearrow$.

As follows from the preceding conclusions, increased workload $W \nearrow$ causes an increase in lung ventilation $\dot{V} \nearrow$ and remarkable breathing resistance $\Delta p \nearrow$ with the square exponent. An increase in breathing resistance exerts an additional effect on ventilation increase. It causes dyspnea.

As follows from the second conclusion, as the diving depth increases $H \nearrow$, breathing resistance increases as well $\Delta p \nearrow$. In order to decrease breathing resistance $\Delta p \searrow$ of the SCR, and to avoid nitrogen narcosis, replacement of nitrogen with the light gas component helium is, for example, applied. Finally, it can be stated that when SCR prototypes are tested, the breathing resistance at the maximum diving depth H_{max} and working load W of the diver should also be tested.

7.2 THE RELIEF VALVE

Except for the scrubber construction, the relief valve setup exerts a great effect on the breathing resistance and stability of the SCR. The basic functions of the relief valve are as follows:

- To maintain the requested filling of the breathing bags
- To maintain the requested overpressure in the breathing space
- To protect the breathing space as safety valve against damage

The analysis of the relief valve functioning is based on the model shown in Figure 7.3.

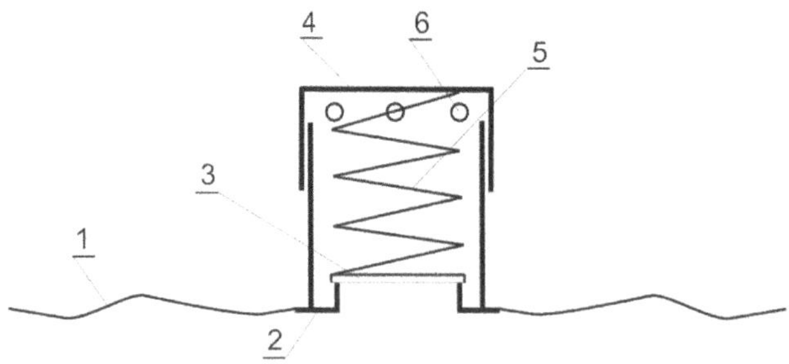

FIGURE 7.3 The scheme of the relief valve: (1) the breathing bag; (2) the valve housing; (3) the valve plate; (4) the mobile housing, for example, screw-in; (5) the pressure spring; (6) outlet holes.

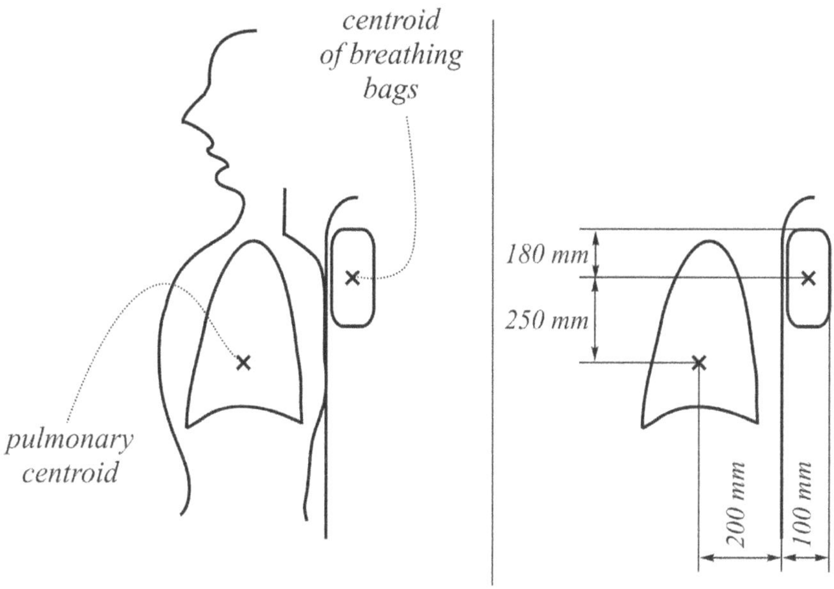

FIGURE 7.4 The anthropometric projection; dimensions needed for relief valve work analysis are given in centimeters.

Anthropometry projection is shown in Figure 7.4, with the following assumptions:

- The breathing bag is perfectly flexible.
- There is no additional resistance of the relief valve except the elasticity force of the pressure spring. It is possible to regulate the elasticity force by screwing in and screwing out the mobile relief valve housing (Figure 7.3).
- The elasticity of the compression spring of the relief valve is set up in such position that for swimming horizontal position, the breathing gas pressure in the breathing bag is the same as the pressure in the middle of the lungs.

In order to analyze the breathing resistance, according to the assumptions made, it is necessary to do thus:

- To calculate the pressure force of the relief valve spring that will exert pressure p_s on its plate necessary to maintain the pressure in the breathing space of the *SCR*, the same as the pressure in the middle of the lungs for the swimming position
- To calculate the gas overpressure in the breathing bag, with respect to the pressure in the middle of the lungs, for the other positions than the swimming one

An example of the calculation of the pressure exerted by the relief valve spring pressure p_s is shown in Figure 7.5.

According to the assumptions, the resistance of the breathing space is neglected. Figure 7.6 shows an example of the calculation of the gas overpressure related to the pressure in the middle of the diver's lungs in vertical position.

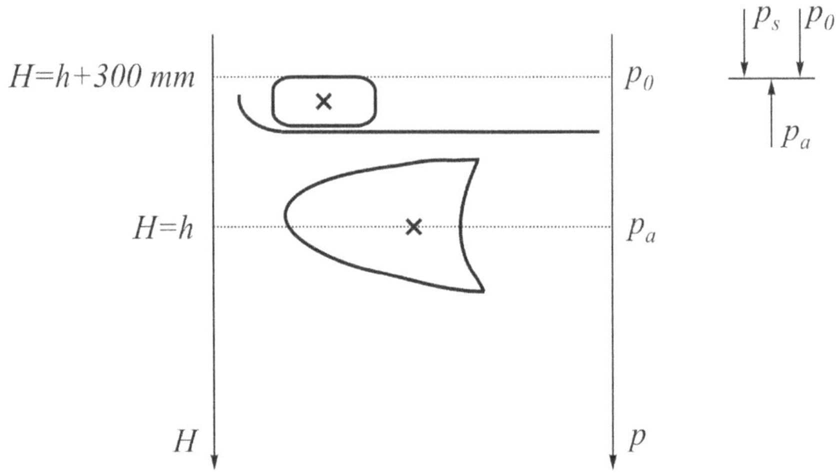

FIGURE 7.5 Diagram illustrating the calculation procedure of the relief valve spring pressure p_s set point for the swimming position of the diver: ↓ – arrow indicates relief valve position; p_s – is the pressure exerted by the relief valve spring; p_0 –is the hydrostatic pressure at the relief valve level; p_a –is the hydrostatic pressure at the diver's middle lung level; p – is hydrostatic pressure; and H – is depth $p_a = p_s + p_0 \rightarrow p_s = p_0 - (p_0 - 30cm\ H_2O) = 30cm\ H_2O$.

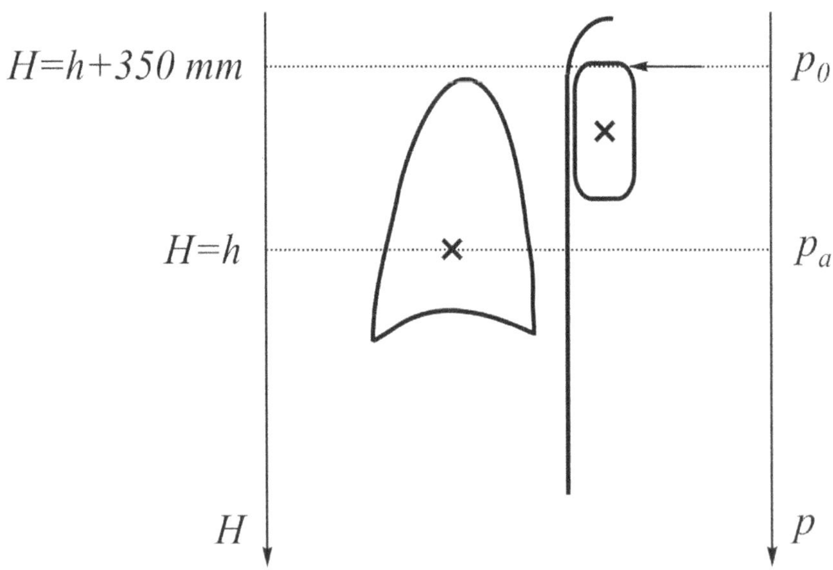

FIGURE 7.6 Diagram showing an example of calculation of the gas overpressure related to the pressure in the interior of the diver's lungs in vertical position: $\Delta p = p_0 + p_s - p_a$ $\Delta p = p_a - 35cmH_2O + 30cmH_2O - p_a = -5cmH_2O \,|\, p_s = 30cmH_2O$.

Table 7.2 shows the gas overpressure in the breathing space calculated for the different places of the relief valve assembly and the different positions of the diver. If the values are positive, the breathing gas will be pushed into the lungs at overpressure, whereas the negative value shows that inspiration is made difficult. As follows from the analysis of the values given in Table 7.2, the theoretical gas overpressure Δp in the breathing space is related to the pressure in the centroid of the diver's lungs calculated for different places of the relief valve assembly and the different positions of the diver and the dimensions in accordance with Figure 7.4

The head-downward position is most frequent during the ascent of the diver. In this case, as a rule, breathing gas deficit occurs in the breathing space. This occurs particularly in small breathing spaces and when gas doses are small. As a result, the gas must be complemented by means of a bypass valve. The situation presented in Table 7.2 deals with work done by the diver in the head-downward position at a constant depth.

The head-upward position is most frequent during the descent of the diver. Due to the risk of lung injury, the breathing gas overpressure is not recommended in this case. The overpressure values given in Table 7.2 are not high, but in real conditions, this overpressure in the breathing space is higher as compared to the calculated one. This is due to the resistance of the relief valve and a rapid increase in gas volume in the breathing space, particularly at small depths.

TABLE 7.2
Position 5 Is the Best Place for the Relief Valve Assembly

No.	Relief valve position	Pressure exerted by the spring *(mmH_20)*	The gas overpressure Δp in the breathing space related to the pressure in the centroid of the diver's lungs for different positions of the diver *(mmH_20)*			
					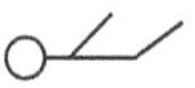	
1	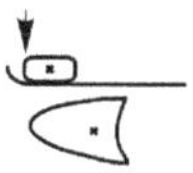	300	0	–50	600	650
2	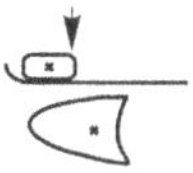	300	0	150	600	450
3	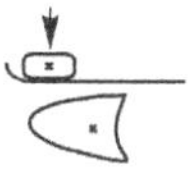	300	0	50	600	550
4	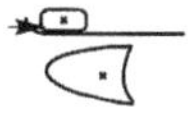	200	0	–150	400	550
5		200	0	50	400	350

The diver works in the back position under the hull of a ship. As it follows from Table 7.2, during this kind of work, high overpressure of the breathing gas always occurs. This unfavorable effect can be decreased by the use of a weight that discharges the valve compression spring pressure. The principle of the discharging weight operation is shown in Figure 7.7.

Theoretically, the highest efficiency of the discharging weight performance will be recorded when the weight exerts pressure equal to the pressure of the compression spring. Practically, this pressure should be twice less due to the necessity of maintaining the breathing gas in the breathing space: $p_s / p_w \geq 2$.

The use of the discharging weight would decrease the overpressure from $400\,mmH_2O$ to $300\,mmH_2O$ in the case of construction 4 presented in Table 7.2 for the back position of the diver.

7.3 STABILITY

A significant local buoyancy caused by Archimedes force has an effect on *SCRs*. This effect is a result of the large breathing space volume of the *SCR* as compared to the open circuit *UBA*. In addition, torque can be observed if there is a shift between the center of buoyancy and the center of gravity appearing in the SCR. It makes it difficult to keep a vertical swimming position. The bending moment must be compensated by means of local weights. An example of calculation of weight needed to compensate local buoyancy is showed in Figure 7.8.

In order to avoid displacing the breathing bag, the static equilibrium of forces shown in Figure 7.8 must exist:

$$\left.\begin{aligned} F_2 &= F_p \\ F_1 + R_x &- F_p \end{aligned}\right\} \Rightarrow R_x = F_2 - F_1 \tag{7.7}$$

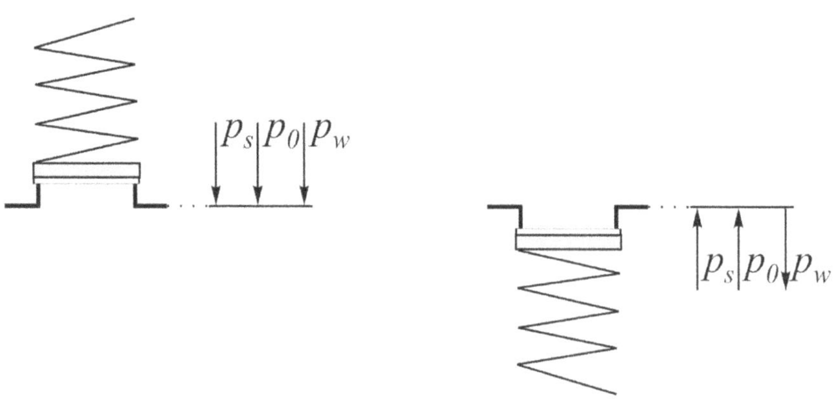

FIGURE 7.7 Pressure p_w exerted on the valve plate by the relieving load depending on the valve position.

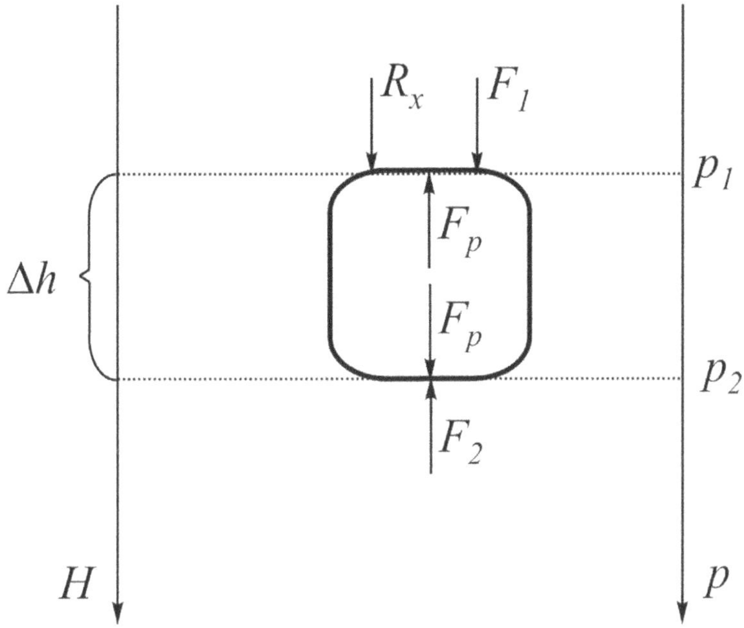

FIGURE 7.8 Distribution of forces having an effect on the breathing bag in water: R_x – is the reaction force of an *SCR* upper cover; F_p – is the force connected with pressure p inside the respiratory bag; H – is depth; Δh – is the height of the respiratory bag; F_i – is the force connected with pressure corresponding to depths for level i.

where R_x is the reaction force of an *SCR* upper cover; F_p – is force connected with pressure p inside the respiratory bag; and F_i – is the force connected with pressure corresponding to depths for the level i.

Let the surface area of the breathing bag be S. As the areas of the upper and lower breathing bag are the same, the following can be written:

$$\begin{cases} F_1 = p_1 \cdot S \\ F_2 = \left(p_1 + \dfrac{dp}{dH} \cdot \Delta h \right) \cdot S \end{cases} \tag{7.8}$$

where H – is the depth; Δh – the height of the respiratory bag; and p_i – the pressure connected with depths for the level i.

From (7.7) and (7.8), it follows that $R_x = \left(p_1 + \dfrac{dp}{dH} \cdot \Delta h \right) \cdot S - p_1 \cdot S = \dfrac{dp}{dH} \cdot \Delta h \cdot S = \dfrac{dp}{dH} \cdot V$, where V represents the breathing bag volume. As water in these circumstances is not compressible, it can therefore be written as: $\dfrac{dp}{dH} = const \cong 10^4 \ Pa \cdot m^{-1}$.

As it can be assumed that for terrestrial circumstances, force $1kG$ corresponds to mass $1kg \cong 10 \cdot 1kG$, therefore thus can be written:

$$R_x \cong 10^3 \cdot V \Big| [Rx] = kg \wedge [V] = m^3 \tag{7.9}$$

To compensate the buoyant force, it is necessary to use the mass m expressed in having the value equal to the volume V of the breathing space given in $V = [dm^3]$. The mass should also include the mass of the buoyant element in question. The mass of the compressed breathing gas is another significant factor that affects variations in buoyancy of the UBA. During SCR diving, the mass of the gas contained in flasks decreases due to the loss of gas escaping to water. For an SCR, a change in mass is less than for an open circuit UBA. For example, the filled-up open-air circuit diving apparatus type $PR-27\,SCUBA$ has negative buoyancy, while the empty apparatus has positive buoyancy, enough to serve as a kickboard. The diver using the SCR should take into account this fact and make sure the sufficient ballast for diving is applied. The mass of gas in cylinders can be calculated as follows:

$$m = \frac{p}{p_0} \cdot V_c \cdot \rho(T) \Big| T = idem \tag{7.10}$$

where m – is the mass of gas contained in the gas cylinders; p – the pressure of the contained gas; p_0 – the atmospheric pressure; T – the temperature; V_c – the cylinder volume; and $\rho(T)$ – the gas density at normal pressure and cylinder temperature T.

7.4 METERING

As regards breathing gas consumption, an SCR is more economical as compared to the open circuit UBA. The nitrox SCRs are used to a diving depth of $30\,mH_2O$ and metering $\dot{V}_0$ of the breathing gas maintained at a level of $\dot{V}_0 \in [7,\ 8]\,dm^3 \cdot min^{-1}$ at NTP conditions, independently of the effort taken by the diver. During open circuit air UBA diving at a depth of $30\,mH_2O$, the ventilation of the lungs for medium effort taken by the diver approximately equals $\dot{V} \in [25,\ 30]\,dm^3 \cdot min^{-1}$ (Table 4.15). Thus, gas consumption is about $\dot{V}_0 \sim 120\ dm^3 \cdot min^{-1}$at NTP conditions. As follows from the preceding text, at a diving depth of $30\,mH_2O$, the breathing gas consumption is 15 times higher for an open circuit UBA than for an SCR.

As was earlier mentioned, in the SCR with constant metering of premix, the stream of the mass of fresh breathing gas must be maintained within the precisely determined limits, which has a major impact on divers' safety. The stream of the mass must be constant and independent of the depth of diving and other diving conditions. This explains why these constructions are called constant *premix* metering $SCRs$. The composition of the fresh breathing gas, the depth of diving, and the reduced pressure exert influence on stream of the premix mass.

It was experimentally confirmed that after exceeding a certain limit of the pressure ratio between the inlet p_1 and outlet p_2 of the nozzle, despite pressure variation at the outlet[1] p_2, the stream of the mass $\dot{m}$ of the flowing gas mixture remains at a constant level $\dot{m} = idem$. In other words, variations in back pressure $\downarrow p_2 \uparrow$ for pressure ratios at the inlet and outlet $\beta = \frac{p_2}{p_1}$ falling below the critical limit value $\beta < \beta_{kr}$ will not cause a change in the stream of the mass $\dot{m} = idem$ of the flowing gas mix.

This serves as the basis for calculating the capacity of the nozzles operating in supercritical flows. A properly selected set consisting of a pressure regulator and a dosing nozzle ensures a constant stream of a gas mix mass $\dot{m} = idem$ is supplied. By developing the required mathematical model, we may obtain a differential form of Bernoulli's equation.

Bernoulli's equation is derived from the principle of energy conservation. Assuming that there is no internal friction,[10] we may write down the differential energy balance per 1 *mole* of gas:

$$de = de_k + de_p + de_o + de = q_{el} - w_{el} \tag{7.11}$$

where $e-$ is the total molar energy of the system; e_k – the molar kinetic energy of the system; e_p – the potential molar energy of the system; e_o – the volumetric molar energy of the system; q_{el} – the molar energy exchanged as heat between the system and the environment; and w_{el} – the molar energy exchanged as work, with the exclusion of volumetric work, between the system and the environment

Equation (7.11) utilizes the *el* symbol instead of the complete differential symbol to define small changes in molar energies exchanged as heat q_{el} and nonvolumetric work w_{el}. Such a procedure is used in order to distinguish a state function[2] from other partial functions which do not explicitly describe the system's state. Heat and work[3] do not usually constitute explicit state functions.[4]

Particular energy parts occurring in balance (7.11) may be expressed as $de_k = d\left(\frac{v^2}{2}\right)$, $de_p = g \cdot dh$, $du = di - d(p \cdot \upsilon)$, $q_{el} = du + p \cdot d\upsilon$, and $de_o = d(p \cdot \upsilon)$, where v represents gas flow velocity, g is gravity, h represents height from the surface of the Earth, i is molar enthalpy, p is pressure, υ is molar volume, and u is internal molar energy of the system. Inserting these equations into balance (7.11) and conducting proper transformations, we receive the differential, molar form of Bernoulli's equation:

$$d\left(\frac{v_2}{2}\right) + g \cdot dh + \upsilon \cdot dp = -w_{el} \tag{7.12}$$

For gas which does not perform volumetric work $w_{el} = 0$ and flows through the horizontal pipe $dh = 0$ from section 1 through the flange with aperture area A_0 into section 2, Bernoulli's equation (7.12) will take the following form (Figure 7.9):

$$\int_{v_0}^{0} d\left(\frac{v^2}{2}\right) = -\int_{p_1}^{p_2} \upsilon \cdot dp \rightarrow \frac{v_0^2}{2} = \int_{p_1}^{p_2} \upsilon \cdot dp \tag{7.13}$$

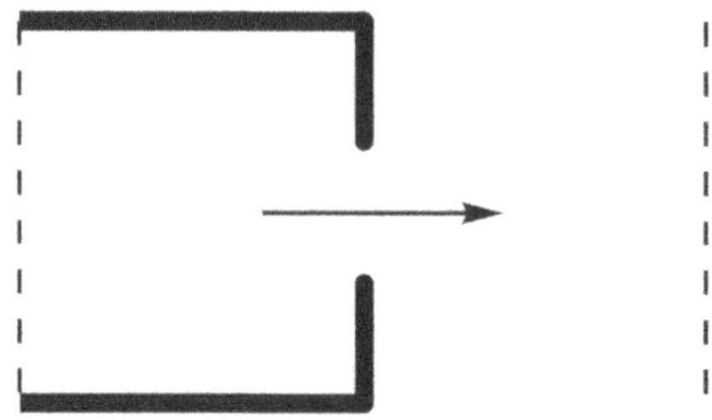

Cross Sections	1	0	2
Preassure	p_1	p_0	p_2
Density	ρ_1		ρ_2
Speed		$v_0=f(\rho)$	$v_2\cong 0$
Temperature	T_1		T_2
Area		A_0	

FIGURE 7.9 The conditions of a gas flow through a horizontal pipe, with a flange of aperture area equal to A_0.

Using the graphic interpretation of an integral illustrated in Figure 7.10, we may write the following:

$$\int_{p_1}^{p_2} \upsilon \cdot dp = \int_{\upsilon_1}^{\upsilon_2} p \cdot d\upsilon - Pole(p_2, 2, \upsilon_2, 0) + Pole(p_1, 1, \upsilon_1, 0)$$
$$= \int_{\upsilon_1}^{\upsilon_2} p \cdot d\upsilon - p_2 \cdot \upsilon_2 + p_1 \cdot \upsilon_1 \tag{7.14}$$

Using the first law of thermodynamics[5] for an isolated system, we may write $p \cdot d\upsilon = c_\upsilon \cdot dT$, and hence, for integral: $\int_{\upsilon_1}^{\upsilon_2} p \cdot d\upsilon = -\int_{T_1}^{T_2} c_\upsilon \cdot dT \equiv c_\upsilon \cdot (T_1 - T_2)$, where c_υ represents the molar thermal volume for isochoric transformation.

By marking the molar thermal volume for isobaric transformation as c_p and using Clapeyron's equation $p \cdot v = R \cdot T$ and general thermodynamic relationships for ideal gas $\frac{c_p}{c_\upsilon} = \kappa \wedge c_p - c_\upsilon = R$, we may write that $c_p - c_\upsilon = R = \frac{p_1 \cdot v_1}{T_1}$, and thus: $c_\upsilon = \frac{p_1 \cdot \upsilon_1}{(\kappa - 1) \cdot T_1}$.

Next, by inserting the last equation into integral $\int_{\upsilon_1}^{\upsilon_2} p \cdot d\upsilon = c_\upsilon \cdot (T_1 - T_2)$, the following could be written: $\int_{\upsilon_1}^{\upsilon_2} p \cdot d\upsilon = \frac{p_1 \cdot \upsilon_1}{(\kappa - 1) \cdot T_1} \cdot (T_1 - T_2)$,. And after transformation, we receive: $\int_{\upsilon_1}^{\upsilon_2} p \cdot d\upsilon = \frac{p_1 \cdot \upsilon_1}{(\kappa - 1) \cdot} \cdot \left(1 - \frac{T_2}{T_1}\right)$.

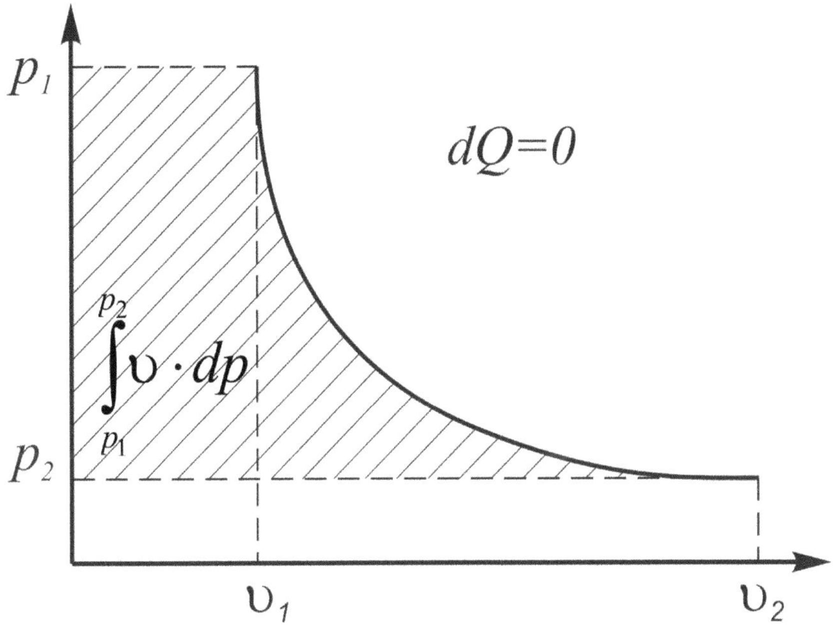

FIGURE 7.10 The graphic interpretation of integral $\int_{p_1}^{p_2} \upsilon \cdot dp$ with coordinates (p,υ) for adiabatic transformation $dQ=0$.

Using Clapeyron's equation $p_1 \cdot v_1 = R \cdot T_1$ and the Poisson formula[6] $p_1 \cdot v_1^{\kappa} = p_2 \cdot v_2^{\kappa}$, it may be written as:

$$\frac{T_2}{T_1} = \frac{p_2 \cdot \upsilon_2}{p_1 \cdot \upsilon_1} = \frac{p_2}{p_1} \cdot \left(\frac{p_1}{p_2}\right)^{1/\kappa} = \left(\frac{p_2}{p_1}\right)^{\kappa - 1/\kappa} \tag{7.15}$$

If equation (7.15) is inserted into integral equation $\int_{\upsilon_1}^{\upsilon_2} p \cdot d\upsilon = \frac{p_1 \cdot \upsilon_1}{(\kappa - 1) \cdot} \cdot \left(1 - \frac{T_2}{T_1}\right)$, the following will be obtained:

$$\int_{\upsilon_1}^{\upsilon_2} p \cdot d\upsilon = \frac{p_1 \cdot \upsilon_1}{\kappa - 1} \cdot \left[1 - \left(\frac{p_2}{p_1}\right)^{\kappa - 1/\kappa}\right] \tag{7.16}$$

By inserting equation (7.16) into equation (7.14), the following will be written:

$$\int_{p_1}^{p_2} \upsilon \cdot dp = p_1 \cdot \upsilon_1 \cdot \frac{1}{\kappa - 1} \cdot \left[1 - \left(\frac{p_2}{p_1}\right)^{\kappa - 1/\kappa}\right] + p_1 \cdot \upsilon_1 - p_2 \cdot \upsilon_2 \tag{7.17}$$

Again, by using equation (7.15) $\frac{T_2}{T_1} = \frac{p_2 \cdot v_2}{p_1 \cdot v_1} = \frac{p_2}{p_1} \cdot \left(\frac{p_1}{p_2}\right)^{1/\kappa} = \left(\frac{p_2}{p_1}\right)^{\kappa-1/\kappa}$, we may write $p_2 \cdot v_2 = p_1 \cdot v_1 \cdot \left(\frac{p_2}{p_1}\right)^{\kappa-1/\kappa}$. By introducing products $p_1 \cdot v_1$ and $p_2 \cdot v_2$ into bracket equation (7.17), it can be transformed into the following form: $\int_{p_1}^{p_2} v \cdot dp = \frac{p_1 \cdot v_1}{\kappa - 1} \cdot \left[1 - \left(\frac{p_2}{p_1}\right)^{\kappa-1/\kappa} + \kappa - 1 - \left((\frac{p_2}{p_1})^{\kappa-1/\kappa} \cdot (\kappa - 1) \right)\right]$. If adiabatic index is bridged κ before the bracket, it may be written as: $\int_{p_1}^{p_2} v \cdot dp = p_1 \cdot v_1 \cdot \frac{\kappa}{\kappa - 1} \cdot \left[-\frac{1}{\kappa} \cdot \left(\frac{p_2}{p_1}\right)^{\kappa-1/\kappa} + 1 - \left(\frac{p_2}{p_1}\right)' + \frac{1}{\kappa} \cdot \left(\frac{p_2}{p_1})^{\kappa-1/\kappa} \right)\right]$. And upon cancellation, the following will be received[7]:

$$\int_{p_1}^{p_2} v \cdot dp = p_1 \cdot v_1 \cdot \frac{\kappa}{\kappa - 1} \cdot \left[1 - \left(\frac{p_2}{p_1}\right)^{\kappa-1/\kappa}\right] \tag{7.18}$$

By inserting equation (8) into (3), $\frac{v_0^2}{2} = \int_{p_1}^{p_2} v \cdot dp$, the following can be written: $\frac{v_0^2}{2} = \int_{p_1}^{p_2} v \cdot dp = p_1 \cdot v_1 \cdot \frac{\kappa}{\kappa - 1} \cdot \left[1 - \left(\frac{p_2}{p_1}\right)^{\kappa-1/\kappa}\right]$. Thus, obtained can be a formula for gas outflow velocity v_0 through a flange with aperture area A_0 as a function of initial gas state (p_1, v_1) and pressure ratio $\beta = \frac{p_2}{p_1}$:

$$v_0 = \sqrt{2 \cdot \frac{p_1}{\rho_1} \cdot \frac{\kappa}{\kappa - 1} \cdot \left[1 - \beta^{\kappa - 1/\kappa}\right]} \tag{7.19}$$

where ρ_1 – is the initial molar density $\rho_1 = \frac{1}{v_1}$, and β – the pressure ratio $\beta = \frac{p_2}{p_1}$.

Following the flow continuity law, we may write $\dot{m} = A_0 \cdot v_0 \cdot \rho_0$, whereas value $\rho_0 \equiv \rho_2$ may be defined with the use of *Clapeyron's equation*. Following the transformations, the following equation can be obtained: $\rho_2 = \rho_1 \cdot \frac{p_2 \cdot T_2}{p_1 \cdot T_1} = \rho_1 \cdot \beta \cdot \beta^{1-\kappa/\kappa} = \rho_1 \cdot \beta^{1/\kappa}$.

Hence, the flow intensity of the mass $\dot{m}$ can be expressed as:

$$\dot{m} = A_0 \cdot \sqrt{p_1 \cdot \rho_1} \cdot \sqrt{2 \cdot \frac{\kappa}{\kappa - 1} \cdot \left[\beta^{2/\kappa} - \beta^{\kappa+1/\kappa}\right]} \tag{7.20}$$

where $\dot{m}$ – is the flow of mass.

Equation (7.20) is known as the Saint-Venant–Wantzel equation. If we introduce the so-called flow number Ψ, $\Psi = \left[2 \cdot \frac{\kappa}{\kappa - 1} \cdot \left(\beta^{2/\kappa} - \beta^{\kappa+1/\kappa}\right)\right]^{0.5} = f(\beta; \kappa)$, equation (7.20) will be expressed in an abbreviated form (Mittleman, 1989):

$$\dot{m} = A_0 \cdot \Psi \cdot \sqrt{p_1 \cdot \rho_1} \tag{7.21}$$

where Ψ – is the flow number.

A graphic representation of the flow number Ψ vs. pressure ratio $\beta = p_2 / p_1$ is a line having the maximum value for a certain critical value β_{kr} (Figure 7.11).

The critical value for pressure ratio β_{kr} can be found by calculating the derivative $\frac{\partial \Psi}{\partial \beta}$ and equating it to zero. Hence:

$$\forall_{\frac{\partial \Psi}{\partial \beta}=0} \beta_{kr} = \frac{p_2}{p_1} = \left(\frac{2}{\kappa+1}\right)^{\kappa/\kappa-1} \tag{7.22}$$

Ψ_{max} value can be found by inserting equation (7.22) into $\Psi = [2 \cdot \frac{\kappa}{\kappa-1} \cdot \left(\beta^{2/\kappa} - \beta^{\kappa+1/\kappa}\right)]^{0.5}$:

$$\Psi_{max} = \sqrt{\kappa \cdot \left(\frac{2}{\kappa+1}\right)^{\kappa+1/\kappa-1}} \tag{7.23}$$

And pressure ratio $\beta = \frac{p_2}{p_1}$ will be slightly different from the theoretical one (Figure 7.12).

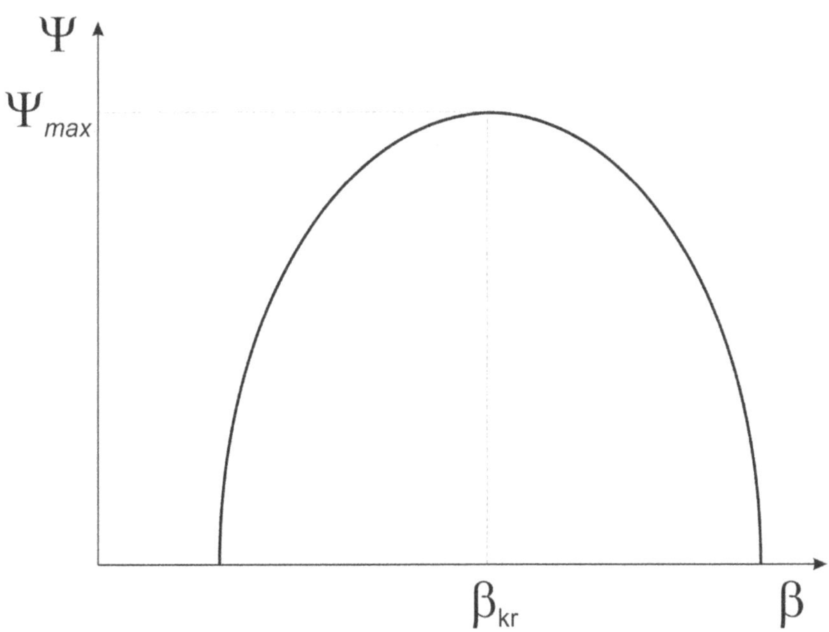

FIGURE 7.11 The theoretical dependence of the flow number Ψ in the function of pressure ratio β.

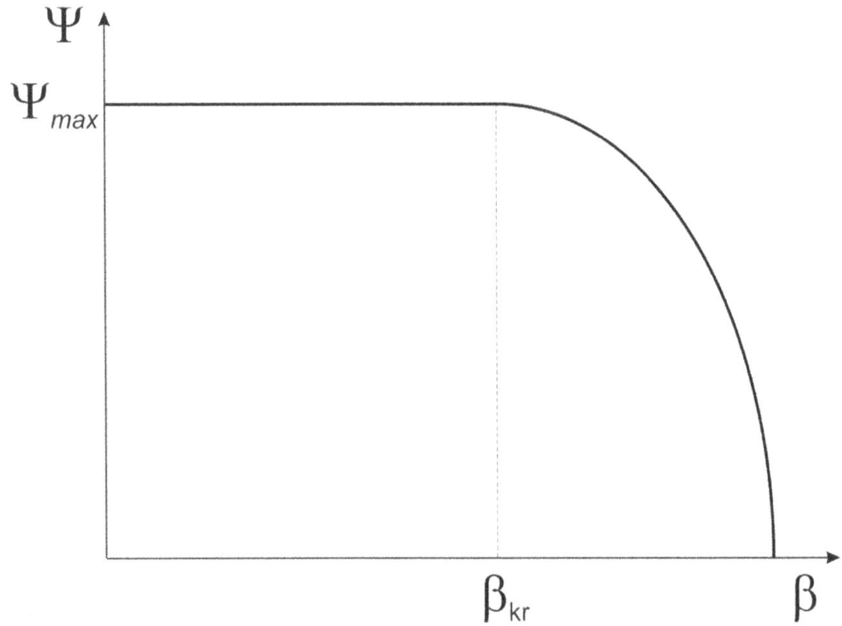

FIGURE 7.12 The practical dependence of flow number Ψ in the function of pressure ratio β.

TABLE 7.3
Values of Selected Parameters for Typical Constituents of Breathing Mixes

	κ	β_{kr}	Ψ_{max}	$\rho(p_0;T_0)$ $(g \cdot dcm^{-3})$	$\bar{M}$ $(g \cdot mol^{-1})$
Air				1.293	28.96
Oxygen	1.4	0.528	0.685	1.429	31.999
Nitrogen				1.251	28.01
Hydrogen				0.0899	2.016
Helium	1.677	0.486	0.728	0.1784	4.003

The values of constants k, β_{kr}, and Ψ_{max} and mean molar mass $\bar{M}$ of some gases are shown in Table 7.3.

The minimal air pressure supplying the nozzle for ultrasonic flow is equal to:

$$p_1^{min} \equiv p_z^{kr}(air) = \frac{p_0}{\beta_{kr}} = \frac{101.325\ kPa}{0.528} \cong 0.2\ MPa \tag{7.24}$$

where $p_z^{kr}(air)$ – is the minimal pressure supply value for the nozzle to ensure supercritical airflow, and β_{kr} – is the critical pressure ratio $\beta = \frac{p_2}{p_1}$ for the nozzle.

As measurements refer to the volumetric flow, to therefore calculate mass flow, equation (7.21) should be transformed to the following equation: $\dot{V} = \frac{\dot{m}}{\rho_1} = \Psi \cdot A_0 \cdot \sqrt{\frac{p_1 \cdot \rho_1}{\rho_1^2}} = \Psi \cdot A_0 \cdot \sqrt{\frac{p_1}{\rho_1}} \bigg| T = idem$, where $\dot{V}$ represents the volumetric flow of the gas mix that is dose-related to the measurement conditions, ρ_1 is gas mix density before the nozzle, and A_0 is nozzle aperture area.

Using Clapeyron's equation $\rho_1 = p_1 \cdot \frac{\bar{M}}{R \cdot T}$, it is possible to transform the preceding relation into the following form: $\dot{V} = \Psi \cdot A_0 \cdot \sqrt{\frac{R \cdot T}{M}} = idem \big| \beta < \beta_{kr} \wedge T = idem$. By comparing the flow to normal conditions $p_0 = 101.325 kPa \wedge t = 0^\circ C$, following Clapeyron's equation, the following can be written:

$$\dot{V}_0 = \frac{p}{p_0} \cdot \frac{T_0}{T} \cdot \dot{V} = \frac{p}{p_0} \cdot \frac{T_0}{T} \cdot \dot{V} \cdot \Psi \cdot A_0 \cdot \sqrt{\frac{R \cdot T}{M}} \tag{7.25}$$

where $\dot{V}-$ is the measured value of the dosed gas stream, and $p-$ is atmospheric pressure at the time of measurement.

The slope of the straight line, constituting the curve of the dependence of the gas stream, refers to the normal conditions in relation to reduced pressure $\dot{V}_0 - \int (p_z)$ equal to the differential of stream $\dot{V}_0$, from relation (7.25), after applying pressure $p_z : \left(\frac{\partial \dot{V}_0}{\partial p_z} \right)_{\beta < \beta kr} = \frac{T_0}{p_0} \cdot \frac{\Psi_{max}}{p_0} \cdot \sqrt{\frac{R \cdot T}{M}} \cdot A_0$. Hence, the required nozzle section A_0 can be calculated from the formula:

$$A_0 = \frac{\pi \cdot D^2}{4} = \frac{p_0 \cdot T}{T_0 \cdot \Psi_{max} \cdot \sqrt{\frac{R \cdot T}{M}}} \cdot \left(\frac{\partial \dot{V}_0}{\partial p_z} \right)_{\beta < \beta kr} \tag{7.26}$$

where $D-$ is the diameter of the nozzle.

As the temperature of the experiment is not exactly determined, it was assumed to be $t = 20^\circ C$. The other temperatures t in experiments will produce a small error in the value being determined. Establishing the temperature enables preparing a diagram for the rapid design of nozzles. Equation (7.26) can be the basic equation for the design of nozzles for reducing – metering systems for constant premix metering in an *SCR*. In order to design the nozzles, it is necessary to use the following relationship: $\left(\frac{\partial \dot{V}_0}{\partial p_z} \right)_{\beta < \beta kr} = f(D) \cong 676.53 \cdot \frac{\Psi_{max}}{\sqrt{\bar{M}}} \cdot D^2$, where $\bar{M}$ represents mean molar mass for gas supply $\bar{M} = [kg \cdot kmol^{-1}]$, $D = [mm]$ is the nozzle diameter, and p_z represents pressure supply value for the nozzle necessary to ensure supercritical airflow $p_z = [MPa]$ (Table 7.4).

TABLE 7.4

Function Calculations $\left(\frac{\partial \dot{V}_0}{\partial p_z}\right)_{20^\circ C}$ for Particular Pure Gases and Mixtures

$D = [mm] \rightarrow$	0.10	0.20	0.30	0.40	0.50	0.60	Equation $\left(\frac{\partial \dot{V}_0}{\partial p_z}\right)_{20^\circ C}$
Air	0.86	3.44	7.75	13.77	21.53	31.00	$86.11 \cdot D^2$
Helium	2.46	9.85	22.15	39.39	61.54	88.62	$246.16 \cdot D^2$
$50\%_v He / air$	1.18	4.71	10.60	18.85	29.45	42.41	$117.82 \cdot D^2$

Note: The temperature of experiments is equal to 20°C; volumetric flow is related to NTP conditions.

The possibility of simple conversion of the gas metering for different gases, for example, air, is an additional advantage of the measuring process. Table 7.4 shows the values obtained from function $\left(\frac{\partial \dot{V}_0}{\partial p_z}\right)_{20^\circ C}$ for the chosen particular pure gas and mix for the temperature of experiments equal to 20°C; volumetric flow is related to NTP conditions.

7.5 WORK OF BREATHING

Work of breathing *WOB* represents volume change work – volume work. The *volume work* is performed, by a system, against external pressure. To illustrate the problem of work, let us consider Figure 7.13.

Movement of a piston to the right causes a change in the gas volume. According to the laws of physics, the work done by the gas in this process is given by:

$$W_{el} = -F_z \cdot d_x = -p_z \cdot S \cdot d_x = -p_z \cdot dV \tag{7.27}$$

where W_{el} is elementary volume work; F_x is external force; d_x is elementary movement; p_z – is external pressure; dV – is change in infinitesimal gas volume related to infinitesimal piston displacement; and S – is the cross-section area of the piston.

And integration form:

$$W = -\int_V p_z \cdot \frac{dV}{dt} \cdot dt = -\int_t p_z \cdot \dot{V} \cdot dt \tag{7.28}$$

where $\dot{V}$ – is the stream of gas flow, W – is volume work, and t – is time.

The opposite direction of a piston movement to that presented in Figure 7.13 will be expressed by the same formula, but with an opposite sign, and p_z will represent the internal pressure in the system. In other words, the work done by gas is expressed by the same equation as before; however, the work direction is opposite. Most often, in thermodynamics, work done by a system is considered positive, and work done on a system is negative – it depends on the accepted convention.

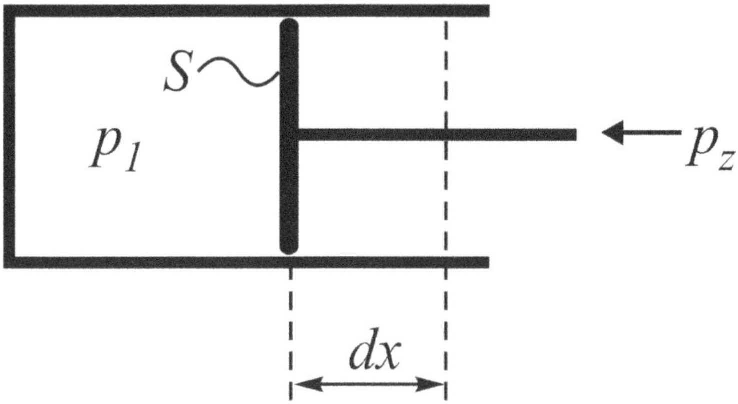

FIGURE 7.13 A diagram illustrating the infinitesimal pushing work done by gas pressured in a piston-cylinder system; the piston is assumed to be weightless, where $p_1 =$ is the pressure of compressed gas; $S =$ is the piston cross-section area; $p_z =$ is external pressure; and $d_x =$ is infinitesimal displacement of the piston.

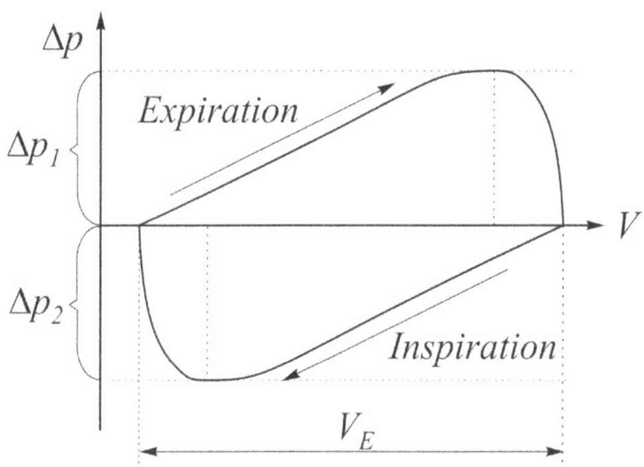

FIGURE 7.14 Example of changes in pressure in the mouthpiece and in the breathing space volume for the *SCR*, where V_E is the volume of inspired gas, Δp_1 is maximal expiration pressure jump, and Δp_2 is maximal inspiration pressure jump.

According to equation (7.28), the relationship between the pressure into the mouthpiece and the volume of inspired gas will be the basis for calculating the *USB* work of breathing. An example of the relationship is shown in Figure 7.14.

The work done by the system is represented by the area between the inspiration and the expiration curves on the $p-V$ diagram. Particularly, the changes in the pressure in the mouthpiece assembly and the changes that accompany them in the breathing space volume can be more complex than those shown in Figure 7.14.

Therefore, two kinds of the mean work of breathing can be distinguished as follows (Reimers, 1989):

- Resistive work of breathing is the first kind of work done against breathing resistance. The resistive work is represented by the area between the mean inspiration and expiration curves that are calculated only when the expiration pressure is higher than the inspiration pressure. This is shown in Figure 7.15a.
- Total work of breathing is the second kind of work. It is the sum of the work done during inspiration and expiration. The work done in each of the two phases is calculated separately with respect to the line for $\Delta p = 0$ – (Figure 7.15b).

Sometimes, due to the lack of equipment, it is not possible to obtain the diagrams $p - V$, which enable calculating *WOB*. However, it is possible, provided that sinusoidal approximation is assumed. The periodic changes in the pressure and the volume are shown in Figure 7.16.

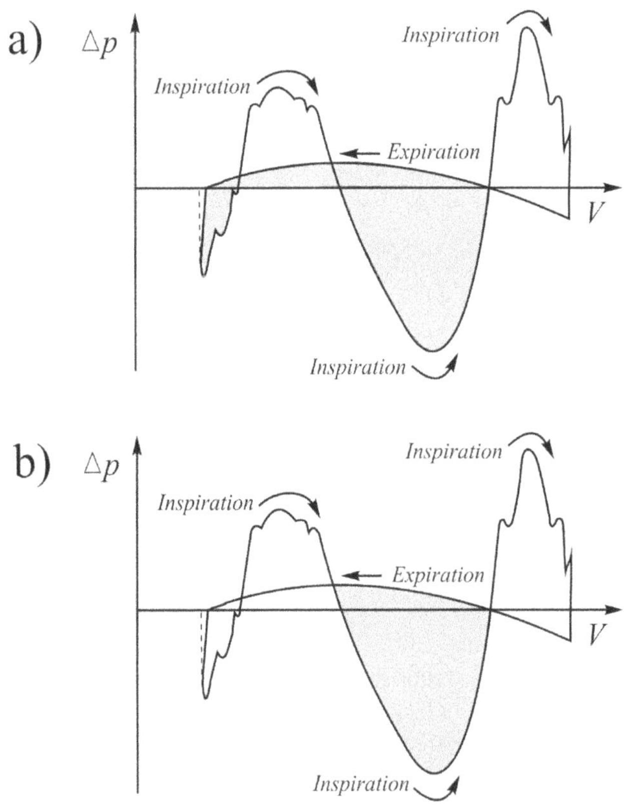

FIGURE 7.15 The kinds of *WOB*: (a) resistive work; (b) total *WOB*.

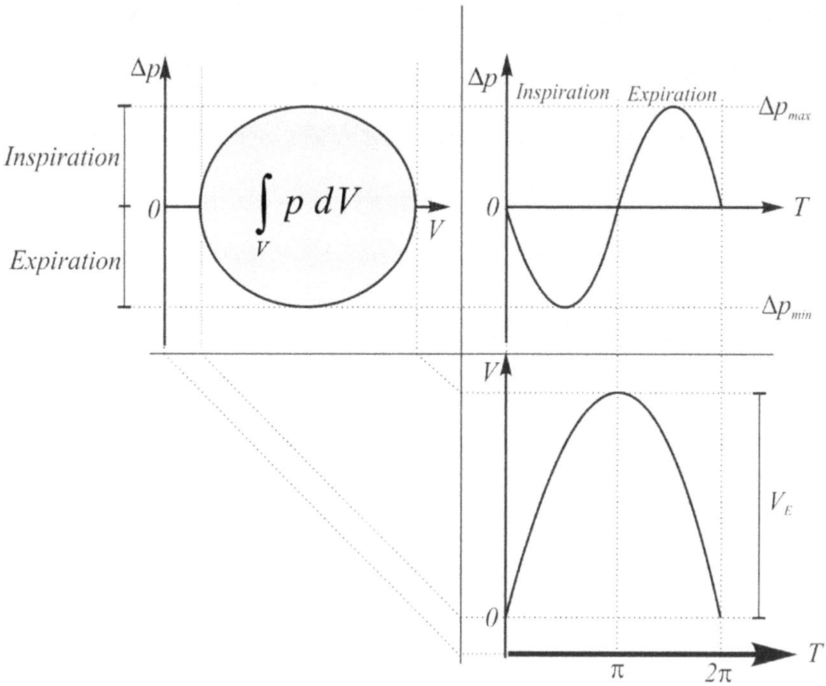

FIGURE 7.16 The variations in the pressure and the breathing space during respiration, sinusoidal approximation, where $\dot{V}$ is ventilation, Δp is variation of pressure, V – is volume, V_E is the breathing space volume, and T – is period.

The work of breathing for sinusoidal approximation can be calculated from the following equation (7.28): $W = -\int_V p_z \cdot \frac{dV}{dt} \cdot dt = -\int_{t=0}^{t=T} \Delta p(t) \cdot \dot{V}(t)$, where T represents a period. The sinusoidal approximation of the relationship between the variation in the pressure in the mouthpiece and the time is expressed as:

$$\Delta p(t) = -\Delta p_{max} \cdot \sin(\varpi \cdot t) \tag{7.29}$$

where Δp_{max} is the maximal/minimum pressure in the mouthpiece during respiratory action, and f – is breathing rate ϖ angular frequency[8] $\varpi = 2 \cdot \pi \cdot f = 2 \cdot \pi \cdot \frac{1}{T}$.

The relationship between the variation in the respiratory volume and the time can be expressed as $V(t) = -V_E \cdot \sin^2(0.5 \cdot \varpi \cdot t) \Rightarrow \frac{dV(t)}{dt} = \varpi \cdot V_E \cdot \sin(0.5 \cdot t) \cdot \cos(0.5 \cdot t) = 0.5\varpi \cdot V_E \cdot \sin(\varpi \cdot t)$; as for function $\dot{V} = f(t)$, period is $T = 2 \cdot \pi$, and the function amplitude is equal to $0.5 \cdot V_E : \dot{V} = 0.5 \cdot V_E \cdot \sin(\varpi \cdot t)$. Hence, work of breathing W is given as:

$$\begin{aligned} W &= 0.5 \cdot \Delta p_{max} \cdot V_E \cdot \int_{t=0}^{t=T} \sin^2(0.5 \cdot \varpi \cdot t) \cdot dt = \\ &= 0.5 \cdot \Delta p_{max} \cdot V_E \cdot \left[0.5 \cdot \varpi \cdot t - \frac{\sin^2(2 \cdot \varpi \cdot t)}{4} \right] \Big|_0^{\frac{2 \cdot \pi}{\varpi}} = \\ &= 0.5 \cdot \pi \cdot \Delta p_{max} \cdot V_E \end{aligned} \tag{7.30}$$

Sinusoidal approximation is one of the methods used for preliminary estimation of work of breathing W. The cyclic variations in the pressure and the volume are presented in Figure 7.16. Usually, the calculated work of breathing W is converted[9] to $V_E = 1\,dm^3$ of the breathing volume V_E corresponding to the assumed depth H of diving and ventilation. In this case, work of breathing $\breve{W}$ calculated for sinusoidal approximation is given as:

$$\breve{W} = 0.5 \cdot \pi \cdot \Delta p_{max} \tag{7.31}$$

where $\breve{W}$ – is standardized resistive effort, and $[J \cdot dm^{-3}]$ Δp_{max} the maximal/minimum pressure in the mouthpiece during respiratory action (kPa).

The work calculated from equation (7.31) is referred to as resistive effort $\breve{W}$ and is given in the following units[10]: $\breve{W} = [J \cdot dm^{-3}]$.

To calculate a sinusoidal approximation of WOS, it is necessary to measure, at constant ventilation, only the maximal pressure Δp_{max} during inspiration and expiration and to calculate inspiratory and expiratory work W. The total resistive work is the sum of inspiratory and expiratory work. In order to establish the same conditions for measurements, a decision was made that the breathing effort expressed by equation (7.31) should be determined for the typical ventilation of lungs $\dot{V}(t)$, depending upon the work performed (Table 4.15).

To establish the same measurement conditions, it was assumed that the resistive effort should be determined for typical pulmonary ventilation dependent upon the activity being performed. In order to act in accordance with the preceding assumption, the model of a standard person has been introduced (Table 7.5; Segedal and Morrison, 1985).

There are a lot of different criteria used to evaluate breathing systems. In principle, all significant institutions dealing with the hyperbaric techniques have established their own standards for evaluating technical facilities aiding respiration. According to NUTEC[11] standard, the maximal value of the total work of breathing is described with the experimental equation (Segedal and Morrison, 1985): $\breve{W} = [J \cdot dm^{-3}] = 0.5 + 0.02 \cdot \dot{V}_E[dm^3 \cdot min^{-1}]$. Apart from that, the maximal overpressure/negative pressure in the mouthpiece should not exceed $2.5\,kPa$.

According to Det Norske Veritas (DNV) standard, the mean work of breathing should not exceed $1.7\,J \cdot dm^{-3}$ for ventilation at a level of $62.5\,dm^3 \cdot min^{-1}$ at the breathing rate 25 $breathes \cdot min^{-1}$.

TABLE 7.5
The Relationship between Lung Ventilation, Inspiration Volume, Carbon Dioxide Emission vs. the Kind of Work Rate for the Standard Person

Lung Ventilation $\dot{V}$ $[dm^3 \cdot min^{-1}]$	Respiratory Rate n $[S^{-1}]$	Inspiration Volume V_E $[dm^3]$	Carbon Dioxide Emission $\dot{V}(CO_2)$ $[dm^3 \cdot min^{-1}]$	Kind of Work Performed
15.0	15	1.0	0.6	Rest
22.5	15	1.5	0.9	Light
40.0	20	2.0	1.6	Medium
62.5	25	2.5	2.5	Moderately heavy
75.0	25	3.0	3.0	Heavy
90.0	30	3.0	3.6	Severely heavy

Source: Segedal and Morrison (1985).

TABLE 7.6
The Maximal Overpressures/Negative Pressures in the Mouthpiece and the Maximal Permissible Values of Work Breathing According to US NEDU Standard

Lung Ventilation $\dot{V}$ $[dm^3 \cdot min^{-1}]$	The Maximal WOB Δp_{max} $[kPa]$	The Maximal WOB $\breve{W}_{max}$ $[J \cdot dm^{-3}]$
22.5	0.10	0.13
40.0	0.45	0.60
62.5	1.00	1.33

Source: Clarke (2015).

According to US Navy Experimental Diving Unit standard for the proper ventilation, the maximal overpressure/negative pressure in the mouthpiece should not exceed the values given in Table 7.6 (Elliott, 1992; Clarke, 2015).

7.6 SCRUBBER

The carbon dioxide scrubber resistance was described in Section 6.2.

7.7 CONCLUSIONS

Flow resistance is the basic parameter for assessing the usefulness elements of life-support systems. It is especially breathing resistance that can lead to excessive fatigue or limit the range of effort to be taken. A considerable breathing resistance may cause respiratory failure. Excessive respiratory support[12] is also dangerous, especially in a hyperbaric environment. For this reason, when designing the apparatuses, it is important to precisely define the function of changes in breathing resistance, depending on the environmental parameters, especially for emergency cases.

NOTES

1 Back pressure.
2 In thermodynamics, this function depends only on the system condition, that is, its current parameter values.
3 As opposed to, for instance, internal system energy.
4 For example, the size of volumetric work performed by the system is defined as an external pressure value and not the pressure of gas which performs work.
5 $du = c_v \cdot dT - \rho \cdot dv \mid du \equiv 0$
6 Adiabatic equation: $\rho \cdot u^{\kappa} = const.$
7 An interesting relationship is revealed if we compare equations (**7.18**) and (**7.16**): $\int_{v^1}^{v^2} \mathrm{p} \cdot \mathrm{d}v = \mathrm{k} \int_{\mathrm{p}^1}^{\mathrm{p}^2} v \cdot \mathrm{dp}.$
8 Pulsatance.
9 Standardized.
10 The absolute values of the work expressed in ($[J \cdot dm^{-3}]$) and (kPa) are equal, but these units have a different physical meaning and are not interchangeable.
11 Norwegian Underwater Technology Center.
12 Zero or negative breathing resistance.

REFERENCES

Clarke J.R. 2015. *The history and implications of design standards for underwater breathing apparatus—1945 to 2015.* Panama City: Navy Experimental Diving Unit. NEDU TR 15-03.

Elliott D.E. 1992. New breathing apparatus performance standard for the diving industry—report on seminar. *Underwater Technology*. 18, 39–41.

Mittleman J. 1989. Computer modeling of underwater breathing systems. In *Physiological and human engineering aspects of underwater breathing apparatus*, ed. Warkander, D.E., Lundgren C.E.G. Bethesda: Undersea and Hyperbaric Medical Society.

Reimers S.D. 1989. Work of breathing calculations. In *Physiological and human engineering aspects of underwater apparatus*, ed. Lundgren C.E.G., Warkander D.E., 223–224. Bethesda: Undersea and Hyperbaric Medical Society, Inc., Appendix A.

Segedal K., Morrison J.B. 1985. *Acceptable criteria and unmanned test procedures for underwater breathing apparatus*. Bergen: Norwegian Underwater Technical Centre. Report NUTEC No. 17-85.

Solomon E.P., Martin C.E., Martin D.W., Berg L.R. 2015. *Biology*. Stamford: College Edition, Cengage Learning. ISBN-13: 978-1-285-42358-6; ISBN-10: 1-285-42358-5.

8 Remarks

This book is the result of many years of research work and a lot of research projects. A new mathematical model of the process of ventilation of an *SCR* was proposed as the base of any investigation on the semi-closed rebreather (*SCR*). Its validation required making a special simulator of gas exchange in the breathing process. The use of this device made the experimental validation of the proposed model possible. The model was adjusted to the process of ventilation in hyperbaric chambers. The validation process required developing a new type of carbon dioxide emission simulator. The generalization of the adopted method for the submarine ventilation process was only an obvious consequence of earlier considerations. However, the validation process required undertaking extensive research based on a real object, which confirmed the validity of the modelling method adopted. The research on ventilation of the mining excavation constituted the validation of the adopted research approach (Kłos, 2021).

The remaining problems discussed in the book are additional or derivative problems, closely related to the adopted ventilation model.

The aim of the work was to develop and verify some mathematical ventilation models for the semi–closed circuit rebreather (SCR;Kłos, 2002b) and, in the future, generalize the presented calculation models to other hyperbaric objects, for example, hyperbaric chambers, submarines, and those used in mining excavations. In most situations, in a submarine or a mining excavation, there exists a slightly higher pressure as compared to the atmospheric pressure, but further, for the purpose of simplification, these facilities will be classified as normobaric. The mathematical modelling enables combining the micro- and macro-objects of the hyperbaric and normobaric technique (Kłos, 2021). The developed models have an analytical character. Therefore, they are subject to physical interpretations. For this reason, neither semianalytical nor analytic models were searched for. For the sake of the nature of the phenomena analyzed, the statistic models of phenomena were not searched for either.

The experiments concerning the mathematical ventilation models were performed during manned and unmanned experimental diving aimed at verification of the decompression procedure proposed. The experiments were carried out for semi-closed rebreathers *SCRs*. The tested *SCRs* should feature the same parameters as the parameters assumed to derive the mathematical model. It is the main condition necessary to perform the experiments by means of a metabolic simulator. A series of experimental dives was carried out with such a simulator.

The adequacy of the assumed mathematical model at limits of oxygen partial pressures was tested during pressure tests with the use of diving apparatuses. The experiments were carried out in parallel with the verification experiments of the decompression procedure based on the experiments carried out in the hyperbaric swimming simulator (Kłos, 2000, 2016, 2021; Kłos, Nishi and Olszański, 2002) (Figure 1.13).

DOI: 10.1201/9781032649207-8

There were more similar trials carried out, but the system was modified. That is why the results obtained are not sufficient to draw conclusions concerning the adequacy of the mathematical ventilation model. The probability of the proposed mathematical model's acceptance was not further maximized because of the high costs of experiments.

It seems that the best method for experimentation is to apply a breathing machine simulating a breathing action with simultaneous oxygen sampling from the breathing space and carbon dioxide emission for unmanned tests. As, in this case, carbon dioxide emission is maintained at the proper level, there is no necessity to carry out manned experiments, and the experimental results are repeatable with known accuracy of reproduction. Within the framework of the financially supported project, a special laboratory was built. It can be used to carry out less-expensive experiments and get the results without the necessity to carry out manned experiments. The mathematical model was verified during unmanned tests with the use of *SCRs* (Figure 1.12).

In order to simulate the breathing process, it is necessary to solve two of some technical problems, that is, the mechanical representation of pulmonary ventilation and the simulation of gas exchange in the course of breathing. The processes should be simulated with sufficient accuracy and repeatability. There is no need for representation of each detail of the processes. It is not necessary to simulate the real ventilation process. Usually, it is assumed that the respiration shape should be approximated to sinusoidal character. As a matter of fact, each of the hyperbaric research centers has their own evaluation standards.

A series of standards concerning the respiration volume and the respiration rate has been established. In fact, they are similar as regards most of the parameters.

During a simulation of the breathing action and gas exchange, the exact representation of the inhalation and exhalation shape is not essential. Attention should be focused on the oxygen consumption and carbon dioxide emission to that place.

A reactor with a catalyst for burning chemical compounds can be used as a metabolic simulator. The combusted chemical substance is added to the volume drawn from the breathing loop of *SCR* through the inhale hose using a breathing simulator. After the mixture thus formed has passed through the reactor, it is returned together with the combustion products to the breathing loop of the diving apparatus through the exhale hose (Kłos, 2002a, 2021). A tested attachment that enables oxygen uptake and carbon dioxide emission was connected to the breathing machine.

That simulator was developed and used to verify the mathematical model of ventilation developed for *SCRs*. Owing to the experiments carried out, the exactness of the theoretical ventilation model in *SCRs* could be tested (Kłos, 2000, 2016, 2021).

The developed experimental laboratory stand enables continuing research on ventilation in various diving apparatuses. Further research focused on developing reactors that could be used for other research purposes, such as regeneration of an ecologically confined space, for example, during saturation dives or in submarines, is recommended (Kłos, 2008, 2014, 2021).

8.1 HYPERBARIC CHAMBERS

Mathematical models for hyperbaric chamber ventilation and regeneration can be derived with the use of a similar procedure, as in the case of *SCR* ventilation. The investigations were performed to check whether the mathematical models of ventilation derived for *SCR* can also be applied to hyperbaric chambers. The experimental results confirmed such a possibility.

The procedures proposed to be employed to obtain mathematical models of ventilation can be used in both a micro scale and a macro scale. The macro scale is concerned with the diving technique. It is important that the presented analytical models, compared to the empirical and semi-empirical models, be related to the physical phenomena occurring in the hyperbaric environment.

The consistency of the assumed theoretical model can be tested using the carbon dioxide simulator (Kłos, 2009, 2021; Figure 8.1).

The comparison of the experimental and theoretical results is satisfactory.

The results of the preliminary experiments confirmed the usability of the applied mathematical models for the interrupted and continuous ventilation parameters. The earlier assumptions concerning the necessity to use an undetermined ventilating medium excess were confirmed as well. As it follows from earlier text, in order to find the mathematical model that can be used to calculate the maximum excess of ventilating medium, it is necessary to carry out further experiments (Kłos, 2000, 2008, 2009, 2021; Kłos and Kłos, 2004).

The aim of the presented experiments was to find out whether the ventilation in the mathematical models applied to the SCRs can be applied also to hyperbaric chambers.

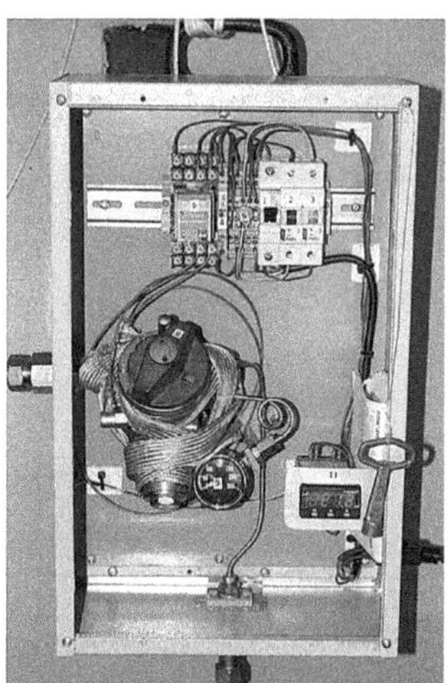

FIGURE 8.1 The carbon dioxide simulator.

Source: Kłos (2021).

The procedures proposed to develop mathematical models of ventilation can be applied on a micro scale and macro scale as well. The macro scale is concerned with the diving methodology. It is important that the empirical and semi-empirical models, when compared to the presented analytical models, be related to the physical phenomena occurring in the hyperbaric environment.

8.2 SUBMARINES

In 2000, a Polish KILO-class submarine was equipped with an experimental submarine atmospheric monitoring system (Figure 8.2).

This system was developed in the Naval Academy. The monitoring system consisted of oxygen, hydrogen, carbon dioxide gas analyzers, and temperature, moisture, and pressure sensors. The sensors and analyzers were connected to a central industrial computer by data acquisition modules into *RS* 485 industrial-standard net. Readouts could be recorded every second or a longer period of time.

The submarine measuring net was connected by a deck-mounted telephone connector and a waterproof electric cable comprising a twisted pair of wires connected to the portable laboratory. And via an *RS* 485/*RS* 232 converter to a typical *PC* situated in a laboratory container, capable of monitoring atmosphere in the submarine with a special computer program (Figure 8.3).

The test involved the simulation of submarine atmosphere carbon dioxide CO_2 contamination, stabilization of CO_2 content, and ventilation with fresh air.

The submarine rescue compartment consisted of two decks which divide the compartment into three subcompartments of equal volume (Figure 8.4). There were stairs between decks, and a cargo hatchway between the lower subcompartments. Preliminary CO_2 concentration was created by releasing CO_2 from four bottles – two on the first and two on the second deck. There were two additional analyzers used for measuring CO_2 concentration by the submarine atmosphere monitoring system, on top of the upper and on the lower bilges of the lowermost compartment.

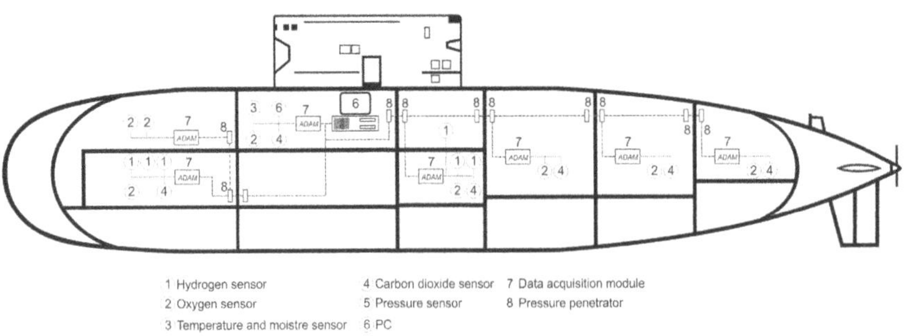

FIGURE 8.2 Submarine atmospheric monitoring system.

Source: Kłos (2021).

FIGURE 8.3 *Top*, arrangement of the main elements for the experiment; *bottom*, a containerized, portable laboratory.

Source: Kłos (2021).

8.3 SEALED MINING EXCAVATION

In mining plants, there exists a hazard of occurrence of unhealthy atmosphere (Kłos, 2017). Harmful pollutants can come from various sources. In coal mining, they come from fires of coal and flammable hydrocarbons. In other mines – for example, in copper mines – they can form during fires of machinery or when gases accumulated in a rock mass fall through cracks. During fires or flarebacks, large amounts of toxic gases can be released, which, if spread quickly, can surprise the crew, even in locations far from the fire or flareback. The smokiness of excavations occurring during a fire makes it difficult for the endangered crew to withdraw along the specific escape routes. In extreme cases, the crew will not be able to withdraw from the excavation face.

A plan for various rescue action scenarios is developed in advance to respond to cases of hazards in a mine. The plan specifies, among other things, escape routes, which the crew may use to withdraw from the endangered zone to a safe place. In the case of long escape routes, the respiratory protection equipment used by the crew may not be sufficient to ensure safe exit from the danger zone. The solution may be the construction of refuge chambers, which, in the event of a fire, will secure a safe haven for the crew. The possibility for the crew to survive the period of danger in such chambers gives the management of the rescue operation time necessary to take action to help them.

Due to the installation of an airtight breathing system in the refuge chamber, when an atmosphere hazardous for safe breathing forms, a group of miners can survive for a period assumed in the chamber design. The basic elements of the lifesaving system are sets of inhalers equipped with half-masks using air contained in high-pressure bottles. The chamber should be fitted with the necessary control and measuring equipment, communication means, etc., in accordance with the design assumptions and applicable regulations.

The chamber was placed in the excavation connecting two gates having a length of minimum10 m, width at the backfill of about 5 m, and height of about 2 m. The chamber was isolated from the gates with walled dams, on which the entrance door to the chamber was built in sluice box arrangement. The main dimensions of the chamber were adopted as guidelines for the construction of the lifesaving system (Figure 8.4).

The dams were made of concrete blocks, the walls were plastered on both sides, and the ground was levelled and hardened. The tightness of the rock mass and walls sealing off the chamber from excavations was provided with generally available means.

The ventilation of the chamber prior to its evacuation was to be provided by one or two racks with six air cylinders designed for ventilation of the atmosphere in the chamber just prior to the planned end of human stay inside it (Figure 8.4). The release of air was to be achieved through the total opening of the tank.

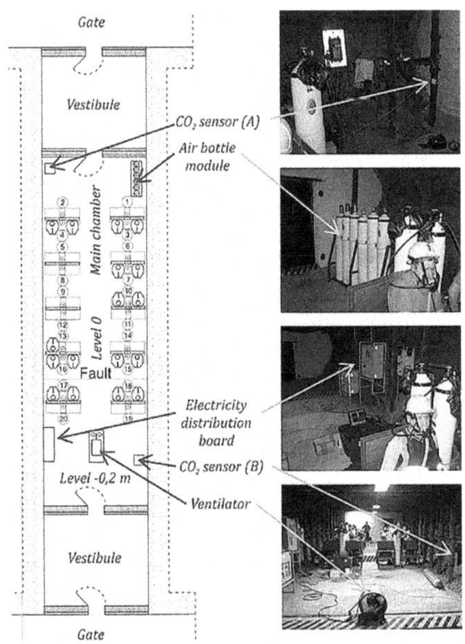

FIGURE 8.4 Arrangement of equipment items during tests.

Source: Kłos (2021).

The ventilation tests were carried out without the participation of people. Carbon dioxide CO_2 was released into the empty and ventilated chamber, stored in two cylinders. The cylinders were located near the sensors of the CO_2 monitoring system.

The complete research results into the chamber ventilation were published earlier and will thus not be referred to here (Kłos, 2010, 2021).

The results of the final test, however, will be discussed here. After $8\,kgCO_2$ was released into the chamber space, ventilation tests were carried out sequentially, based on two schemes:

- Simulation of laminar flow through slow and even release of $90\,m^3$ of air into the chamber space for about $25\,min$
- Simulation of turbulent flow of even air release of 31 m^3 to the space of the chamber for about 1 min

In the initial period, the stratification of the carbon dioxide content in the atmosphere of the chamber after its release from the cylinders was clearly visible. However, after about 10 min, rapid homogenization and the disappearance of the initial stratification took place. Such rapid homogenization could be caused by mechanical movements of the atmosphere of the chamber caused by a stream of released carbon dioxide CO_2. Initially carried-out ventilation was at a lower intensity of the airstream. Then, after the disappearance period of the effect of the first ventilation process, determined for approximately 20 min, a more intensive ventilation process was carried out using a tenfold higher airstream (Kłos, 2010, 2021).

The measurement results for both ventilation processes demonstrated extraordinary compliance with the theoretical model. Previously, several times obtained were the results which did not show less compatibility of the theoretical model with the results of measurements (Kłos, 2010, 2021).

It can be stated that the results of the research on such a large and complex object as a mining excavation were surprisingly consistent with the proposed mathematical model. This suggests that the approach used to model the ventilation process for a mining excavation is sufficiently accurate to predict its course at the assumed level of credibility, thus enabling its practical use for designing mine refuge chambers.

The scientific measures described here were aimed at finding out whether the mathematical ventilation models applied for the $SCRs$, hyperbaric chambers, sealed mining excavation, and submarine ventilation mathematical model.

There is good conformity between the mathematical model and the real measurements from experiments. Only at the end of the ventilation process, a slight deviation was observed, probably caused by weak homogenization into all compartment volumes.

REFERENCES

Kłos I., Kłos R. 2004. *Polish soda lime in military applications*. Oświęcim: Chemical Company DWORY S.A. ISBN: 83-920272-0-5.

Kłos R. 2000. *Aparaty Nurkowe z regeneracją czynnika oddechowego*. Poznań: COOPgraf. ISBN: 83-909187-2-2.

Kłos R. 2002a. Metabolic simulator supports diving apparatus researches. *Sea Technology*. 12, 53–56.

Kłos R. 2002b. Mathematical modelling of the breathing space ventilation for semi-closed circuit diving apparatus. *Biocybernetics and Biomedical Engineering*. 22, 79–94.

Kłos R. 2008. *Systemy podtrzymania życia na okręcie podwodnym*. Gdynia: Polskie Towarzystwo Medycyny i Techniki Hiperbarycznej. ISBN: 978-83-924989-4-0.

Kłos R. 2009. *Wapno sodowane w zastosowaniach wojskowych*. Gdynia: Polskie Towarzystwo Medycyny i Techniki Hiperbarycznej. ISBN: 978-83-924989-5-7.

Kłos R. 2010. Komorowy system zachowania życia zabezpieczający górników w przypadku powstania atmosfery niezdatnej do oddychania. *Polish Hyperbaric Research*. 33, 71–88.

Kłos R. 2014. *Helioksowe nurkowania saturowane—podstawy teoretyczne do prowadzenia nurkowań i szkolenia*. Gdynia: Polskie Towarzystwo Medycyny i Techniki Hipernbarycznej. ISBN: 978-83-938322-1-7.

Kłos R. 2016. *System trymiksowej dekompresji dla aparatu nurkowego typu CRABE*. Gdynia: Polskie Towarzystwo Medycyny i Techniki Hiperbarycznej. ISBN: 978-83-938322-5-5.

Kłos R. 2017. Pollutions of the hyperbaric breathing atmosphere. *Scientific Journal of Polish Naval Academy*. 208, 31–44. DOI: 10.5604/0860889X.1237621.

Kłos R. 2021. *Ventilation of normobaric and hyperbaric objects*. Boca Raton, FL: CRC Press (Taylor & Francis Group, LLC). ISBN: 978-0-367-67523-3 (hbk); ISBN: 978-0-367-67524-0 (pbk); ISBN: 978-1-003-13164-9 (ebk).

Kłos R., Nishi R., Olszański R. 2002. Validation of diving decompression tables. *International Maritime Health*. 53, 1–4.

Index

Note: Page numbers in *italics* indicate a figure and page numbers in **bold** indicate a table on the corresponding page.

U

W

For Product Safety Concerns and Information please contact our EU
representative GPSR@taylorandfrancis.com
Taylor & Francis Verlag GmbH, Kaufingerstraße 24, 80331 München, Germany

www.ingramcontent.com/pod-product-compliance
Ingram Content Group UK Ltd.
Pitfield, Milton Keynes, MK11 3LW, UK
UKHW020022200926
475702UK00014B/2016
* 9 7 8 1 0 3 2 6 4 9 1 9 1 *